Primatology Monographs

Series Editors

Tetsuro Matsuzawa
Inuyama, Japan

Juichi Yamagiwa
Kyoto, Japan

For further volumes:
http://www.springer.com/series/8796

Hirohisa Hirai · Hiroo Imai · Yasuhiro Go
Editors

Post-Genome Biology of Primates

Editors

Hirohisa Hirai
Professor
Primate Research Institute
Kyoto University
41-2 Kanrin, Inuyama
Aichi 484-8506, Japan
hhirai@pri.kyoto-u.ac.jp

Yasuhiro Go
Assistant Professor
Primate Research Institute
Kyoto University
41-2 Kanrin, Inuyama
Aichi 484-8506, Japan
yago@pri.kyoto-u.ac.jp

Hiroo Imai
Associate Professor
Primate Research Institute
Kyoto University
41-2 Kanrin, Inuyama
Aichi 484-8506, Japan
imai@pri.kyoto-u.ac.jp

ISSN 2190-5967   e-ISSN 2190-5975
ISBN 978-4-431-54010-6   e-ISBN 978-4-431-54011-3
DOI 10.1007/978-4-431-54011-3
Springer Tokyo Dordrecht Heidelberg London New York

Library of Congress Control Number: 2011943308

© Springer 2012

This work is subject to copyright. All rights are reserved, whether the whole or part of the material is concerned, specifically the rights of translation, reprinting, reuse of illustrations, recitation, broadcasting, reproduction on microfilm or in any other way, and storage in data banks.

The use of general descriptive names, registered names, trademarks, etc. in this publication does not imply, even in the absence of a specific statement, that such names are exempt from the relevant protective laws and regulations and therefore free for general use.

Cover illustration:

Front cover: *Top*: An adult male white-handed gibbon (*Hylobates lar*) at the Nakhon Ratchasima (Khorat) Zoo, Thailand. Photo by Hirohisa Hirai. *Center left*: Hybridization of a human bacterial artificial chromosome shows split signals (*red*) on two chromosomes of the white-cheeked gibbon, revealing an evolutionary breakpoint. Photo by Roscoe Stanyon. *Center middle*: A proboscis monkey metaphase counterstained in *blue* and hybridized by human chromosome paints: 1 in *green*, 3 in *yellow*, and 19 in *red*. Photo by Roscoe Stanyon. *Center right*: Screenshot of an alignment of short nucleotide reads produced by next-generation sequencing (NGS). Photo by Yasuhiro Go.
Back cover: Grooming by Japanese macaques (*Macaca fuscata fuscata*) at the Primate Research Institute, Kyoto University, Japan. Photo by Hirohisa Hirai.

Printed on acid-free paper

Springer is part of Springer Science+Business Media (www.springer.com)

Foreword

It is a great pleasure and honor to be asked to write a foreword to this volume, which addresses the *Post-Genome Biology of Primates*. It is hard to believe that just over 10 years ago there was a raging debate as to which primate genome should be selected for sequencing next, after the human, mouse, and rat genomes had been completed. The chimpanzee eventually won out, based on feedback from the academic community, but a strong minority believed that a better studied and more experimentally tractable animal such as the rhesus monkey or baboon should have had priority. In hindsight, all these arguments turned out to be meaningless, as the pace of genome sequencing increased so rapidly and the costs fell so dramatically that many primate genomes have been partially or completely sequenced within the past decade.

There is still a long way to go before one can say that we have covered all genomes that would be worth sequencing (one even could argue that all of them are), and population-level genomic information is still very limited for most primate genomes. But I think it is safe to say we are now indeed in an era where the genomic sequences that are already available can be used to explicate the genetic and genomic contributions toward primate evolution and phenotype. Indeed, we are now in a situation in which it is the phenotypic information has become rate limiting. In this volume, the editors have brought together an excellent collection of papers covering a wide variety of topics relevant to primate genomes, including evolution, genome structure, chromosome genomics, bioinformatics, and functions. Although it is impossible to do justice to all possible topics in this huge area of research, this book covers many that should be of interest, not only to those who study primate and primate genomes, but also for those wishing to understand human origins ("anthropogeny") and the remarkable phenotypic diversity of primates. Also included are somewhat more theoretical papers about issues of interest to other readers.

This valuable resource will undoubtedly catalyze further sequencing of primate genomes as well as studies of primate phenotypes. Thus, although we are in a "Post-Genome Era," we will also continue to be in the "Genome Era" for some time yet. Meanwhile, please enjoy reading this timely and informative volume.

Ajit Varki
Distinguished Professor of Medicine and Cellular & Molecular Medicine
Co-Director, Center for Academic Research and Training in Anthropogeny (CARTA)
Co-Director, Glycobiology Research and Training Center
University of California, San Diego, La Jolla, CA, USA

Contents

1 Introduction 1
Yasuhiro Go, Hiroo Imai, and Hirohisa Hirai

Part I Post-Genomic Approaches Toward Phenotype

**2 An Overview of Transcriptome Studies
in Non-Human Primates** 9
Naoki Osada

**3 The Role of Neoteny in Human Evolution:
From Genes to the Phenotype** 23
Mehmet Somel, Lin Tang, and Philipp Khaitovich

**4 Evolution of Chemosensory Receptor Genes
in Primates and Other Mammals** 43
Yoshihito Niimura

5 Functional Evolution of Primate Odorant Receptors 63
Kaylin A. Adipietro, Hiroaki Matsunami, and Hanyi Zhuang

**6 Post-Genome Biology of Primates Focusing
on Taste Perception** 79
Tohru Sugawara and Hiroo Imai

**7 Polymorphic Color Vision in Primates:
Evolutionary Considerations** 93
Shoji Kawamura, Chihiro Hiramatsu, Amanda D. Melin,
Colleen M. Schaffner, Filippo Aureli, and Linda M. Fedigan

Part II Genome Structure and Its Applications

8 Human-Specific Changes in Sialic Acid Biology 123
Toshiyuki Hayakawa and Ajit Varki

vii

viii

Contents

9 Duplicated Gene Evolution of the Primate Alcohol Dehydrogenase Family 149
Hiroki Oota and Kenneth K. Kidd

10 Genome Structure and Primate Evolution 163
Yoko Satta

11 Contribution of DNA-Based Transposable Elements to Genome Evolution: Inferences Drawn from Behavior of an Element Found in Fish 175
Akihiko Koga

12 Application of Phylogenetic Network 181
Takashi Kitano

Part III Chromosome Genomics

13 Comparative Primate Molecular Cytogenetics: Revealing Ancestral Genomes, Marker Order, and Evolutionary New Centromeres 193
Roscoe Stanyon, Nicoletta Archidiacono, and Mariano Rocchi

14 Chromosomal Evolution of Gibbons (Hylobatidae) 217
Stefan Müller and Johannes Wienberg

15 Evolution and Biological Meaning of Genomic Wastelands (RCRO): Proposal of Hypothesis 227
Hirohisa Hirai

Part IV Evolution of Humans and Non-Human Primates

16 Molecular Phylogeny and Evolution in Primates 243
Atsushi Matsui and Masami Hasegawa

17 Origins and Evolution of Early Primates 269
Masanaru Takai

Index 281

Contributors

Kaylin A. Adipietro (*Chapter 5*)
Department of Molecular Genetics and Microbiology, Duke University
Medical Center, Durham, NC, USA

Nicoletta Archidiacono (*Chapter 13*)
Department of Genetics and Microbiology, University of Bari, Bari, Italy

Filippo Aureli (*Chapter 7*)
Research Centre in Evolutionary Anthropology and Palaeoecology,
Liverpool John Moores University, Liverpool, UK
and
Instituto de Neuroetologia, Universidad Veracruzana, Xalapa, Mexico

Linda M. Fedigan (*Chapter 7*)
Department of Anthropology, University of Calgary, Calgary, AB, Canada

Yasuhiro Go (*Chapter 1*)
Primate Research Institute, Kyoto University, Inuyama, Aichi, Japan

Masami Hasegawa (*Chapter 16*)
School of Life Sciences, Fudan University, Shanghai, China
and
The Institute of Statistical Mathematics, Tachikawa, Tokyo, Japan

Toshiyuki Hayakawa (*Chapter 8*)
Center for Human Evolution Modeling Research, Primate Research Institute,
Kyoto University, Inuyama, Aichi, Japan

Hirohisa Hirai (*Chapter 1, 15*)
Primate Research Institute, Kyoto University, Inuyama, Aichi, Japan

Chihiro Hiramatsu (*Chapter 7*)
Department of Integrated Biosciences, Graduate School of Frontier Sciences,
The University of Tokyo, Kashiwa, Chiba, Japan
and
Department of Psychology, Graduate School of Letters, Kyoto University,
Kyoto, Japan

Hiroo Imai (*Chapter 1, 6*)
Primate Research Institute, Kyoto University, Inuyama, Aichi, Japan

Shoji Kawamura (*Chapter 7*)
Department of Integrated Biosciences, Graduate School of Frontier Sciences,
The University of Tokyo, Kashiwa, Chiba, Japan

Philipp Khaitovich (*Chapter 3*)
Partner Institute for Computational Biology, Shanghai Institutes
for Biological Sciences, Chinese Academy of Sciences, Shanghai, China
and
Max Planck Institute for Evolutionary Anthropology, Leipzig, Germany

Kenneth K. Kidd (*Chapter 9*)
Department of Genetics, Yale University School of Medicine,
New Haven, CT, USA

Takashi Kitano (*Chapter 12*)
Department of Biomolecular Functional Engineering, College of Engineering,
Ibaraki University, Hitachi, Japan

Akihiko Koga (*Chapter 11*)
Primate Research Institute, Kyoto University, Inuyama, Aichi, Japan

Stefan Müller (*Chapter 14*)
Institut für Humangenetik, Klinikum der Ludwig-Maximilians-Universität,
Munich, Germany

Atsushi Matsui (*Chapter 16*)
Department of Cellular and Molecular Biology, Primate Research Institute,
Kyoto University, Inuyama, Aichi, Japan

Hiroaki Matsunami (*Chapter 5*)
Department of Molecular Genetics and Microbiology, Duke University
Medical Center, Durham, NC, USA
and
Department of Neurobiology, Duke University Medical Center,
Durham, NC, USA

Amanda D. Melin (*Chapter 7*)
Department of Anthropology, University of Calgary, Calgary, AB, Canada
and
Department of Anthropology, Dartmouth College, Hanover, NH, USA

Contributors xi

Yoshihito Niimura (*Chapter 4*)
Medical Research Institute, Tokyo Medical and Dental University,
Tokyo, Japan

Hiroki Oota (*Chapter 9*)
Graduate School of Frontier Sciences, The University of Tokyo,
Kashiwa, Chiba, Japan
and
Kitasato University School of Medicine, Sagamihara, Kanagawa, Japan

Naoki Osada (*Chapter 2*)
Department of Population Genetics, National Institute of Genetics,
Shizuoka, Japan

Mariano Rocchi (*Chapter 13*)
Department of Genetics and Microbiology, University of Bari, Bari, Italy

Yoko Satta (*Chapter 10*)
Department of Evolutionary Studies of Biosystems, The Graduate University
for Advanced Studies, Sokendai Hayama, Hayama, Kanagawa, Japan

Colleen M. Schaffner (*Chapter 7*)
Psychology Department, University of Chester, Chester, UK
and
Instituto de Neuroetologia, Universidad Veracruzana, Xalapa, Mexico

Mehmet Somel (*Chapter 3*)
Partner Institute for Computational Biology, Shanghai Institutes
for Biological Sciences, Chinese Academy of Sciences, Shanghai, China
and
Max Planck Institute for Evolutionary Anthropology, Leipzig, Germany

Roscoe Stanyon (*Chapter 13*)
Laboratory of Anthropology, Department of Evolutionary Biology,
University of Florence, Florence, Italy

Tohru Sugawara (*Chapter 6*)
Primate Research Institute, Kyoto University, Inuyama, Aichi, Japan
and
Department of Reproductive Biology, National Research Institute
for Child Health and Development, Tokyo, Japan

Masanaru Takai (*Chapter 17*)
Primate Research Institute, Kyoto University, Inuyama, Aichi, Japan

Lin Tang (*Chapter 3*)
Partner Institute for Computational Biology, Shanghai Institutes
for Biological Sciences, Chinese Academy of Sciences, Shanghai, China

Ajit Varki (*Chapter 8*)
Center for Academic Research and Training in Anthropogeny, Glycobiology
Research and Training Center, Departments of Medicine and Cellular & Molecular
Medicine, University of California, San Diego, La Jolla, CA, USA

Johannes Wienberg (*Chapter 14*)
Anthropology and Human Genetics, Department of Biology II,
Ludwig-Maximilians-Universität, Munich, Germany
and
Chrombios GmbH, Raubling, Germany

Hanyi Zhuang (*Chapter 5*)
Department of Pathophysiology, Key Laboratory of Cell Differentiation
and Apoptosis of National Ministry of Education, Shanghai Jiaotong University
School of Medicine, Shanghai, China
and
Institute of Health Sciences, Shanghai Institutes for Biological Sciences
of Chinese Academy of Sciences and Shanghai Jiaotong University School
of Medicine, Shanghai, China

Chapter 1
Introduction

Yasuhiro Go, Hiroo Imai, and Hirohisa Hirai

1.1 Introduction

A decade ago, the first reports of the human draft genome were simultaneously published in *Nature* from the international Human Genome Project (International Human Genome Sequencing Consortium 2001) and in *Science* from the company Celera Genomics (Venter et al. 2001). Since the milestone of the human genome, genome projects of many organisms have been proposed, undertaken, and achieved in the past decade. These organisms include the mouse (Mouse Genome Sequencing Consortium 2002), rat (Rat Genome Sequencing Project Consortium 2004), dog (Lindblad-Toh et al. 2005), chimpanzee (Chimpanzee Sequencing and Analysis Consortium 2005), rhesus macaque (Rhesus Macaque Genome Sequencing and Analysis Consortium 2007), marsupial (Mikkelsen et al. 2007), and, more recently, the Neanderthal (Green et al. 2010). As for primates, besides the chimpanzee and rhesus macaque, many other primate genomes have been sequenced, such as gorilla, orangutan, gibbon, baboon, marmoset, tarsier, galago, and lemur. New insights are thus required to think about how we should use the vast information of genome sequences for post-genome investigations. Now is the best time to establish standpoints for genomic primatology in these early days in several areas of genomic research. Here we introduce the angles from which we investigate primates with the aim of understanding what makes us human.

This book consists of four sections: each has two to six chapters, as listed in the Table of Contents.

The first section is "Post-Genomic Approaches Toward Phenotype," in which we introduce approaches to uncover phenotypic changes, including physiological and behavioral changes, based on the genomic and transcriptome level, especially focusing

---

Y. Go • H. Imai • H. Hirai (✉)
Primate Research Institute, Kyoto University,
41-2 Kanrin, Inuyama, Aichi 484-8506, Japan
e-mail: yago@pri.kyoto-u.ac.jp; imai@pri.kyoto-u.ac.jp; hhirai@pri.kyoto-u.ac.jp

H. Hirai et al. (eds.), *Post-Genome Biology of Primates*, Primatology Monographs,
DOI 10.1007/978-4-431-54011-3_1, © Springer 2012

on the sensory functions. The sensory functions that perceive various external signals, such as light, smell, and taste, play major roles in sensing physical or chemical environmental changes and in taking such information inside the organisms. The necessity for such sensory functions for each organism could be variable and heavily dependent on the environment to which each organism has adapted. This feature thus results in producing functional diversity of the sensory functions from organism to organism, and this diversity has conferred various species-specific phenotypic characters. The first chapter of this section, Chap. 2, is written by Naoki Osada from National Institute of Genetics. He highlights the importance of the transcriptome, a first outcome of the genome and a key component linking the genotype and phenotype of an organism, and introduces the recent advance of transcriptome studies in nonhuman primates and the quantification methodology for the transcriptome. The next chapter, Chap. 3, is a review by Mehmet Somel et al. from the Chinese Academy of Sciences and German Max Planck Society. They draw attention to one of the most distinguishing features that characterizes humans as distinct from the other primates, so-called neoteny. Neoteny is a form of heterochrony that is defined as a developmental change in the timing of events, leading to changes in size and shape. The authors examine whether human-specific changes can be seen and what kind of genes are involved in the molecular basis of neoteny using brain transcriptome results of humans, chimpanzees, and rhesus macaques covering almost all ages. In the third chapter, Chap. 4, Yoshihito Niimura from Tokyo Medical and Dental University reviews chemosensory receptor gene evolution in primates and mammals. Among chemosensory receptor genes, olfactory receptor genes are the largest multigene family in the mammalian genome, and the number of genes differs greatly among species (~1,000 genes in rodents, but fewer than 400 in primates). Niimura argues for a dynamic change of the repertoire of the olfactory receptor genes in the context of trade-off between vision and the olfaction system. The fourth chapter of this section, Chap. 5, is a review by Kaylin Adipietro et al. from Duke University. Although they also examine the evolution of odorant (olfactory) receptors in primates, as in the previous chapter, their studies are based on a more functional or system point of view. Using in vitro functional assays of the ligand sensitivity of odorant receptors and behavioral evaluation of responses to a set of smells, they found different responses to a set of ligands or smells among very closely related species and even between sexes in humans. This discovery in general implies that the response to some smells might be modulated at the transcriptional, metabolic, or epigenetic level. In the fifth chapter, Chap. 6, Tohru Sugawara and Hiroo Imai from Kyoto University highlight another important chemosensory receptor involved in taste perception. Of five taste modalities (sweet, bitter, sour, salty, and umami), the sense of bitter taste is known to be highly polymorphic. For instance, although people can taste some bitter compounds, such as phenylthiocarbamide (PTC), there are people who cannot sense the same bitter compound (called "nontasters"). The genetic basis of this polymorphism has recently been attributed to one of the bitter taste receptor genes (*T2R38*). It is also known that such behavioral polymorphisms are observed not only in humans but also in some other primates. The authors then examined the genetic basis of such polymorphisms in chimpanzees and uncovered the evolutionary origin and significance of such "non-tasters" in humans and chimpanzees. They also give us a nice review of the function and evolution of other taste-related genes. The final chapter of this section, Chap. 7, is a review by Shoji Kawamura et al. from the University of Tokyo on the evolution of the color vision system in primates. Color vision is a crucial cue for object detection, food identification, mate choice, and predator avoidance. Kawamura et al. show an evolutionary significance of the color vision (opsins) system in primates and examine precisely to what extent three-color vision (trichromacy) has an advantage over two-color vision (dichromacy) in the environment of free-ranging living primates. As New World monkeys are unique with respect to the existence of dichromatic and trichromatic individuals in a species or even in a group, they examine the genotypes of opsin genes in free-ranging New World monkeys and record each individual's behavioral data. Connecting the genetic and behavioral data in the wild monkeys, they uncover the advantages and disadvantages of trichromacy in the environmental context. This study gives us one of the best examples of how we can incorporate multidisciplinary approaches in the post-genome era of biology.

The second section, including chapters on "Genome Structure and Its Applications," explores the impact of genomic structural changes on human evolution. The first chapter, Chap. 8, is on the evolution of sialic acids, one of the important components of sugar chains. Toshiyuki Hayakawa from Kyoto University and Ajit Varki from the University of California, San Diego, give an overview of the evolution of sialic acid biology in primates and highlight sialic acid-related human-specific changes and their possible impact on human evolution. The discovery of such human-specific genetic changes is one of the hallmarks of human uniqueness and can be a good clue for thinking about a longstanding question: What makes us human? The second chapter of this section, Chap. 9, describes the evolution of the genes involved in alcohol metabolism in primates. Hiroki Oota from Kitasato University and Kenneth Kidd from Yale University show copy number variation of the alcohol dehydrogenase (*ADH*) gene in detail and disclose the independent origin of each *ADH* gene between apes and Old World monkeys. Based on the findings, they hypothesize that frugivorous feeding behavior facilitates the maintenance of taxon-specific duplicated genes because of the necessity of digesting ethanol generated by the fermentation of fruit sugar. In the third chapter, Chap. 10, Yoko Satta from The Graduate University for Advanced Studies (Sokendai) gives us a review of genome structure evolution especially focused on sex chromosomes. Because the genome sequencing projects covered a wide range of organisms, many types of structural changes, such as segmental duplications, copy number variations, and insertions and deletions, have been discerned and quantified. Among the genome (chromosomes), the Y chromosome is exceptional because it exists as a hemizygous chromosome in the genome and most of the mutations that could be deleterious are not then eliminated as a result of the arrest of recombination with the X chromosome. She shows the discontinuous structure of the human Y chromosome with respect to the evolutionary relationships of gametologous (homologous relationships between sex chromosomes) genes on the X chromosome and discusses the evolutionary origin and biological significance of sex chromosomes in the light of human evolution. Akihiko Koga from Kyoto University reviews the impact of DNA-based transposed elements (DTEs) on the genome and their evolution in the fourth chapter, Chap. 11. Although most DTEs are thought to be dead in mammals, DTEs can trigger chromosomal rearrangements such as inversions, deletions, duplications, and translocations because of their repetitive nature. Koga discusses the potential contribution of DTEs to mammalian genome evolution. The last chapter of this section, Chap. 12, is a review by Takashi Kitano from Ibaraki University. He argues the possibility and extensibility of the phylogenetic network, an extended framework of the phylogenetic tree. Phylogenetic network methods have advantages of describing the genes with complex evolutionary genealogies resulting from processes such as recombination, hybridization, and gene conversion. He also introduces some practical applications using the phylogenetic network method.

The third section, "Chromosome Genomics," concerns molecular cytogenetics and chromosome evolution in primates. Classical comparative cytogenetics has a long history dating back to the 1950s, and since then it has used the information of the number of chromosomes (karyotypes), and chromosome banding such as Q-banding and G-banding. During the past 20 years, the introduction of molecular methods has made it possible to examine precise chromosome rearrangements among species, as revealed by the fluorescent in situ hybridization (FISH) method. Moreover, cytogenetic studies have also uncovered the mechanisms and biological meaning of the essential components of chromosomes, such as the centromere and telomere. Even now, these components are difficult to sequence by the ordinary genomic approaches as a consequence of the highly repetitive nature of their sequences. The first chapter of this section, Chap. 13, is a review by Roscoe Stanyon et al. from University of Florence. Focusing on the cytogenetic level of primate genome organization as shown by the chromosome painting method, they reveal the complex chromosome rearrangements that occurred during primate evolution and reconstruct the ancestral genome organizations. Moreover, they show intriguing phenomena of neocentromeres, which are newly formed in ectopic chromosomal regions and are even heritable in some cases, and the meaning of such neocentromeres from an evolutionary point of view. They also give a perspective of the future and possibilities of cytogenetics in the high-throughput genomic era. In the second chapter, Chap. 14, Stefan Müller and Johannes Wienberg from The Ludwig Maximilian University of Munich highlight chromosomal evolution in gibbons, one of the organisms with the highest rate of chromosome rearrangements in mammals. Although gibbons are classified as lesser or smaller apes and are phylogenetically close relatives, they diverged from humans and great apes 15–20 million years ago, and the rates of rearrangements in gibbons are 10–20 times higher than the mammalian default rate. They examine the evolutionary relationship of the highly differentiated chromosomal organizations in the four genera of gibbons (chromosome numbers from 2n=38 to 2n=52) based on chromosome painting methods such as those covered in the previous chapter (Chap. 13). In addition, they summarize the recent progress of elucidating the cause of the higher rate of rearrangements with respect to the epigenomic changes occurring in this group. The final chapter, Chap. 15, is by Hirohisa Hirai from Kyoto University, who discusses the evolution and biological meaning of so-called genomic wastelands, mainly constructed from repetitive sequences in heterochromatic regions. Although humans and chimpanzees are reported to share approximately 99% of their genome sequences, this was calculated from alignable sequences of the genome. When one considers unalignable regions such as insertions and deletions, the difference of the genome between humans and chimpanzees is estimated to be about 3%. One such genomic component is heterochromatic regions that are usually enriched in centromere, telomere, and subtelomeric regions. He then argues how such heterochromatic regions (genomic wastelands) contribute to make us humans.

The fourth and last part of this book, "Evolution of Humans and Non-Human Primates," addresses the topic of primate evolution from the molecular and fossil points of view. In the first chapter, Chap. 16, Atsushi Matsui from Kyoto University and Masami Hasegawa from Fudan University summarize the recent advances of molecular primate phylogeny and point out remaining unsolved problems of molecular phylogenetic studies. Although the phylogenetic relationships of living primates are relatively well established, the divergence times among them are still controversial. They comprehensively examine the divergence time in each taxon and discuss the evolutionary scenario of primate evolution with reference to the geographic and fossil records. The last chapter, Chap. 17, is a review of primate fossil studies contributed by Masanaru Takai from Kyoto University. He especially focuses on the early time of primate evolution and draws conclusions about the place of primate origin. The North America origin hypothesis has long been accepted and is widespread among primatologists and paleontologists because of the rich fossil records of early primates. However, the author advocates that this view should be reconsidered and can be replaced by the southern continent origin hypothesis involving the Indian Continent or East Asia, based on incorporating the results from geographic evidence and recent molecular phylogenetic studies.

It is clear that this book does not completely cover the comprehensive fields of genome and post-genome biology in primates. Instead, we intended to organize the contents of the book to show front-line research for broadening one's insights and extending one's research interests incorporating various methods, technologies, and knowledge, as shown in this volume.

We are truly grateful to all the authors of this book for devoting their time to write the chapters and contributing to several refinements of the book. Thanks to all the effort, we are proud of publishing this book. We would also like to thank the series editors, Tetsuro Matsuzawa and Juichi Yamagiwa, for their special leadership and continuous support, and we thank Aiko Hiraguchi and Kaoru Hashimoro of Springer Japan for their dedicated assistance with the editing of this book. Finally, we give special thanks to numerous colleagues, postdoctoral researchers, and students. Without them, this book could never have been accomplished.

References

Chimpanzee Sequencing and Analysis Consortium (2005) Initial sequence of the chimpanzee genome and comparison with the human genome. Nature (Lond) 437:69–87

Green RE, Krause J, Briggs AW et al (2010) A draft sequence of the Neandertal genome. Science 328:710–722

International Human Genome Sequencing Consortium (2001) Initial sequencing and analysis of the human genome. Nature (Lond) 409:860–921

Lindblad-Toh K, Wade CM, Mikkelsen TS et al (2005) Genome sequence, comparative analysis and haplotype structure of the domestic dog. Nature (Lond) 438:803–819

Mikkelsen TS, Mikkelsen TS, Wakefield MJ et al (2007) Genome of the marsupial *Monodelphis domestica* reveals innovation in non-coding sequences. Nature (Lond) 447:167–177

Mouse Genome Sequencing Consortium (2002) Initial sequencing and comparative analysis of the mouse genome. Nature (Lond) 420:520–562

Rat Genome Sequencing Project Consortium (2004) Genome sequence of the brown Norway rat yields insights into mammalian evolution. Nature (Lond) 428:493–521

Rhesus Macaque Genome Sequencing and Analysis Consortium (2007) Evolutionary and biomedical insights from the rhesus macaque genome. Science 316:222–234

Venter JC, Venter JC, Adams MD et al (2001) The sequence of the human genome. Science 291:1304–1351

Part I
Post-Genomic Approaches
Toward Phenotype

Chapter 2
An Overview of Transcriptome Studies in Non-Human Primates

Naoki Osada

Abbreviations

cDNA   Complementary DNA
EST   Expressed sequence tag
SAGE   Serial analysis of gene expression

2.1 What Is a Transcriptome?

The word *transcriptome* is a combination of *transcript* and *genome*, which refers to the whole set of transcripts expressed in a cell or tissue. The word *genome* itself is a blend of *gene* and *chromosome*, which refers to the whole set of genes in an organism. Now the ending *-ome* has been applied somewhat excessively to represent any kind of massive biological dataset, for example, proteome, metabolome, phenome, interactome, and phylome. In a classical view, transcripts are equivalent to messenger RNAs (mRNAs) that encode functional proteins. However, recent progress in transcriptome analysis has demonstrated that a large number of noncoding sequences are transcribed to RNA, more than previously estimated. For example, a very deep sequencing of RNA expressed in mice revealed that more than 70% of the genome is actually transcribed to RNA, if introns are included (Carninci et al. 2005). The functions of some transcribed RNAs, such as micro-RNA and small interfering (si)RNA, have been extensively studied, whereas those of most of the noncoding RNAs remain unclear.

N. Osada (✉)
Department of Population Genetics, National Institute of Genetics,
1111 Yata, Mishima, Shizuoka 411-8540, Japan
e-mail: nosada@lab.nig.ac.jp

H. Hirai et al. (eds.), *Post-Genome Biology of Primates*, Primatology Monographs,
DOI 10.1007/978-4-431-54011-3_2, © Springer 2012

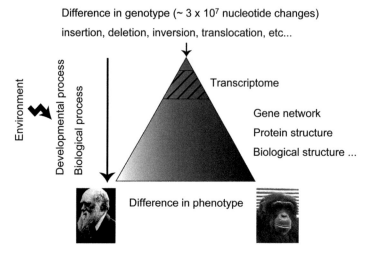

Fig. 2.1 A conceptual picture of the genotypic and phenotypic differences between humans and chimpanzees

Transcriptome data enable us to link the genotype to the phenotype of an organism. Figure 2.1 represents a conceptual picture of the genotypic and phenotypic differences between humans and chimpanzees. The human and chimpanzee genomes differ by about 1% at the nucleotide substitution level, and more at other levels including insertion, deletion, inversion, and translocation. Although not all these changes have been thoroughly cataloged, they are finite and countable features. These small genotypic differences increase according to their developmental and biological processes and by interactions with the environment of the organism, which leads to large phenotypic differences such as morphology and behavior that can be easily recognized. Compared to genotypic differences, phenotypic differences are far more complex. One can measure and compare some phenotypes, but the number of measurable phenotypes is very large (probably an infinite number). In addition, there are many cryptic phenotypes in organisms, which make the comparison of whole sets of phenotypes (phenomes) almost impossible. Transcriptome studies can be more complex than genomic studies, because a transcriptome may be a quantitative measurement and may vary in space and time (see also Chap. 3). However, compared to a phenotypic measurement, a transcriptome measurement could be a more neutral measurement of the feature of organisms. The transcriptome is sometimes referred to as an endophenotype. In other words, the transcriptome represents the very beginning of genotypic development into a complex of phenotypes.

For a long time, transcriptome studies were restricted to well-known model organisms. However, recent advances in molecular genetic techniques enable us to extend transcriptome studies to many non-model organisms, including non-human primates. Transcriptome studies in non-human primates have been strongly promoted for two major purposes: for conducting evolutionary and biomedical studies. In this review, the methodologies used in transcriptome studies and some results obtained from previous studies have been summarized.

2.2 Different Methods for Analyzing Transcriptome

2.2.1 Qualitative/Quantitative Studies

According to the type of data analyzed, transcriptome studies are classified into qualitative and quantitative studies. Qualitative studies analyze the sequence and structural differences among genes or the repertoire of expressed genes in samples, whereas quantitative studies evaluate the level of gene expression of many genes. Two major methods are widely used to analyze these different aspects of transcriptome studies. The underlying technologies, advantages, and disadvantages of these methods are briefly discussed in the following sections.

2.2.2 cDNA Sequencing

The first method is classical complementary DNA (cDNA) sequencing. mRNA in a sample is reverse transcribed to cDNA and cloned into plasmid vectors. A set of hosts (usually *Escherichia coli*) containing the vectors are called cDNA libraries. Because gene expression patterns differ by species, individual, sex, tissue, time course, and experimental treatment, many different types of cDNA libraries can be established.

In a practical transcriptome study by cDNA sequencing, hundreds to thousands of clones are randomly obtained from cDNA libraries, and their sequences are determined with a DNA sequencer to catalog many genes expressed in a target tissue. Usually, cDNA sequences at either the 5′-end or 3′-end are determined. These one-pass sequences are called expressed sequence tag (EST) sequences. EST sequences provide relatively limited information because the currently used Sanger sequencing method reads less than 1,000 bp and the average length of primate genes is greater than 2,000 bp. After EST sequencing has been performed, one may determine a full insert of clones by further sequencing with the primer-walking or shotgun methods. With standard cDNA library construction methods, obtained EST sequences are highly redundant, which signifies that many EST sequences represent the same gene. Therefore, the normalization method, which subtracts highly redundant mRNAs from a sample, is sometimes performed to construct cDNA libraries. Because most protein-coding genes in primates are highly conserved, one can investigate whether the obtained EST sequences are homologous to known human protein-coding genes by a homology search of public databases using a program such as BLAST.

There is a modified form of the conventional EST sequencing method, called serial analysis of gene expression (SAGE), in which short-sequence fragments cleaved by restriction enzymes (~15–30 bp) from the 5′- or 3′-end of cDNAs are concatenated and analyzed. The concatenated tags are sequenced with a DNA sequencer. Because a single one-pass sequence can simultaneously identify several tags, this method can determine sequence tags more efficiently than the conventional

EST sequencing method. Because the number of tags in SAGE analysis outnumbers that in ESTs, the tag count can be treated as a quantitative measure of the gene expression level. Unfortunately, this method is not as useful for studies involving non-human primate samples. Because the sequence tags of the SAGE method are much shorter than the EST sequencing, changes in even a few nucleotides may cause the misidentification of tags. The link between tags and genes is mainly based on the human genome data. Therefore, SAGE is not efficient in non-human primate studies unless there is prior information about which tag represents which gene.

Several cDNA libraries derived from non-human primates have been constructed and analyzed; for example, from chimpanzees (*Pan troglodytes*) (Hellmann et al. 2003; Sakate et al. 2003), orangutans (*Pongo pygmaeus*) (Mewes et al. 2004), rhesus macaques (*Macaca mulatta*) (Magness et al. 2005; Spindel et al. 2005), cynomolgus macaques (*Macaca fascicularis*) (Hida et al. 2000; Magness et al. 2005; Chen et al. 2006; Osada et al. 2008, 2009), pigtail macaques (*Macaca nemestrina*) (Magness et al. 2005), and common marmosets (*Callithrix jacchus*) (Datson et al. 2007). The EST sequence data of cynomolgus monkeys obtained by our research group is one of the largest non-human primate transcriptome datasets, containing 112,587 EST sequences (Osada et al. 2009). These EST sequences or cDNA clones can be used for the generation of DNA microarrays. Unfortunately, the EST data are not suitable for quantitative measurement of the gene expression level because a limited number of tags are counted. These one-pass sequences were error prone, but can be used for comparative studies of primate genes if many sequences per gene are obtained. Before the advent of non-human primate genome sequences, cDNA sequence comparison was the only method to compare many genes between primate species. Some interesting results of comparative genomics using transcriptome data are discussed later.

2.2.3 DNA Microarray

The second method uses DNA microarrays. In DNA microarray experiments, mRNAs are reverse transcribed to cDNAs, labeled, and hybridized to cDNAs or synthesized oligo-DNAs that are arrayed densely on a glass surface. The number of mRNA molecules in a sample can be measured from the relative signal intensity of each labeled probe.

cDNA microarrays are constructed by spotting cDNA clones, whose sequences are known by EST or full-length sequencing. On the other hand, oligo-DNA microarrays are constructed by spotting or synthesizing short oligonucleotides on a glass slide. Typical oligonucleotide probes for microarrays are 30–120-mers that are complementary to transcript sequences. In oligo-DNA microarrays, recent techniques can design multiple probes for a single transcript. If different probes are designed for different exons, then the microarray can detect a different expression level among alternatively spliced transcripts. For non-human primate microarrays, sequences from EST data are frequently used to design oligo-DNA microarrays.

In addition, there is a special type of DNA microarray called a tiling array. Oligo-DNA probes on tiling arrays are designed to cover a wide range of the whole genome sequence. Indeed, genome sequences of target regions must be determined to design a tiling array. Although a microarray based on transcript sequences cannot detect the expression of unknown genes, a tiling array can detect the expression of unknown transcripts in the genome, including that of noncoding RNAs and micro-RNAs. Studies using tiling arrays have established a very complex expression pattern in the human transcriptome (Bertone et al. 2004; Kampa et al. 2004).

Although a DNA microarray is a powerful tool to quantitatively analyze the transcriptome, there are some drawbacks of DNA microarray experiments in non-human primates. A species-specific DNA microarray should be designed in advance, because the DNA–DNA hybridization accuracy depends on sequence similarity between sample species and species used to design microarrays. This effect may cause serious problems, particularly cross-species expression comparison, because it is difficult to identify whether the observed changes in signal intensity are caused by sequence mismatches or changes in actual gene expression. If sequence divergence between species is relatively small, as between humans and chimpanzees, then its effect may be negligible. However, the effect becomes more pronounced when the target species are distantly related to microarray species. Because the effect of hybridization mismatches is much stronger in shorter probes, cDNA microarrays are supposed to be more robust to sequence mismatches than oligo-DNA microarrays (Walker et al. 2006; Jacquelin et al. 2007). Before the draft genome sequence of the rhesus macaque was determined, many studies used commercially available or custom-made human microarrays. Therefore, early studies of gene expression in non-human primates were based on human-specific microarrays (Zou et al. 2002; Marvanova et al. 2003; Sui et al. 2003; Vahey et al. 2003; Baskin et al. 2004; Rubins et al. 2004; Dillman and Phillips 2005; Ylostalo et al. 2005; Kothapalli et al. 2007; Nijland et al. 2007; Djavani et al. 2009). Although these studies obtained satisfactory results at some level, microarrays designed specifically for the particular non-human primate species would produce a much more accurate estimation of the gene expression level (Gilad et al. 2005). Up to the present, several microarrays specific to non-human primates have been developed (Osada et al. 2002; Gilad et al. 2005; Spindel et al. 2005; Datson et al. 2007; Kobasa et al. 2007; Wallace et al. 2007; Osada et al. 2008). Most of them are intended for biomedical research in non-human primates. At present, several DNA microarrays for humans and macaques are commercially available, and bioinformatics methods that mitigate the effect of sequence mismatches in oligo-DNA microarrays have been developed (Wang et al. 2004; Royce et al. 2007; Lin et al. 2009; Lu et al. 2009).

2.2.4 Next-Generation Sequencer

Recently, new DNA sequencing technologies that can identify a large number of short DNA fragments have been developed. Three different platforms are currently available: FLX (Roche), Solexa GA (Illumina), and AB SOLiD (Life Sciences). The technologies used by these platforms differ, but each can identify millions to billions of transcript fragments in a single run. Because the number of detectable fragments is enormous, the counted fragments would correlate with the level of gene expression, as in SAGE analysis. Many studies have suggested that these new methods are capable of quantifying very low gene expression levels. For example, mRNAs that are expressed at less than one copy per cell were detectable (Hashimoto et al. 2009). Initial versions of GA and SOLiD produced only 25-bp-long sequences, and thus were not as efficient as the SAGE method for analyzing non-human primate data. The sequencing length, however, has been increased to 50–100 bp, which enables us to overcome the sequence mismatch problem. Therefore, these methods can identify a non-human primate transcriptome both qualitatively and quantitatively. If we have known genome sequences of target organisms, fragments can be easily mapped on the genome sequences; otherwise, the assembly of fragments would be more challenging. Such an assembly is designated as de novo transcriptome sequencing. Although only a limited number of non-human primate research studies have been performed using these methods, the technique has a great potential for investigating the transcriptome of non-human primates.

2.3 Subjects of Transcriptome Studies

2.3.1 *Application to Biomedical Research*

Many non-human primates are used as a model for humans in biomedical research. Biomedical studies include studies on infectious diseases, tissue transplantation, neurology, toxicology, and many other human diseases. In particular, pharmaceutical studies using genome-wide gene expression data are referred to as toxicogenomics and are of interest to many pharmaceutical researchers. The most popular non-human primates for biomedical research are the Old World monkeys such as macaques and baboons. Among New World monkeys, the marmoset is the most popular animal because it has a small body size and grows relatively fast. Biomedical research using invasive treatments in apes is strongly restricted because of ethical reasons. Even in other non-human primates, the investigational use of monkeys has been a debatable issue for a long time from the point of view of animal rights (Editorial 2008), but that question is not discussed in this review.

In typical biomedical studies, differences in gene expression after certain treatments are measured using DNA microarrays to identify the gene relevant to the biological response. A sampling point may be a time course for measuring temporal changes in gene expression. Because most biomedical studies try to detect differences in gene expression patterns between experimental and control samples, the sequencing mismatch problem, which was described in the previous section, is less problematic. Sequence mismatches may reduce the probe detection efficiency, but will have a lesser effect on reproducibility in intraspecies comparison. Therefore, in many biomedical studies, non-human primate cDNA was hybridized to human-specific DNA microarrays (Zou et al. 2002; Marvanova et al. 2003; Sui et al. 2003; Vahey et al. 2003; Baskin et al. 2004; Rubins et al. 2004; Dillman and Phillips 2005; Ylostalo et al. 2005; Kothapalli et al. 2007; Nijland et al. 2007; Djavani et al. 2009).

2.3.2 Comparative Studies

Besides biomedical research, research comparing humans and non-human primates has attracted much attention of evolutionary biologists. It is not clear as to what kind of genetic components make humans phenotypically distinct from other non-human primates, especially in their high cognitive ability. Understanding humanity from a genomic perspective is a challenging but tantalizing issue in human evolutionary biology studies. In 1967, Sarich and Wilson used immunological reactions to investigate the similarity in protein structures of albumin (Sarich and Wilson 1967). They used these data to date the divergence between gorillas, chimpanzees, and humans, and concluded that their divergence was much more recent (~5 Mya) than was previously thought from morphological and fossil evidence (~15 Mya) (see also Chap. 16). At present, the very close relationship between humans and chimpanzees is supported by enormous amounts of DNA sequence data. It is now known that only approximately 1% of their genomes differ at the DNA sequence level (The Chimpanzee Sequencing and Analysis Consortium 2005).

2.3.2.1 Molecular Evolution Rate of the Primate Transcriptome

Using a comparative transcriptome analysis, which contrasts whole sets of genes among genomes, genes responsible for human-specific traits could be identified. A comparison between human and chimpanzee genome sequences revealed a difference of about 40,000 amino acids in their protein sequences (The Chimpanzee Sequencing and Analysis Consortium 2005). Many of these differences are probably neutral, that is, have no phenotypic effect, but some of them may have been affected by positive or negative selection. Here, positive and negative selection means natural selection on beneficial and deleterious mutations, respectively. It is intuitive that mutations causing beneficial phenotypic changes quickly spread within populations, whereas bad mutations are easily removed from populations. Mutations that are neither good nor bad are assumed to be selectively neutral.

To estimate the mode of protein evolution, the relative rate of protein evolution was measured by observing synonymous and nonsynonymous changes between species. Synonymous substitutions are nucleotide changes that do not affect encoded protein sequences, whereas nonsynonymous substitutions are nucleotide changes that alter encoded proteins. The rate of nucleotide substitution per site for synonymous (K_S) and nonsynonymous (K_A) substitutions can be estimated using several statistical methods. It can be assumed that synonymous substitutions are mostly selectively neutral, although it is known that weak negative selection caused by translational efficiency may act on synonymous substitutions (Akashi 1994). K_A and K_S are also referred as d_N and d_S, respectively (see also Chap. 4).

If synonymous substitutions are selectively neutral, K_S equals the mutation rate under neutral theory of molecular evolution (Kimura 1968). Neutral theory of molecular evolution also predicts that if the changes in the protein are largely deleterious, K_A becomes much smaller than K_S. Comparison between human and chimpanzee genomes revealed that the average K_A/K_S was around 0.20–0.25, indicating that about 70–80% of amino acid changes in humans and chimpanzees are deleterious. However, the intensity of natural selection (estimated by K_A/K_S) is different for different genes. If some gene is biologically important and the negative selection intensity of the gene is very severe, the K_A/K_S value of the gene becomes extremely small. For example, humans and mice share identical amino acid sequences for one of the histone proteins, H4, which is a fundamental protein constituting chromosome structure. On the other hand, if many new mutations in the protein are beneficial to the organism, those mutations spread rapidly and become fixed in the species. Then, it is assumed that K_A exceeds K_S ($K_A/K_S > 1$) for such a gene. Thus, comparison of human and chimpanzee transcriptome data at a nucleotide level may be useful to identify which genes have the greatest impact on the human–chimpanzee divergence.

Before the genomic era, when no non-human primate genome sequences had been determined, the only way to catalog a large number of genes in non-human primate genomes was by the analysis of cDNA libraries. Hellmann et al. (2003) constructed cDNA libraries derived from chimpanzee brain and testis, sequenced about 5,000 EST sequences from the libraries, and compared the sequences with those of humans. In conjunction with the human polymorphism data, they estimated the level of negative selection on genes after the human–chimpanzee divergence. Similarly, Sakate et al. (2003) constructed cDNA libraries derived from chimpanzee brain, skin, and liver, and estimated the molecular evolution rate of hundreds of genes. These two studies were the first large-scale comparisons of human–chimp genes.

Using the Old World monkeys, Osada et al. (2002) attempted to discover genes that have evolved rapidly after the human–macaque divergence, with about 10,000 EST sequences of cynomolgus macaques. They identified eight candidate genes that showed $K_A/K_S > 1$. Interestingly, four of these are nuclear genes that encode mitochondrial components. By analyzing many other mitochondrial genes, Goodman and colleagues hypothesized that the rapid evolution of mitochondrial component genes may be responsible for the development of brains in the ape lineage, which consume much more energy than other organs (Grossman et al. 2004).

2.3.2.2 Finding Rapidly Evolving Genes Between Humans and Chimpanzees

After the chimpanzee and macaque genome sequences were published (The Chimpanzee Sequencing and Analysis Consortium 2005; Gibbs et al. 2007), genome-wide comparison of transcript sequences became easier and much more popular. Combined with human polymorphism data, recent studies have established that several genes related to brain function evolved rapidly under positive selection after the human–chimpanzee divergence. These genes include the forkhead box P2 (*FOXP2*) (Enard et al. 2002b), abnormal spindle homologue, microcephaly associated (*ASPM*) (Zhang 2003; Mekel-Bobrov et al. 2005), and microcephalin 1 (*MCPH1*) (Evans et al. 2005). *FOXP2* is known to be involved in human genetic diseases related to vocalization. The other two genes are involved in brain development. Laboratory experiments in mice carrying genetically modified *FOXP2* revealed that disruption or mutations in *FOXP2* changed the ultrasonic vocalization and behavior of mice (Shu et al. 2005; Fujita et al. 2008; Groszer et al. 2008). The functions of these genes and their molecular evolution pattern suggest that they are good candidates for contributing to human-specific cognitive abilities.

These studies demonstrated that the function of some genes in the brain may have rapidly evolved under positive selection in humans. Then, what about the other brain-expressed genes? Dorus et al. (2004) reported that hundreds of genes related to the nervous system have evolved more rapidly in the human lineage than in macaques. This finding suggests that not a few neuronal genes have evolved under positive selection. On analyzing brain-expressed genes more extensively, however, researchers observed an opposite trend and concluded that brain-expressed genes may have evolved more slowly in the human lineage than in chimpanzees (Wang et al. 2007). The nearly neutral theory of molecular evolution predicts that any set of genes could have evolved more rapidly in the human lineage than in chimpanzees simply because of the smaller effective size of the human population, in which the elimination of slightly deleterious mutations is less effective (Ohta 1973). Therefore, to find a relative evolution rate in brain-expressed genes, it is necessary to calibrate rate with respect to the genome-wide average. The results of Wang et al. (2007) showed that although some functionally important genes have evolved rapidly because of positive selection, most brain-expressed genes have evolved rather slowly in the human lineage. This idea has been explained by the hypothesis that in tissues that have a more complex gene network, functional constraints on genes are stronger than in tissues which have a simpler gene network. As a result, protein sequences of brain-expressed genes in humans have evolved at an extremely slow rate. A similar trend has been reported by Shi et al. (2006). These findings, in turn, suggest that the changes in gene expression may be more important than those in protein sequences for the evolution of human brain. The pattern of gene expression evolution in the human brain is summarized later.

Although the brain is one of the most fascinating organs in humans, another class of genes – the immune genes – have been found to evolve rapidly according to many genome-wide comparisons between human and chimpanzee genomes (Clark et al. 2003; The Chimpanzee Sequencing and Analysis Consortium 2005; Nielsen et al. 2005). Pathogens evolve to adapt to the host immune system, and the host immune systems evolve to defend against pathogens. This process is sometimes referred to as an arms race. Indeed, many pathogens are known to specifically infect humans and not chimpanzees. Some pathogens, such as human immunodeficiency virus

(HIV), infect both humans and chimpanzees, but the symptoms of these infections differ between the species. Therefore, immune system genes are likely to evolve very rapidly under positive selection. One example is the rapid evolution of glycophorin proteins, which are surface proteins of red blood cells and are subject to infection by malaria. Wang et al. (2003) compared cDNA sequences of cynomolgus macaques with human genes and demonstrated very rapid molecular evolution of the glycophorin C (*GYPC*) gene after the divergence of humans and Old World monkeys.

It is necessary to mention that there are several limitations in using K_A/K_S-based methods to identify rapidly evolving genes. Because K_A/K_S may be elevated by relaxation of selective constraints, one must show that K_A/K_S is greater than 1 with statistical significance. In addition, because the observed numbers of substitutions between humans and chimpanzees are often very small, K_A/K_S frequently exceeds 1 by statistical fluctuation without any positive selective force. Many sophisticated statistical methods based on probabilistic nucleotide substitution models that could estimate selection intensity at a particular codon site and evolutionary lineage have been developed (Zhang et al. 2005). Model-based methods are usually preferred because they can be evaluated with a rigorous statistical framework. However, the model-based methods are appropriate only when the assumption in the model has not deviated from what is actually occurring. Indeed, the violation of assumption is problematic in any statistical test. For example, many methods assume an equilibrium state of a nucleotide substation process. That is, the composition of the four nucleotides in a genome and the mutation rates of each nucleotide are constant over a long time. If this assumption is violated, K_A/K_S may exceed 1 by many other factors without invoking a positive selection such as by biased gene conversion with relaxation of the selection constraint (Berglund et al. 2009).

2.3.2.3 Comparative Analysis of Gene Expression Using Microarrays

The comparison of human and chimpanzee transcriptome sequences demonstrated that although some genes may have evolved rapidly under positive selection, the vast majority of genes evolved in a conservative manner. In 1975, King and Wilson proposed that differences in gene expression are more important than those in protein structure for determining phenotypic differences between humans and chimpanzees (King and Wilson 1975). Their insight was quite reasonable, but their hypothesis had to wait a long time before advances in molecular biology techniques made it possible to support it with experimental data. DNA microarrays had a great impact on this research area. These comparative gene expression studies are not described in detail here, but some of the interesting findings are briefly summarized.

Several independent experiments and data analyses have demonstrated that (1) gene expression is more conservative in the brain than in other tissues; (2) the divergence rate of gene expression in the brain is higher in the human lineage than in the chimpanzee lineage; and (3) more genes are upregulated in the human lineage than in the chimpanzee lineage (Enard et al. 2002a; Caceres et al. 2003; Gu and Gu 2003; Khaitovich et al. 2005). It is important to note that these trends were not observed in other tissues such as liver or testis. In addition, Oldham et al. (2006) performed a network analysis of expression data across six brain regions in humans and chimpanzees. They identified several coexpressed gene network modules in brains. Interestingly, the network connectivity of the cerebral cortex module showed the weakest conservation between humans and chimpanzees. These network changes in the human cerebral cortex might be associated with the expansion of the region and might be responsible for the greater cognitive ability of humans. By looking at gene expression patterns in many other tissues, Gilad et al. (2006) established that there was an overrepresentation of transcription factors in the upregulated genes in the human lineage. All these findings support the original idea of King and Wilson that gene expression divergence is important for phenotypic divergence between humans and chimpanzees.

2.4 Future Direction

Transcriptome analysis is an important subject in non-human primate research because of the availability of many new molecular biology and computational tools. The next generation of sequencers will definitely give us many novel findings for non-human transcriptome studies. Although transcriptome data consist of huge dimensions (space and time), they are still simpler than complex phenotype comparisons and easier to quantify. Therefore, transcriptome analysis could help us link the genotype and phenotype for an improved understanding of primate evolution. As previous studies have reported, transcriptome studies are much more powerful with genome sequences of individual non-human primates. Indeed, the next-generation sequencers will be useful for determining novel genome sequences of non-human primates, which enable us to deeply understand primate evolution and promote biomedical research.

References

Akashi H (1994) Synonymous codon usage in *Drosophila melanogaster*: natural selection and translational accuracy. Genetics 136:927–935

Baskin CR, Garcia-Sastre A, Tumpey TM et al (2004) Integration of clinical data, pathology, and cDNA microarrays in influenza virus-infected pigtailed macaques (*Macaca nemestrina*). J Virol 78:10420–10432

Berglund J, Pollard KS, Webster MT (2009) Hotspots of biased nucleotide substitutions in human genes. PLoS Biol 7:e26

Bertone P, Stolc V, Royce TE et al (2004) Global identification of human transcribed sequences with genome tiling arrays. Science 306:2242–2246

Caceres M, Lachuer J, Zapala MA et al (2003) Elevated gene expression levels distinguish human from non-human primate brains. Proc Natl Acad Sci USA 100:13030–13035

Carninci P, Kasukawa T, Katayama S et al (2005) The transcriptional landscape of the mammalian genome. Science 309:1559–1563

Chen WH, Wang XX, Lin W et al (2006) Analysis of 10,000 ESTs from lymphocytes of the cynomolgus monkey to improve our understanding of its immune system. BMC Genomics 7:82

Clark AG, Glanowski S, Nielsen R et al (2003) Inferring nonneutral evolution from human-chimp-mouse orthologous gene trios. Science 302:1960–1963

Consortium TCSaA (2005) Initial sequence of the chimpanzee genome and comparison with the human genome. Nature (Lond) 437:69–87

Datson NA, Morsink MC, Atanasova S et al (2007) Development of the first marmoset-specific DNA microarray (EUMAMA): a new genetic tool for large-scale expression profiling in a non-human primate. BMC Genomics 8:190

Dillman JF 3rd, Phillips CS (2005) Comparison of non-human primate and human whole blood tissue gene expression profiles. Toxicol Sci 87:306–314

Djavani M, Crasta O, Zhang Y et al (2009) Gene expression in primate liver during viral hemorrhagic fever. Virol J 6:20

Dorus S, Vallender EJ, Evans PD et al (2004) Accelerated evolution of nervous system genes in the origin of *Homo sapiens*. Cell 119:1027–1040

Editorial (2008) When less is not more. Nat Med 14:791–792

Enard W, Khaitovich P, Klose J et al (2002a) Intra- and interspecific variation in primate gene expression patterns. Science 296:340–343

Enard W, Przeworski M, Fisher SE et al (2002b) Molecular evolution of FOXP2, a gene involved in speech and language. Nature (Lond) 418:869–872

Evans PD, Gilbert SL, Mekel Bobrov N et al (2005) Microcephalin, a gene regulating brain size, continues to evolve adaptively in humans. Science 309:1717–1720

Fujita E, Tanabe Y, Shiota A et al (2008) Ultrasonic vocalization impairment of FOXP2 (R552H) knockin mice related to speech-language disorder and abnormality of purkinje cells. Proc Natl Acad Sci USA 105:3117–3122

Gibbs RA, Rogers J, Katze MG et al (2007) Evolutionary and biomedical insights from the rhesus macaque genome. Science 316:222–234

Gilad Y, Rifkin SA, Bertone P et al (2005) Multi-species microarrays reveal the effect of sequence divergence on gene expression profiles. Genome Res 15:674–680

Gilad Y, Oshlack A, Smyth GK et al (2006) Expression profiling in primates reveals a rapid evolution of human transcription factors. Nature (Lond) 440:242–245

Grossman LI, Wildman DE, Schmidt TR et al (2004) Accelerated evolution of the electron transport chain in anthropoid primates. Trends Genet 20:578–585

Groszer M, Keays DA, Deacon RMJ et al (2008) Impaired synaptic plasticity and motor learning in mice with a point mutation implicated in human speech deficits. Curr Biol 18:354–362

Gu J, Gu X (2003) Induced gene expression in human brain after the split from chimpanzee. Trends Genet 19:63–65

Hashimoto S, Qu W, Ahsan B et al (2009) High-resolution analysis of the 5′-end transcriptome using a next generation DNA sequencer. PLoS ONE 4:e4108

Hellmann I, Zollner S, Enard W et al (2003) Selection on human genes as revealed by comparisons to chimpanzee cDNA. Genome Res 13:831–837

Hida M, Suzuki Y, Sugano S et al (2000) Construction and preliminary characterization of full-length enriched cDNA libraries for nonhuman primates (in Japanese). Primate Res 16:95–110

Jacquelin B, Mayau V, Brysbaert G et al (2007) Long oligonucleotide microarrays for African green monkey gene expression profile analysis. FASEB J 21:3262–3271

Kampa D, Cheng J, Kapranov P et al (2004) Novel RNAs identified from an in-depth analysis of the transcriptome of human chromosomes 21 and 22. Genome Res 14:331–342

Khaitovich P, Hellmann I, Enard W et al (2005) Parallel patterns of evolution in the genomes and transcriptomes of humans and chimpanzees. Science 309:1850–1854

Kimura M (1968) Evolutionary rate at the molecular level. Nature (Lond) 217:624–626

King MC, Wilson AC (1975) Evolution at two levels in humans and chimpanzees. Science 188:107–116

Kobasa D, Jones SM, Shinya K et al (2007) Aberrant innate immune response in lethal infection of macaques with the 1918 influenza virus. Nature (Lond) 445:319–323

Kothapalli KS, Anthony JC, Pan BS et al (2007) Differential cerebral cortex transcriptomes of baboon neonates consuming moderate and high docosahexaenoic acid formulas. PLoS ONE 2:e370

Lin L, Liu S, Brockway H et al (2009) Using high-density exon arrays to profile gene expression in closely related species. Nucleic Acids Res 37:e90

Lu Y, Huggins P, Bar-Joseph Z (2009) Cross species analysis of microarray expression data. Bioinformatics 25:1476–1483

Magness CL, Fellin PC, Thomas MJ et al (2005) Analysis of the *Macaca mulatta* transcriptome and the sequence divergence between *Macaca* and human. Genome Biol 6:R60

Marvanova M, Ménager J, Bezard E et al (2003) Microarray analysis of nonhuman primates: validation of experimental models in neurological disorders. FASEB J 17:929–931

Mekel-Bobrov N, Gilbert SL, Evans PD et al (2005) Ongoing adaptive evolution of ASPM, a brain size determinant in *Homo sapiens*. Science 309:1720–1722

Mewes HW, Amid C, Arnold R et al (2004) MIPS: analysis and annotation of proteins from whole genomes. Nucleic Acids Res 32:D41–D44

Nielsen R, Bustamante C, Clark AG et al (2005) A scan for positively selected genes in the genomes of humans and chimpanzees. PLoS Biol 3:e170

Nijland MJ, Schlabritz-Loutsevitch NE, Hubbard GB et al (2007) Non-human primate fetal kidney transcriptome analysis indicates mammalian target of rapamycin (MTOR) is a central nutrient-responsive pathway. J Physiol 579:643–656

Ohta T (1973) Slightly deleterious mutant substitutions in evolution. Nature (Lond) 246:96–98

Oldham MC, Horvath S, Geschwind DH (2006) Conservation and evolution of gene coexpression networks in human and chimpanzee brains. Proc Natl Acad Sci USA 103:17973–17978

Osada N, Hida M, Kusuda J et al (2002) Cynomolgus monkey testicular cDNAs for discovery of novel human genes in the human genome sequence. BMC Genomics 3:36

Osada N, Hashimoto K, Kameoka Y et al (2008) Large-scale analysis of *Macaca fascicularis* transcripts and inference of genetic divergence between *M. fascicularis* and *M. mulatta*. BMC Genomics 9:90

Osada N, Hirata M, Tanuma R et al (2009) Collection of *Macaca fascicularis* cDNAs derived from bone marrow, kidney, liver, pancreas, spleen, and thymus. BMC Res Notes 2:199

Royce TE, Rozowsky JS, Gerstein MB (2007) Toward a universal microarray: prediction of gene expression through nearest-neighbor probe sequence identification. Nucleic Acids Res 35:e99

Rubins KH, Hensley LE, Jahrling PB et al (2004) The host response to smallpox: analysis of the gene expression program in peripheral blood cells in a nonhuman primate model. Proc Natl Acad Sci USA 101:15190–15195

Sakate R, Osada N, Hida M et al (2003) Analysis of 5′-end sequences of chimpanzee cDNAs. Genome Res 13:1022–1026

Sarich VM, Wilson AC (1967) Rates of albumin evolution in primates. Proc Natl Acad Sci USA 58:142–148

Shi P, Bakewell MA, Zhang J (2006) Did brain-specific genes evolve faster in humans than in chimpanzees? Trends Genet 22:608–613

Shu W, Cho JY, Jiang Y et al (2005) Altered ultrasonic vocalization in mice with a disruption in the FOXP2 gene. Proc Natl Acad Sci USA 102:9643–9648

Spindel ER, Pauley MA, Jia Y et al (2005) Leveraging human genomic information to identify nonhuman primate sequences for expression array development. BMC Genomics 6:160

Sui Y, Potula R, Pinson D et al (2003) Microarray analysis of cytokine and chemokine genes in the brains of macaques with SHIV-encephalitis. J Med Primatol 32:229–239

Vahey MT, Nau ME, Taubman M et al (2003) Patterns of gene expression in peripheral blood mononuclear cells of rhesus macaques infected with SIVmac251 and exhibiting differential rates of disease progression. AIDS Res Hum Retroviruses 19:369–387

Walker SJ, Wang Y, Grant KA et al (2006) Long versus short oligonucleotide microarrays for the study of gene expression in nonhuman primates. J Neurosci Methods 152:179–189

Wallace JC, Korth MJ, Paeper B et al (2007) High-density rhesus macaque oligonucleotide microarray design using early-stage rhesus genome sequence information and human genome annotations. BMC Genomics 8:28

Wang HY, Tang H, Shen CK et al (2003) Rapidly evolving genes in human. I. The glycophorins and their possible role in evading malaria parasites. Mol Biol Evol 20:1795–1804

Wang Z, Lewis MG, Nau ME et al (2004) Identification and utilization of inter-species conserved (ISC) probesets on affymetrix human genechip platforms for the optimization of the assessment of expression patterns in non human primate (NHP) samples. BMC Bioinformatics 5:165

Wang HY, Chien HC, Osada N et al (2007) Rate of evolution in brain-expressed genes in humans and other primates. PLoS Biol 5:e13

Ylostalo J, Randall AC, Myers TA et al (2005) Transcriptome profiles of host gene expression in a monkey model of human malaria. J Infect Dis 191:400–409

Zhang J (2003) Evolution of the human ASPM gene, a major determinant of brain size. Genetics 165:2063–2070

Zhang J, Nielsen R, Yang Z (2005) Evaluation of an improved branch-site likelihood method for detecting positive selection at the molecular level. Mol Biol Evol 22:2472–2479

Zou J, Young S, Zhu F et al (2002) Microarray profile of differentially expressed genes in a monkey model of allergic asthma. Genome Biol 3(5):research0020

Chapter 3
The Role of Neoteny in Human Evolution: From Genes to the Phenotype

Mehmet Somel, Lin Tang, and Philipp Khaitovich

Abbreviations

miRNA   Micro-RNA
mRNA   Messenger RNA
ROS   Reactive oxygen species

3.1 The Paradox of the Human Phenotype

When studying human evolution, one encounters a major paradox: genetically, humans are remarkably similar to other primate species. For instance, humans share close to 99% of their genetic material with chimpanzees. This similarity reflects the fact that divergence between the human and the chimpanzee evolutionary lineages occurred somewhere in Africa only 6–7 million years ago, that is, very recently (Carroll 2003). Such genetic similarity is striking, especially when considering differences between other species that appear very much alike to our eyes. For instance, the genetic difference between two species of mice, one living in Europe (*Mus musculus*) and one living in East Asia (*Mus caroli*), is approximately five times greater than the difference between humans and chimpanzees (Enard et al. 2002).

M. Somel • P. Khaitovich (✉)
Partner Institute for Computational Biology, Shanghai Institutes for Biological Sciences, Chinese Academy of Sciences, 320 Yue Yang Road, Shanghai 200031, China

Max Planck Institute for Evolutionary Anthropology, Deutscher Platz 6, Leipzig, Germany
e-mail: khaitovich@eva.mpg.de

L. Tang
Partner Institute for Computational Biology, Shanghai Institutes for Biological Sciences, Chinese Academy of Sciences, 320 Yue Yang Road, Shanghai 200031, China

H. Hirai et al. (eds.), *Post-Genome Biology of Primates*, Primatology Monographs,
DOI 10.1007/978-4-431-54011-3_3, © Springer 2012

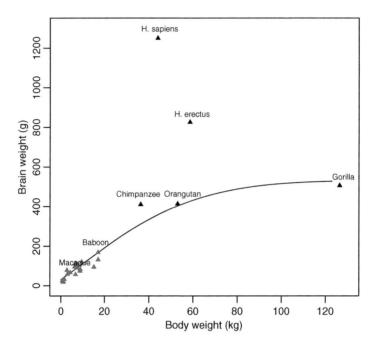

Fig. 3.1 Encephalization among primates. Graph shows brain and body weight among modern humans and great apes (*black triangles*), and gibbons and monkeys (*gray triangles*). The spline curve describes the brain to body weight relationship for non-human primates. (Data from Aiello and Dean 1990)

Still, phenotypically, the two mice species are so similar that one needs to be a qualified specialist to distinguish them. Humans and chimpanzees, on the other hand, can be readily distinguished based on their respective phenotypes. With respect to anatomy and morphology, human-specific phenotypes include bipedalism, the opposable thumb, and loss of body hair. Most striking, however, are human-specific features related to the brain. Human brains have grown three times larger than those of chimpanzees, and humans are outside the primate encephalization range (Gould 1977; Aiello and Dean 1990) (Fig. 3.1). What makes this change more amazing is that the brain is an expensive tissue, that is, it demands large amounts of energy to function (Aiello and Wheeler 1995). Along with large brains, humans have evolved unique cognitive abilities, especially communication skills and the use of abstract symbols, which have dramatically increased the pace of cultural accumulation and niche construction (Smith and Szathmary 1998; Laland et al. 2001). The question is: how could humans evolve into such a phenotypically distinct species over such a short evolutionary time?

When addressing this question, we have to consider several points. First, when examining the human phenotype, it is common to pool together biological features, such as the ability to learn language, with the products of such features, such as writing, and books created by humans over the course of history. As biologists, however, we have to distinguish clearly between the two, as genetic changes that took place on the human lineage would underlie the former but not the latter. Simply speaking, we have to consider humans outside of all artifacts that accumulated during the course of human history as a consequence of cultural transmission (Tomasello 2008). Instead, we need to identify the genetic background of cognitive features enabling humans to transmit knowledge between generations.

Second, even the small difference between the human and the chimpanzee genomes still accounts for millions of point mutations, thousands of insertions and deletions, as well as several large-scale genetic rearrangements (Consortium CSaA 2005). Most of these differences do not affect the phenotype and are evolutionarily neutral (Kimura 1968). Still, even a few mutations can lead to a strong phenotypic effect. For instance, certain single nucleotide substitutions can enable human individuals to digest milk as adults (Bersaglieri et al. 2004), whereas others disable the ability to process alcohol (Han et al. 2007). Both types of mutations could have substantial effects on reproductive success under certain environments and thereby affect evolutionary fitness. Note, however, that the examples mentioned here refer to differences among humans, while the main challenge in the field of human evolutionary genetics is to identify human–chimpanzee genetic differences responsible for the emergence of the human phenotype, among millions of phenotypically neutral mutations (Carroll 2003; Vallender and Lahn 2004; Sabeti et al. 2006). This is not a simple task, exacerbated by the impossibility of conducting genetic experiments on humans and apes. Nevertheless, recent studies have started to provide examples of human genetic features associated with certain human-specific phenotypes, such as a frameshift mutation in the myosin heavy chain (MYH) gene causing a reduction in masticatory muscles in humans (Stedman et al. 2004), or two amino acid substitutions in the transcription factor FOXP2 associated with speech and language abilities (Enard et al. 2009).

Finally, the emergence of human-specific cognitive traits does not necessarily require multiple independent adaptations that give rise to completely novel biological functions. Evolutionary biologists have long realized that novelty is rarely created from scratch. Instead, positive selection usually acts on already existing features by exploiting them in new contexts or remodeling them during ontogenesis. Human-specific cognitive features are probably not an exception. Indeed, despite the substantial change in brain size between humans and chimpanzees, the brain structures, as well as general functionality, is conserved between the two species. For instance, when communicating, chimpanzees use the same brain area (the Broca's area) as humans, even though they use gestures rather than spoken language (Taglialatela et al. 2008). Therefore, there may have been a limited number of evolutionary changes that lead to a global shift in brain functionality and to the recruitment of already existing neural circuitry to create novel cognitive tasks. One of the possible mechanisms producing such a shift could be neoteny.

3.2 The Neoteny Hypothesis of Human Evolution

Development in multicellular organisms involves a precise sequence of events to build a functional and reproductively capable adult form. Even small changes in the timing or magnitude of these events can result in large alterations in adult phenotype (e.g., gigantism can arise in humans because of excess production of growth hormones during development). The term heterochrony defines such changes in developmental rate and timing between two organisms or species, causing a change in shape or size.

A specific type of heterochrony is called neoteny: it describes slowed or delayed development of a species compared to its ancestral form, which phenotypically leads to adult individuals resembling the immature form of the ancestor (Gould 1977; Alberch et al. 1979; Shea 1989; McKinney and McNamara 1991; McNamara 1997; Klingenberg 1998; Horder 2006). The classical example of neoteny is the axolotl, a species of salamander. Although most salamanders loose their gills during metamorphosis, this developmental process is blocked in axolotls (Brown 1997). So, adult axolotls retain the gills and resemble immature, juvenile salamanders. Coming back to humans, morphological studies dating back to the nineteenth century had already noted visual similarities between adult humans and juvenile chimpanzees. Thus, many human-specific phenotypic features were thought to result from neotenic or delayed development (Montagu 1955; Gould 1977) (reviewed in Klingenberg 1998). For instance, chimpanzees have large jaws compared to humans, while humans retain small, underdeveloped jaws, small teeth, and a flat face. The result is that the adult human face is similar to that of young chimpanzees (Fig. 3.2). Gould argued that this resemblance was the result of retarded jaw development in humans and that this was a neotenic feature (Gould 1977). The list of traits classified as neotenic by various authors is extensive, including a large brain-to-body ratio, sparse body hair growth, and various skull features (reviewed in Gould 1977). Human neoteny has also been suggested to involve psychological traits: for example, the extension of childhood and fast learning periods in humans has been considered a form of neoteny (Montagu 1955; Gould 1977). This is an important point regarding the evolution of human intelligence. Human infants are recognized to have innate social skills, superior to those of chimpanzees (Tomasello 2008). This ability to communicate, coupled with a longer period of brain plasticity, would mean that human brains mature by assimilating cultural information already existing in the population. Indeed, recent research indicates that in human infants, many cognitive abilities may not be distinct from other great apes, but that these abilities develop gradually through social interactions, facilitated by the aforementioned social skills (Herrmann et al. 2007). Putting together these two lines of evidence – extended human development and the phenotypic similarity between adult humans and juvenile chimpanzees – S.J. Gould and others suggested that neoteny could be a major mechanism in human evolution.

Despite the popularity of the human neoteny hypothesis, many anthropologists and paleontologists have received both the evidence and methodology with skepticism

3 The Role of Neoteny in Human Evolution...

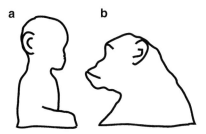

Fig. 3.2 A cartoon of juvenile and adult chimpanzee profiles. The drawings are based on (Gould 1977), citing Naef (Naef 1926). Gould uses these pictures to emphasize the similarity between the relatively flat face of the juvenile chimpanzee (**a**) with that of adult humans. Note the large jaws of the adult chimpanzee (**b**), who lacks a prominent chin or nose

(Shea 1989; McKinney and McNamara 1991; Rice 2002; Vinicius 2005) (reviewed in Klingenberg 1998), for a number of reasons. First, in many morphometric studies, sample sizes, especially for ape samples, have been limited (DeSilva and Lesnik 2006). Second, detailed models of morphological development (Kavanagh et al. 2007) are rare. Without large sample collections and lacking knowledge on developmental mechanisms, it is difficult to distinguish between changes in developmental timing (i.e., when a developmental process starts) and the rate of development (i.e., how fast development proceeds), or even to discern cause and effect. Third, estimating the direction of an ontogenetic shift requires estimating extinct ancestral forms or using outgroup species, which is not always feasible (the hominid fossil record is notoriously weak). Finally, morphometric studies analyze changes in size and shape, and these two sets of parameters are usually not independent (Alberch et al. 1979). Shape itself is a multidimensional object, such that a change pattern observed in one part (e.g., overgrowth) can be coupled with an opposite pattern in a neighboring part (e.g., undergrowth). Critically, depending on which dimensions are taken as reference, one may reach opposing conclusions on the type of heterochrony observed. The prominent human nose and chin can illustrate this point. These features could be considered overgrowths, or derived features, as apes lack both (Fig. 3.2). In contrast, the human nose and chin can also been seen as side-products of the general reduction of the human face (Lieberman 1998), and Gould argued that they were thus tied to neoteny (Gould 1977). Another such example is human encephalization. Adult humans have large brain-to-body ratios, and thus resemble juvenile chimpanzees, as a consequence of faster brain growth in human infants in the first years of life (Leigh 2004). Some researchers have considered high encephalization a neotenic trait, assuming that fast fetal growth rates have extended into human infancy (Gould 1977; Shea 1989). However, others regarded human encephalization as an example for overgrowth (McKinney and McNamara 1991; Vrba 1998; Rice 2002). All in all, a substantial number of researchers have dismissed the implication of neoteny as an explanation for novel human traits. Some recent studies have further suggested that morphological traits such as skull and face shape, or life history traits such as childhood, result from independent developmental trajectories in each

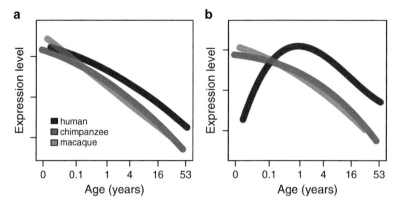

Fig. 3.3 Two types of human neotenic expression. (**a**) An example of difference in the rate of change, in which changes in humans (*black line*) are delayed relative to chimpanzees (*gray line*) and macaques (*light gray line*). (**b**) An example of difference in the timing, in which expression downregulation initiates later in humans (post displacement)

species and therefore are not amenable to categorization as simple heterochronies (Bogin 1997; Penin et al. 2002; Mitteroecker et al. 2004; Vinicius 2005).

The confusion over methods and terminology in heterochrony analysis has lead to the opinion that, to understand the evolution of developmental patterns in depth, we need an improved understanding of the underlying processes (Klingenberg 1998; Laurie 1999; Mitteroecker et al. 2004; Horder 2006). To obtain insight into such processes, monitoring gene expression changes with age is a powerful approach (see Osada, Chap. 2, this book). The expression level of a single gene is a one-dimensional variable, thus readily comparable across species (Fig. 3.3). Furthermore, microarrays and high-throughput sequencing technologies developed within the past two decades have allowed simultaneous assessment of messenger RNA (mRNA) expression levels across thousands of genes. Using these methods, research in molecular evolution and development has intensified, while work on model organisms has provided a range of examples in which simple genetic changes cause heterochrony, leading to novel phenotypes (e.g., Ambros and Horvitz 1984). Given that many developmental switches rely on changes in gene expression level, shifting the timing of expression changes during ontogenesis might be one of the most potent mechanisms of heterochrony (Zakany et al. 1997; Kim et al. 2000; Caygill and Johnston 2008).

3.3 Human Gene Expression Heterochrony

More than 30 years ago, M.C. King and A. Wilson had already proposed that identifying differences in the timing of gene expression during brain development between humans and apes would be crucial for understanding human evolution

(King and Wilson 1975). Why would changes in the timing of brain development be so relevant to the evolution of human cognitive abilities? As we know, both human brain and human cognitive abilities develop to large extent postnatally. In contrast to other primates, humans are born with relatively small brains, approximately 25–30% of the adult brain size. By contrast, the chimpanzee brain is 40% of the adult brain size at birth, and the macaque brain is close to 70% (Coqueugniot et al. 2004; DeSilva and Lesnik 2006). Thus, the main proportion of human brain growth takes place after birth. Evolutionarily, this phenomenon is attributed to the conflict between the size of the baby head, requiring a large birth canal, and the constraint on birth canal size imposed by bipedal locomotion (Rosenberg and Trevathan 2002). In other words, giving birth to a baby with a bigger head would necessitate human women to have a broader pelvis. This, in turn, would compromise their ability to run, an important quality in pre-modern human societies. This evolutionary contradiction was solved by giving birth to neonates at an earlier stage of development, compared to other primate species. An important consequence of this solution is the baby's early exposure to environmental stimuli, a factor suggested to influence human brain development, facilitating the emergence of human-specific cognitive abilities (Coqueugniot et al. 2004).

More importantly, even though human brain growth proceeds rapidly and finalizes by mid-childhood (Leigh 2004), brain restructuring and maturation continue well into adulthood. For instance, synaptic pruning extends into adolescence, whereas the myelinization of neuronal connections in the prefrontal cortex, evolutionarily the most recent brain region, only finalizes by approximately 30–35 years of age (Sowell et al. 2004; de Graaf-Peters and Hadders-Algra 2006; Marsh et al. 2008). During this time, the maturing brain is constantly exposed to a complex social environment; this is crucial to normal cognitive development, as illustrated by unfortunate cases when infants are deprived of social contact (Cavalli 2007; Veenema 2009). In parallel with earlier exposure of the developing brain to environmental clues, the prolonged period of human postnatal brain development was proposed to provide a critical condition for the evolution of human cognitive abilities (Montagu 1955; Gould 1977; Johnson 2001).

Importantly, the extension of human brain development and maturation might be a very recent evolutionary event. Even though the brain size of human ancestors, such as a *Homo erectus* specimen dating back 1.8 million years, was almost three times as large as that of chimpanzees or other apes, its development still followed the rapid trajectory characteristic of extant ape species (Coqueugniot et al. 2004). Even the Neanderthals, the evolutionary "cousins" of modern humans who separated from the human evolutionary branch just about 400,000 years ago, apparently grew and developed at a more ape-like pace compared to modern humans (Smith et al. 2007). Hence, neotenic delay in brain development and maturation could be at least partly specific to our species. If so, it is highly possible that this delay has contributed substantially to the emergence of cognitive abilities specific to modern humans. Remarkably, symbolic behavior, reflected in cave drawings and similar artifacts, also appears very recently in the archeological record (~100,000 years ago) and is thought to be unique to *Homo sapiens* (Powell et al. 2009).

Finally, the attractiveness of the neoteny hypothesis is to a large extent based on its relative simplicity. According to this hypothesis, the extension of brain development and maturation periods in humans may not have required multiple complex adaptations. Instead, this can be caused by one or several regulatory events shifting or extending the timing of existing developmental programs (Gould 1977). An example for such an ontogenic master switch is the reduction of a human skull bone, the sphenoid, which was suggested to shape the modern human cranium to a large extent (Lieberman 1998).

Despite all the attractiveness and visible simplicity of the human neoteny hypothesis, the molecular comparison of human and chimpanzee development has visibly lagged behind those in model organisms, mainly because of limited access to experimental samples (Khaitovich et al. 2006a). The first few studies comparing the two species showed that the adult brain, relative to other tissues, underwent substantial remodeling of expression patterns in humans (Enard et al. 2002; Caceres et al. 2003; Khaitovich et al. 2005), implying brain-specific positive selection on gene expression patterns (Khaitovich et al. 2006b). However, adult gene expression divergence is largely an outcome of differences in ontogeny; therefore, to understand how adult differences arise, one has to analyze the two species developmental trajectories.

The first study to compare human and chimpanzee development at the molecular level was published in 2009 (Somel et al. 2009). This study focused on gene expression changes in the frontal cortex region of the brain during postnatal development, from birth and into adulthood, until approximately 40 years of age in both species. The results indicated that gene expression changes during brain development are generally conserved between humans and chimpanzees. A substantial number of genes, however, did show differences in timing and rate of expression changes with age between the species. Using rhesus macaque expression as outgroup, many of these expression differences were attributed to the human lineage (Fig. 3.3). Interestingly, although at the organism level human development is delayed compared to the chimpanzee (humans reach sexual maturity at ~13 years of age, whereas chimpanzees do so at ~8), at the level of gene expression in the brain cortex, there was no uniform delay. Instead, many genes showed no difference in ontogenetic timing between the two species, whereas some genes showed neotenic patterns, and some, accelerated expression patterns in humans (Fig. 3.4); this implies that human transcriptome evolution may follow a mosaic model, rather than a uniform neotenic shift.

Nonetheless, the study did find that genes exhibiting neotenic expression patterns in the human brain were much more numerous than genes with other patterns (Fig. 3.4). Further, in contrast to other gene groups that showed no functional specificity, genes following neotenic expression trajectories were enriched in the gray matter of the brain, a histological structure containing neuronal cell bodies. Interestingly, more detailed analysis of these genes showed that they too could be separated into three distinct groups based on the shift in timing between the human and the chimpanzee expression profiles (Yuan et al. 2011). Notably, genes specifically expressed in gray matter and associated with neurons showed a distinct type of shift, with expression levels of young adult humans matching those of juvenile chimpanzees (Fig. 3.3). This point is interesting, as in the human prefrontal cortex this

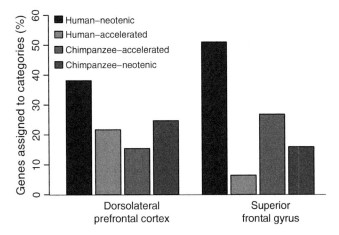

Fig. 3.4 Proportion of genes with different phylo-ontogenetic expression profiles in the brain. (1) Human neoteny: expression changes occurred in the human lineage, and human expression corresponds to that in younger chimpanzees; (2) human acceleration: expression changes occurred in the human lineage, and human expression corresponds to that in older chimpanzees; (3) chimpanzee neoteny; or (4) chimpanzee acceleration. The two prefrontal cortex regions analyzed both show excess of human neotenic genes, although other patterns also can be detected. (Results are from Somel et al. 2009)

period corresponds to the end of maturation processes, including axon myelinization (Sowell et al. 2004). Notably, alterations in the normal pace of cortical maturation can affect mental functions in humans (Shaw et al. 2007). It is thus possible that in the chimpanzee brain, cortical maturation finishes substantially earlier, leading to limited plasticity and, potentially, limiting the chimpanzee's learning abilities relative to the human. Conversely, this finding further supports the notion that neoteny might have played an essential role in evolution of human cognitive abilities by extending the period of neuronal plasticity in the human brain.

The mosaic pattern of human-specific gene expression changes, consisting of distinct groups of genes following shared ontogenetic trajectories, indicates that these groups may share the same regulation. Among common regulators of gene expression are transcription factors and micro-RNA (miRNA). Transcription factors are proteins binding to DNA in the vicinity of a gene's transcription start site, activating or inhibiting the transcription of that gene. miRNAs are a specific class of small inhibitory RNA involved in posttranscriptional gene regulation. Typical miRNA are single-stranded RNA molecules 21–23 nucleotides in length, derived from a double-stranded hairpin precursor, and inhibit gene expression by targeting mRNA for degradation or inhibiting mRNA translation (Bartel 2004; Baek et al. 2008; Selbach et al. 2008). Recent results obtained in our laboratory indicate that many miRNA and transcription factors expressed in the human brain show mosaic patterns of heterochrony, dominated by neotenic shifts. Furthermore, these regulators target dozens of genes that show corresponding heterochronic expression patterns

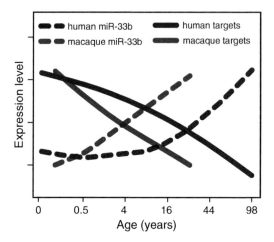

Fig. 3.5 Human and rhesus macaque expression profiles for a micro-RNA (miRNA; miR-33b) and 12 of its target genes (Somel et al. 2010). The *y*-axis represents normalized miRNA and mean mRNA expression levels. The *solid and dashed lines* show spline curves of gene and miRNA expression changes with age, respectively. The timing shift between the two species among the target genes corresponds to an opposite shift in miRNA expression levels. Because miRNAs can act as repressors of gene expression, the human–macaque difference in the 12 target genes is potentially regulated by the timing difference in miR-33b expression

(Fig. 3.5). Thus, diverse neotenic expression patterns found in the human brain might indeed be caused by neotenic expression of a limited number of key gene expression regulators. This finding fits well into the framework of the human neoteny hypothesis, postulating that broad changes in human brain ontogenesis might result from a few evolutionary events that changed the rate and the timing of existing developmental processes.

3.4 Neoteny and Human Longevity

Although development constitutes the initial phase of lifespan, molecular and morphological changes in the body do not cease with adulthood, but continue during aging. Despite this continuity, development and aging are commonly viewed as two independent processes. Indeed, in development, robust regulatory programs propel an organism from a single-cell stage of the zygote into a fully developed reproductively capable individual. By contrast, the body deteriorates during aging, leading to an ever-increasing probability of death. The mechanisms underlying this age-related deterioration, or senescence, are still poorly understood.

Human aging presents a particularly interesting problem: Despite their extremely close genetic similarity, humans live substantially longer than apes (Rose and Mueller 1998). For instance, in a modern hunter-gatherer society, the Ache, 42% of individuals can reach 50 years of age, compared to only 9% among wild chimpanzees (Hill et al. 2001). In terms of maximum lifespan, humans can live about two times longer than chimpanzees and three to four times longer than rhesus macaques (122, 66, and 36 years, respectively) (de Magalhães 2006; Walker et al. 2006; Hill et al. 2007; Viegas 2008). Such an extension of lifespan is even more intriguing given that human females reach menopause by around 45 years, and thus spend a substantial amount of their lives in the postreproductive state, whereas chimpanzee females reach menopause toward the end of their life, as is discussed below (Atsalis and Videan 2009; Herndon and Lacreuse 2009).

There are a number of hypotheses addressing the deceleration of human aging. The "grandmother hypothesis" postulates a kin selection mechanism, wherein long-living nonreproductive grandmothers care for their grandchildren, thus increasing their fitness (Hawkes et al. 1998). As grandmothers and their grandchildren share a quarter of their genetic material, such a mechanism could lead to indirect selection of long-life genes. Recent studies have found indication for such a "grandmother" effect in pre-modern societies (e.g., Lahdenpera et al. 2004). Another possibility is that, as a result of technological innovation, as well as social support motifs, extrinsic mortality decreased in human ancestors, allowing older individuals to reproduce and leading to selection for old age (Hill et al. 2007). The rapid pace of cultural innovation could also have favored the increased longevity, if elder, experienced hunters/gatherers increased the fitness of their kin. Notably, these hypotheses imply direct selection for longevity. An alternative idea builds on the observation that brain size is strongly correlated with lifespan among primates (Walker et al. 2006), and it tries to connect human encephalization with longevity through a hypothetical physiological clock (Comfort 1979; Rose and Mueller 1998). As a consequence of the clock, prolonged development would automatically lead to prolonged lifespan. For instance, the "embodied capital hypothesis" suggests that the complex hunting and gathering niche of *Homo* selected for individuals with larger brains and longer lifespan, allowing for an extensive learning period and older age (Kaplan and Robson 2002; Gurven et al. 2006). Based on observations from our own work, described in detail later in this chapter, we offer a refined version of this hypothesis. Specifically, we propose that the neotenic shift in human development, leading to a substantial delay in timing of sexual and tissue maturation, might have played a role in delaying molecular changes associated with the aging phenotype. This delay, in turn, would lead to an extension of the human lifespan. In other words, we propose that extension of the human lifespan was a by-product of neotenic development. Such a model is supported by observations in other natural populations, as well as laboratory models. For instance, in *Drosophila*, early reproduction is shown to dramatically shorten lifespan (Sgro and Partridge 1999). Similarly, across animals, there is a general correlation between rates of maturation and rates of aging (Charnov 1993). Thus, if development and aging are also connected on the molecular level, delayed maturation in humans (neoteny) can explain increased longevity, along with human-specific cognitive abilities.

To gauge the feasibility of these various models of human longevity, we need to understand the molecular mechanisms underlying aging, as well as the relationship between aging and development. Although our knowledge on the physiological mechanisms of aging and their evolution lags behind our knowledge on developmental processes, studies within the twentieth century have already identified a number of molecular processes associated with aging. The most widely recognized hypothesis describes aging as a consequence of accumulating oxidative damage on somatic cell DNA and other biomolecules (reviewed in Beckman and Ames 1998). Such accumulation, induced by products of cellular catabolism called reactive oxygen species (ROS), eventually leads to cellular dysfunction. This oxidative damage (or free radical) theory of aging has remained the leading mechanistic explanation of aging in humans and other species since its original formulation in 1956 (Harman 1956). Recent experiments in mice and worms, however, suggest that oxidative damage on DNA may not be the main mechanism of aging (Vermulst et al. 2007; Gems and Doonan 2009). Other related sources of somatic damage also exist, such as spontaneous misfolding of proteins, or telomere attrition (Bell and Sharpless 2007; Cohen and Dillin 2008). Intriguingly, the animal soma does harbor mechanisms to protect itself from external or intrinsic damage, such as antioxidants or autophagy of altered proteins (Schriner et al. 2005; Zhang and Cuervo 2008). However, such maintenance pathways are not fully activated under normal conditions. The "disposable soma hypothesis" suggests that this is because investing resources in reproduction, rather than somatic maintenance, is a superior evolutionary strategy (Kirkwood 2005). Damage is thus left unrepaired, causing aging.

Finally, according to the antagonistic pleitropy hypothesis, there may exist developmental processes that improve survival and boost reproduction at young ages but are detrimental at later ages (Williams 1957; de Magalhaes and Church 2005). Such processes can evolve in the wild because, as a result of extrinsic factors such as predation or starvation, too few individuals survive to an old age to enable efficient purifying selection at that stage. For example, mice rarely live past their first birthday in the wild (Kirkwood 2005). Thus, in mice, developmental processes that are beneficial during the first year of life, but detrimental past this age, can contribute to senescence. Interestingly, recent gene expression studies in *Caenorhabditis elegans* have identified several developmental regulatory patterns persisting into aging and effectively limiting the animal's lifespan (Boehm and Slack 2005; Budovskaya et al. 2008). If such mechanisms also exist in humans, shifting developmental timing could lead to a concomitant shift in lifespan.

During the past decade, microarray-based mRNA expression profiling has been widely used to investigate gene expression changes in human and mammalian aging, with a special emphasis on the brain (Lee et al. 2000; Blalock et al. 2003; Lu et al. 2004; Erraji-Benchekroun et al. 2005; Zahn et al. 2007; Loerch et al. 2008). These studies have identified a number of mRNA expression patterns consistently found in aging individuals, including an upregulation of stress and immune response pathways, as well as a decrease in expression of genes involved in energy metabolism and neuronal functions in the brain. Many of the observed expression changes were attributed to ROS-related accumulation of DNA damage (Lu et al. 2004). Intriguingly, in the human brain, some of the gene expression changes observed in aging are already detectable as early as 13 years of age (Erraji-Benchekroun et al. 2005); this might either reflect very early effects of DNA damage, which could be surprising in the pre-reproductive period, or it might indicate that some of the expression changes

3 The Role of Neoteny in Human Evolution... 35

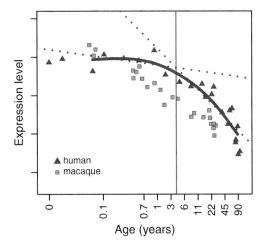

Fig. 3.6 Expression changes in neural-related genes across lifespan. The *triangles* represent the mean expression levels of a group of genes with similar expression profiles among 23 humans (Somel et al. 2010). These genes are predominantly involved in neural development and function, such as axon guidance and long-term depression. It was previously recognized that, during aging, expression levels of these genes decrease dramatically (Lu et al. 2004), which was associated with cognitive decline during aging. However, this decrease apparently already starts early in childhood, around 4 years of age (*vertical line*), implying a connection between developmental processes and aging. The rhesus macaque (*gray squares*) also shows a parallel decrease in expression levels with age, although the trend starts earlier than in humans

seen in aging reflect regulatory patterns established in development. As already mentioned, such senescence-driving developmental regulation, recently observed in model organisms, involved both miRNA and transcription factors (Boehm and Slack 2005; Budovskaya et al. 2008).

Studies conducted by our group on human and rhesus macaque brains throughout each species lifespan indicate that, similar to development, gene expression patterns during aging are conserved across primates (Somel et al. 2010). The main difference between humans and macaques in aging is the timing of age-related expression changes, including both mRNA expression, as well as the expression of its regulators, miRNA, and transcription factors. Furthermore, two observations suggest a connection between developmental and postdevelopmental expression changes. First, several miRNA and transcription factors that regulate gene expression in development continue their regulation well into the postdevelopmental period: at least until 98 years of age in humans and 28 years of age in macaques (the oldest individuals used in our study). Second, our results show that gene expression changes observed during aging are either reversals or continuations of developmental patterns, rather than changes arising de novo and being specific to old age. A striking pattern found here is that many gene expression changes observed in previous aging studies, and frequently explained by accumulation of oxidative damage with age, in fact begin at as early as 4 years of age in humans, and much earlier in macaques (Fig. 3.6). Specifically, these changes include downregulation of genes involved in neural development and function. Thus, many gene expression patterns observed at old age represent the continuation of developmental expression trajectories established in early childhood. These developmental changes themselves could be related to brain maturation processes, such as synaptic pruning (Sowell et al. 2004). Thus, it is plausible that these lifelong expression patterns are beneficial during development but detrimental at older ages, as proposed by the antagonistic pleiotropy hypothesis.

Our results suggest that in humans, and in other primates as well, gene expression patterns found in development may have direct effects on gene expression patterns observed in aging. It is then possible to suggest that the neotenic gene expression shift found in the human brain and, possibly, present in other tissues, may have caused further delay of expression changes later in life, thus leading to an extension of human lifespan. As some of the neotenic gene expression shifts found in the human brain are quite large, from 5 to 7 years in the chimpanzee to 30–35 years in the human (Yuan et al. 2011), they could conceivably cause an approximately twofold lifespan extension in humans.

Even in the brain, however, not all genes show human-specific neotenic shifts. Similarly, on the phenotypic level, not all human life history traits are extended relative to chimpanzees. Specifically, while the maximum human lifespan is twofold greater than that of chimpanzees, the length of the female reproductive period differs little between the two species. Although chimpanzee females also experience menopause, it occurs relatively late in life, frequently after 50 years of age (Atsalis and Videan 2009; Herndon and Lacreuse 2009). In all studied wild chimpanzee populations, female fertility persist at least through to 40–44 years of age, such that chimpanzee males tend to favor elderly females for mating (Muller et al. 2006). Thus, analogous to the diversity of gene expression patterns seen in brain, it is possible that some of the human phenotypic features, such as the maximum lifespan, are delayed compared to chimpanzees, whereas others, such as reproductive decline in females, have been retained, or specifically remodeled, in humans (Hawkes et al. 1998). The analogy between mosaic patterns of heterochrony in gene expression and phenotypic features is not far fetched, as gene expression largely determines the functional phenotype of cells and tissues, thus shaping the phenotype of an organism.

The notion that human longevity is linked to human neoteny and childhood is not new. Hill et al. argue that, given current life history models, a decrease in extrinsic mortality could lead to delays both in sexual maturity and in aging (Hill et al. 2001, 2007). Our results indicate that developmental and postdevelopmental delays could also be linked at the molecular level. Compared to a model with independent selection events for extended childhood and for extended lifespan, this might present a more parsimonious model of human life history evolution. Meanwhile, we do not dispute the possible importance of a "grandmother effect" or a similar role of elderly individuals in human life history evolution. Indeed, multiple features of human development, including the difficulty of human childbirth (Rosenberg and Trevathan 2002), the immature state of human neonates (Coqueugniot et al. 2004), and the extended period of childhood, which demands intense social input (Johnson 2001; Herrmann et al. 2007; Veenema 2009), all suggest that support from elder kin could have been particularly critical in the shaping of modern human life history. Human neoteny and longevity might thus have evolved synergistically.

3.5 Conclusions

Recent studies indicate that the molecular basis of human-specific phenotypic features, such as human cognition and human longevity, might be understood by studying molecular changes over the course of the human lifespan and comparing them to the molecular changes found in other primate species. Although this work has just begun, the results yield optimism regarding the possibility of discovering key evolutionary changes leading to the emergence of the human phenotype. In fact, it is possible, as we speculate in the preceding paragraphs, that the very same changes responsible for human cognitive abilities also resulted in the extension of human lifespan. This is not implausible, as only a short evolutionary time passed since the divergence of the human and the chimpanzee evolutionary branches, 6–7 million years, which does not provide much room for complex multimodular adaptations. Instead, the emergence of both human phenotypes could have been caused by just a few events changing the timing or rate of the ontogenetic changes of several key gene expression regulators. Such changes would affect the expression of hundreds of genes, leading to profound modifications in the functionality of preexisting brain structures. In humans, this change might have been facilitated further by extended exposure of the developing brain to environmental and social stimuli. Further, as gene expression changes found late in life commonly represent extensions of developmental trends, delay in development would also postpone the commencement of aging. If proven, this notion would indicate that manipulating the regulation of human ontogenetic changes by modifying the expression profiles of a few key regulatory genes could lead to further extensions in human brain maturation and, possibly, of the human lifespan as a whole.

Acknowledgments M.S. was supported by Chinese Academy of Sciences young scientist fellowship (no. 2009Y2BS12) and a Natural Science Foundation of China research grant (no. 31010022).

References

Aiello L, Dean C (1990) An introduction to human evolutionary anatomy. Academic, London
Aiello LC, Wheeler P (1995) The expensive-tissue hypothesis: the brain and the digestive system in human and primate evolution. Curr Anthropol 36:199–221
Alberch P, Gould SJ, Oster GF et al (1979) Size and shape in ontogeny and phylogeny. Paleobiology 5:296–317
Ambros V, Horvitz HR (1984) Heterochronic mutants of the nematode *Caenorhabditis elegans*. Science 226:409–416

Atsalis S, Videan E (2009) Reproductive aging in captive and wild common chimpanzees: factors influencing the rate of follicular depletion. Am J Primatol 71:271–282

Baek D, Villén J, Shin C et al (2008) The impact of microRNAs on protein output. Nature (Lond) 455:64–71

Bartel DP (2004) MicroRNAs: genomics, biogenesis, mechanism, and function. Cell 116:281–297

Beckman K, Ames BN (1998) The free radical theory of aging matures. Physiol Rev 78:547–581

Bell JF, Sharpless NE (2007) Telomeres, p21 and the cancer-aging hypothesis. Nat Genet 39:11–12

Bersaglieri T, Sabeti P, Patterson N et al (2004) Genetic signatures of strong recent positive selection at the lactase gene. Am J Hum Genet 74:1111–1120

Blalock E, Chen K, Sharrow K et al (2003) Gene microarrays in hippocampal aging: statistical profiling identifies novel processes correlated with cognitive impairment. J Neurosci 23:3807–3819

Boehm M, Slack F (2005) A developmental timing microRNA and its target regulate life span in *C. elegans*. Science 310:1954–1957

Bogin B (1997) Evolutionary hypotheses for human childhood. Yearb Phys Anthropol Am J Phys Anthropol 104:63–89

Brown DD (1997) The role of thyroid hormone in zebrafish and axolotl development. Proc Natl Acad Sci USA 94:13011–13016

Budovskaya YV, Wu K, Southworth LK et al (2008) An elt-3/elt-5/elt-6 GATA transcription circuit guides aging in *C. elegans*. Cell 134:291–303

Caceres M, Lachuer J, Zapala MA et al (2003) Elevated gene expression levels distinguish human from non-human primate brains. Proc Natl Acad Sci USA 100:13030–13035

Carroll SB (2003) Genetics and the making of *Homo sapiens*. Nature (Lond) 422:849–857

Cavalli A (2007) Casper or "the cabinet of horrors". J Anal Psychol 52:607–623

Caygill EE, Johnston LA (2008) Temporal regulation of metamorphic processes in *Drosophila* by the let-7 and mir-125 heterochronic microRNAs. Curr Biol 18:943–950

Charnov EL (1993) Life history invariants. Oxford University Press, New York

Cohen E, Dillin A (2008) The insulin paradox: aging, proteotoxicity and neurodegeneration. Nat Rev Neurosci 9:759–767

Comfort A (1979) The biology of senescence. Churchill Livingstone, Edinburgh

Consortium CSaA (2005) Initial sequence of the chimpanzee genome and comparison with the human genome. Nature (Lond) 437:69–87

Coqueugniot H, Hublin JJ, Veillon F et al (2004) Early brain growth in *Homo erectus* and implications for cognitive ability. Nature (Lond) 431:299–302

de Graaf-Peters VB, Hadders-Algra M (2006) Ontogeny of the human central nervous system: what is happening when? Early Hum Dev 82:257–266

de Magalhães JP (2006) Anage database, build 9. http://genomics.senescence.info/species

de Magalhaes JP, Church GM (2005) Genomes optimize reproduction: aging as a consequence of the developmental program. Physiology 20:252–259

DeSilva J, Lesnik J (2006) Chimpanzee neonatal brain size: implications for brain growth in *Homo erectus*. J Hum Evol 51:207–212

Enard W, Khaitovich P, Klose J et al (2002) Intra- and interspecific variation in primate gene expression patterns. Science 296:340–343

Enard W, Gehre S, Hammerschmidt K et al (2009) A humanized version of FoxP2 affects cortico-basal ganglia circuits in mice. Cell 137:961–971

Erraji-Benchekroun L, Underwood MD, Arango V et al (2005) Molecular aging in human pre-frontal cortex is selective and continuous throughout adult life. Biol Psychiatry 57:549

Gems D, Doonan R (2009) Antioxidant defense and aging in *C. elegans*: is the oxidative damage theory of aging wrong? Cell Cycle 8:1681–1688

Gould SJ (1977) Ontogeny and phylogeny. Harvard University Press, Cambridge, MA

Gurven M, Kaplan H, Gutierrez M (2006) How long does it take to become a proficient hunter? Implications for the evolution of extended development and long life span. J Hum Evol 51:454–470

Han Y, Gu S, Oota H et al (2007) Evidence of positive selection on a class I ADH locus. Am J Hum Genet 80:441–456

Harman D (1956) Aging: a theory based on free radical and radiation chemistry. J Gerontol 11:298–300

Hawkes K, O'Connell JF, Jones NG et al (1998) Grandmothering, menopause, and the evolution of human life histories. Proc Natl Acad Sci USA 95:1336–1339

Herndon JG, Lacreuse A (2009) Commentary: "Reproductive aging in captive and wild common chimpanzees: factors influencing the rate of follicular depletion". Am J Primatol 71:891–892

Herrmann E, Call J, Hernandez-Lloreda MV et al (2007) Humans have evolved specialized skills of social cognition: the cultural intelligence hypothesis. Science 317:1360–1366

Hill K, Boesch C, Goodall J et al (2001) Mortality rates among wild chimpanzees. J Hum Evol 40:437–450

Hill K, Hurtado AM, Walker RS (2007) High adult mortality among Hiwi hunter-gatherers: implications for human evolution. J Hum Evol 52:443–454

Horder T (2006) Heterochrony. In: Encyclopedia of Life Sciences. http://www.els.net

Johnson MH (2001) Functional brain development in humans. Nat Rev Neurosci 2:475–483

Kaplan HS, Robson AJ (2002) The emergence of humans: the coevolution of intelligence and longevity with intergenerational transfers. Proc Natl Acad Sci USA 99:10221–10226

Kavanagh KD, Evans AR, Jernvall J (2007) Predicting evolutionary patterns of mammalian teeth from development. Nature (Lond) 449:427–432

Khaitovich P, Hellmann I, Enard W et al (2005) Parallel patterns of evolution in the genomes and transcriptomes of humans and chimpanzees. Science 309:1850–1854

Khaitovich P, Enard W, Lachmann M et al (2006a) Evolution of primate gene expression. Nat Rev Genet 7:693–702

Khaitovich P, Tang K, Franz H et al (2006b) Positive selection on gene expression in the human brain. Curr Biol 16:R356–R358

Kim J, Kerr JQ, Min G (2000) Molecular heterochrony in the early development of *Drosophila*. Proc Natl Acad Sci USA 97:212–216

Kimura M (1968) Evolutionary rate at the molecular level. Nature (Lond) 217:624–626

King MC, Wilson AC (1975) Evolution at two levels in humans and chimpanzees. Science 188:107–116

Kirkwood TBL (2005) Understanding the odd science of aging. Cell 120:437–447

Klingenberg CP (1998) Heterochrony and allometry: the analysis of evolutionary change in ontogeny. Biol Rev 73:79–123

Lahdenpera M, Lummaa V, Helle S et al (2004) Fitness benefits of prolonged post-reproductive lifespan in women. Nature (Lond) 428:178–181

Laland KN, Odling-Smee J, Feldman MW (2001) Cultural niche construction and human evolution. J Evol Biol 14:22–33

Laurie G (1999) What is heterochrony? Evol Anthropol 7:186–188

Lee C, Weindruch R, Prolla TA (2000) Gene-expression profile of the ageing brain in mice. Nat Genet 25:294

Leigh S (2004) Brain growth, life history, and cognition in primate and human evolution. Am J Primatol 62:139–164

Lieberman DE (1998) Sphenoid shortening and the evolution of modern human cranial shape. Nature (Lond) 393:158–162

Loerch PM, Lu T, Dakin KA et al (2008) Evolution of the aging brain transcriptome and synaptic regulation. PLoS One 3:e3329

Lu T, Pan Y, Kao S et al (2004) Gene regulation and DNA damage in the ageing human brain. Nature (Lond) 429:883

Marsh R, Gerber AJ, Peterson BS (2008) Neuroimaging studies of normal brain development and their relevance for understanding childhood neuropsychiatric disorders. J Am Acad Child Adolesc Psychiatry 47:1233–1251

McKinney ML, McNamara KJ (1991) Heterochrony: the evolution of ontogeny. Plenum Press, New York

McNamara KJ (1997) Shapes of time: the evolution of growth and development. Johns Hopkins University Press, Baltimore

Mitteroecker P, Gunz P, Bernhard M et al (2004) Comparison of cranial ontogenetic trajectories among great apes and humans. J Hum Evol 46:679–698

Montagu MFA (1955) Time, morphology, and neoteny in the evolution of man. Am Anthropol 57:13–27

Muller MN, Thompson ME, Wrangham RW (2006) Male chimpanzees prefer mating with old females. Curr Biol 16:2234–2238

Naef A (1926) Über die urformen der anthropomorphen und die stammesgeschichte des menschenschädels. Naturwissenschaften 14:472–477

Penin X, Berge C, Baylac M (2002) Ontogenetic study of the skull in modern humans and the common chimpanzees: neotenic hypothesis reconsidered with a tridimensional Procrustes analysis. Am J Phys Anthropol 118:50–62

Powell A, Shennan S, Thomas MG (2009) Late Pleistocene demography and the appearance of modern human behavior. Science 324:1298–1301

Rice SH (2002) The role of heterochrony in primate brain evolution. In: Minugh-Purvis N, McNamara KJ (eds) Human evolution through developmental change. Johns Hopkins University Press, Baltimore

Rose MR, Mueller LD (1998) Evolution of human lifespan: past, future, and present. Am J Hum Biol 10:409–420

Rosenberg K, Trevathan W (2002) Birth, obstetrics and human evolution. Br J Obset Gynaecol 109:1199–1206

Sabeti PC, Schaffner SF, Fry B et al (2006) Positive natural selection in the human lineage. Science 312:1614–1620

Schriner SE, Linford NJ, Martin GM et al (2005) Extension of murine life span by overexpression of catalase targeted to mitochondria. Science 308:1909–1911

Selbach M, Schwanhäusser B, Thierfelder N et al (2008) Widespread changes in protein synthesis induced by microRNAs. Nature (Lond) 455:58–63

Sgro CM, Partridge L (1999) A delayed wave of death from reproduction in *Drosophila*. Science 286:2521–2524

Shaw P, Eckstrand K, Sharp W et al (2007) Attention-deficit/hyperactivity disorder is characterized by a delay in cortical maturation. Proc Natl Acad Sci USA 104(49):19649–19654

Shea BT (1989) Heterochrony in human evolution: the case for neoteny reconsidered. Am J Phys Anthropol 32:69–101

Smith JM, Szathmary E (1998) The major transitions in evolution. Oxford University Press, Oxford

Smith TM, Toussaint M, Reid DJ et al (2007) Rapid dental development in a Middle Paleolithic Belgian Neanderthal. Proc Natl Acad Sci USA 104:20220–20225

Somel M, Franz H, Yan Z et al (2009) Transcriptional neoteny in the human brain. Proc Natl Acad Sci USA 106:5743–5748

Somel M, Guo S, Fu N et al (2010) MicroRNA, mRNA, and protein expression link development and aging in human and macaque brain. Genome Res 20:1207–1218

Sowell ER, Thompson PM, Toga AW (2004) Mapping changes in the human cortex throughout the span of life. Neuroscientist 10:372–392

Stedman HH, Kozyak BW, Nelson A et al (2004) Myosin gene mutation correlates with anatomical changes in the human lineage. Nature (Lond) 428:415–418

Taglialatela JP, Russell JL, Schaeffer JA et al (2008) Communicative signaling activates Broca's homolog in chimpanzees. Curr Biol 18:343–348

Tomasello M (2008) Origins of human communication. MIT Press, Cambridge

Vallender EJ, Lahn BT (2004) Positive selection on the human genome. Hum Mol Genet 13:R245–R254

Veenema AH (2009) Early life stress, the development of aggression and neuroendocrine and neurobiological correlates: what can we learn from animal models? Front Neuroendocrinol 30:497–518

Vermulst M, Bielas JH, Kujoth GC et al (2007) Mitochondrial point mutations do not limit the natural lifespan of mice. Nat Genet 39:540–543

Viegas J (2008) Africa's oldest chimp, a conservation icon, dies. http://dsc.discovery.com/news/2008/12/23/gregoire-oldest-chimp.html

Vinicius L (2005) Human encephalization and developmental timing. J Hum Evol 49:762–776

Vrba ES (1998) Multiphasic growth models and the evolution of prolonged growth exemplified by human brain evolution. J Theor Biol 190:227–239

Walker R, Burger O, Wagner J et al (2006) Evolution of brain size and juvenile periods in primates. J Hum Evol 51:480–489

Williams GC (1957) Pleiotropy, natural selection, and the evolution of senescence. Evolution 11:398–411

Yuan Y, Yi-Ping PC, Shengyu N et al (2011) Development and application of a modified dynamic time warping algorithm (DTW-S) to analyses of primate brain expression time series. BMC Bioinformatics 12:347

Zahn JM, Poosala S, Owen AB et al (2007) AGEMAP: a gene expression database for aging in mice. PLoS Genet 3:e201

Zakany J, Gerard M, Favier B et al (1997) Deletion of a HoxD enhancer induces transcriptional heterochrony leading to transposition of the sacrum. EMBO J 16:4393–4402

Zhang C, Cuervo AM (2008) Restoration of chaperone-mediated autophagy in aging liver improves cellular maintenance and hepatic function. Nat Med 14:959–965

Chapter 4
Evolution of Chemosensory Receptor Genes in Primates and Other Mammals

Yoshihito Niimura

Abbreviations

CNV	Copy number variation
FPR	Formyl peptide receptor
GPCR	G-protein-coupled receptor
MOE	Main olfactory epithelium
MOS	Main olfactory system
MRCA	Most recent common ancestor
NWM	New World monkey
OR	Olfactory receptor
OWM	Old World monkey
SNP	Single nucleotide polymorphism
TAAR	Trace amine-associated receptor
VNO	Vomeronasal organ
VNS	Vomeronasal system

4.1 Introduction

Chemosensation is the most "primitive" sense in organisms. Even bacteria detect chemical molecules in the environment. They swim toward higher concentrations of attractants such as sugars or amino acids and away from repellents of toxic substances. The chemical senses – olfaction and taste – are also essential to the survival of mammals. The olfactory information is used for finding foods, avoiding danger,

---

Y. Niimura (✉)
Medical Research Institute, Tokyo Medical and Dental University,
1-5-45 Yushima, Bunkyo-ku, Tokyo 113-8510, Japan
e-mail: niimura@bioinfo.tmd.ac.jp

H. Hirai et al. (eds.), *Post-Genome Biology of Primates*, Primatology Monographs,
DOI 10.1007/978-4-431-54011-3_4, © Springer 2012

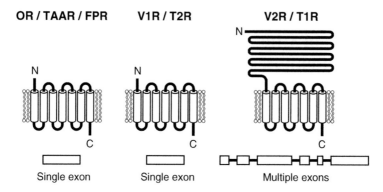

Fig. 4.1 Structure of seven chemosensory receptor gene families, showing membrane topologies of the receptors and their exon–intron structures. *OR* olfactory receptor, *TAAR* trace amine-associated receptor, *FPR* formyl peptide receptor, *V1R* vomeronasal receptor type 1, *T2R* taste receptor type 2, *V2R* vomeronasal receptor type 2, *T1R* taste receptor type 1

recognizing territories, and identifying offspring and mates; taste is used for evaluating the quality, quantity, and the safety of ingested food.

Chemical molecules are detected by chemosensory receptors. Although the chemosensory system is present essentially in all motile organisms, chemosensory receptor genes are different among bacteria, worms, insects, and vertebrates (Niimura 2009b). In vertebrates, genes involved in chemosensation belong to at least seven different multigene families: olfactory receptors (ORs), vomeronasal receptors type 1 and 2 (V1Rs and V2Rs), trace amine-associated receptors (TAARs), formyl peptide receptors (FPRs), and taste receptors type 1 and 2 (T1Rs and T2Rs). ORs, V1Rs, V2Rs, TAARs, and FPRs are olfactory or pheromone receptors, and T1Rs and T2Rs are taste receptors. All these gene families encode G-protein-coupled receptors (GPCRs), membrane proteins having seven transmembrane α-helical regions (Fig. 4.1). Binding of a ligand to a receptor activates a G protein and a subsequent signaling cascade.

Among them, the OR multigene family is by far the largest (Fig. 4.2a). OR genes were first identified from rats by Linda Buck and Richard Axel (Buck and Axel 1991). They discovered a huge multigene family of GPCRs, the expression of which is restricted to the main olfactory epithelium (MOE) in the nasal cavity, and suggested that there are approximately 1,000 different OR genes in mammalian genomes. The discovery opened the door to the molecular study of chemosensation, and they received the Novel Prize in 2004 for this achievement. OR genes typically constitute 3–5% of the entire proteome in mammals, forming the largest multigene family in vertebrates.

Thanks to the recent advancement of genome sequencing technologies, the whole genome sequences of diverse organisms have become available and are freely accessible by anyone via the Internet. This situation allowed us to identify nearly complete repertoires of chemosensory receptor genes in each species with the aid of bioinformatics. These studies revealed that the numbers of chemosensory receptor

4 Evolution of Chemosensory Receptor Genes in Primates and Other Mammals 45

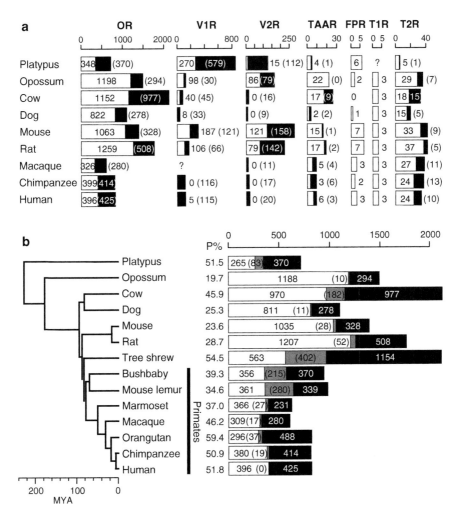

Fig. 4.2 (**a**) Numbers of genes in seven chemosensory receptor gene families. Numbers of functional genes and pseudogenes (*in parentheses*) are depicted by *white and black bars*, respectively. The numbers of OR genes are from Niimura and Nei (2007), Go and Niimura (2008), and Matsui et al. (2010). The numbers of FPR genes were obtained from Liberles et al. (2009); those of T2R genes are from Nei et al. (2008) and Dong et al. (2009). Data for other gene families are from Nei et al. (2008). (**b**) Numbers of olfactory receptor (OR) genes in various mammalian species with their phylogeny. The numbers of intact genes, truncated genes (*in parentheses*), and disrupted genes are indicated by *white*, *gray*, and *black bars*, respectively. The fractions of pseudogenes (*P%*) are calculated by assuming that all intact and truncated genes are functional while all disrupted genes are pseudogenes. From Niimura and Nei (2007), Go and Niimura (2008), and Matsui et al. (2010). *MYA* million years ago

genes and the fractions of pseudogenes vary enormously among different species. For example, in silico analyses using the whole human genome sequences showed that the number of functional OR genes is about 400, which is considerably smaller than that in mice, but the fraction of pseudogenes is as high as 50% or more (Fig. 4.2b) (Glusman et al. 2001; Zozulya et al. 2001; Niimura and Nei 2003).

By investigating the evolution of these huge multigene families of chemosensory receptors, we could expect to obtain some insights into the interaction of genomic contents with the environment, because olfaction and taste are sensors to the external world. In this chapter, I would like to present a recent progress on the evolution of chemosensory receptor genes from the point of view of comparative genomics. Because OR genes are the largest and the most thoroughly studied among the seven chemosensory receptor gene families, I explain the evolution of OR genes in some detail. I mention non-primate mammals as well as primates, because the characteristics of primates will be highlighted by the comparison with other organisms. However, regardless of the presence of chemosensory receptor genes in all vertebrate and some invertebrate species, I confine my argument to mammals because of limitations of space. For the evolution of chemosensory receptor genes across a broader range of animals, see Niimura and Nei (2005a), Nei et al. (2008), and Niimura (2009a, b).

4.2 Olfactory Receptor Genes

4.2.1 *Expression*

OR genes are predominantly expressed in olfactory sensory neurons of the MOE in the nasal cavity. The consensus is that each olfactory neuron expresses only a single functional OR gene among approximately 1,000 genes. This "one neuron–one receptor rule" is hypothesized to be important for olfactory coding, such that only a given population of olfactory neurons responds to a restricted number of odor molecules. However, the detailed molecular mechanism is still unclear, although it has been proposed that one functional OR gene is stochastically chosen in each olfactory neuron and its expression prevents the activation of other OR genes though negative feedback regulation (Serizawa et al. 2003).

A small subset of OR genes are expressed in other chemosensory organs, such as the vomeronasal organ (VNO; see below; Lévai et al. 2006). Some OR genes are also expressed in the testis, and these ORs are apparently involved in sperm chemotaxis (Spehr et al. 2003). Moreover, several researchers have reported that OR genes are expressed in various nonolfactory tissues including the tongue, brain, kidney, and placenta. However, Feldmesser et al. (2006) found no correlation in expression levels in nonolfactory tissues for orthologous OR genes between humans and mice. In contrast, De la Cruz et al. (2009) detected a statistically significant correlation for human–chimpanzee OR orthologues. Therefore, it is controversial whether such "ectopic expression" of OR genes has any functional significance.

4.2.2 Ligands

Mammals detect airborne, low molecular mass compounds as odorants. (In contrast, fishes recognize water-soluble molecules such as amino acids, bile acids, sex steroids, and prostaglandins as odorants.) It is generally believed that the olfactory system uses combinatorial coding (Malnic et al. 1999). The relationships between odor molecules and ORs are multiple to multiple; that is, one OR recognizes multiple odorants, and one odorant is recognized by multiple ORs. Therefore, different odorants are recognized by different combinations of ORs. This system would explain why humans can detect and discriminate tens of thousands of different odors by using several hundreds of OR genes. Moreover, this would be the reason why the classification of odors has long been unsuccessful, because the multiple-to-multiple relationships suggest that "primary odors," from which all other odors could be reproduced by appropriate mixtures, actually do not exist (Niimura 2009b).

Each OR is thought to be specialized to recognize physicochemical features of odor molecules, such as functional groups or molecular sizes. However, the relationships between ORs and odorants are still largely unknown, and ligands have been identified for only about 100 mammalian ORs so far (Saito et al. 2009).

It is known that human olfactory perception greatly differs among individuals (Hasin-Brumshtein et al. 2009). There is a phenomenon called specific anosmia, meaning diminished sensitivity toward specific odorants. Examples are documented for dozens of different odors, such as hydrogen cyanide (an extreme poisonous gas with an almond-like odor), butyl mercaptan (an odor of the skunk; Amoore 1967), the pig pheromone androstenone, and the sweaty odorant isovaleric acid. Androstenone is perceived as offensive ("sweaty, urinous"), pleasant ("sweet, floral"), or odorless. A novel approach employing a combination of in vitro experiments with genetic association studies revealed the genetic basis accounting for the variation in sensitivity and quality of androstenone perception (Keller et al. 2007). The human OR *OR7D4*, which is selectively activated by androstenone, contains two nonsynonymous single nucleotide polymorphisms (SNPs), R88W and T133M, and homozygous or heterozygous subjects having the WM haplotype were less sensitive to androstenone and felt it to be less unpleasant (see also Chap. 5). Moreover, Menashe et al. (2007) reported a significant association between the presence of a nonsense SNP in an OR *OR11H7P* and the detection threshold difference for isovaleric acid.

4.2.3 Gene Structure

As already mentioned, ORs are members of GPCRs (Fig. 4.1). GPCRs can be classified into five groups by sequence similarities, and OR genes belong to the largest one among them, a rhodopsin-like GPCR superfamily. This superfamily includes opsin genes for detecting light and many other receptor genes for neurotransmitters, peptide hormones, chemokines, lipids, nucleotides, etc. (Fredriksson et al. 2003). Several motifs are characteristic of OR genes. In a phylogenetic tree of rhodopsin-like GPCR genes, OR genes form a well-supported monophyletic clade and are clearly distinguishable from other non-OR GPCR genes (Niimura 2009a).

OR genes do not have any introns in their coding regions. The number of exons in the 5'-untranslated region can be variable, and these noncoding exons are often alternatively spliced to generate multiple mRNA isoforms, which, however, results in the same protein (Young et al. 2003). The intronless gene structure is not specific to OR genes but is widely observed among rhodopsin-like GPCR genes. Phylogenetic analyses showed a widespread loss of introns during the evolution of the mammalian rhodopsin-like GPCR genes (Bryson-Richardson et al. 2004).

4.2.4 Genomic Distribution

Mammalian OR genes form many genomic clusters and are dispersed in almost all chromosomes. In the case of humans, they are located on all chromosomes except chromosomes 20 and Y (Niimura and Nei 2003; Nei et al. 2008). Especially, chromosome 11 contains more than 40% of all OR genes. The largest human OR gene cluster contains approximately 100 OR genes occupying a genomic region of about 2 Mb, whereas in mice the largest one includes about 270 OR genes and is approximately 5 Mb long (Niimura and Nei 2003, 2005b). OR genes are generally densely arrayed in a genomic cluster without interspersed non-OR genes. In both humans and mice, the intergenic distances are more or less constant, and its distribution shows a sharp peak at about 1.1 kb (Niimura and Nei 2003, 2005b).

The genes located close to each other on a chromosome tend to be evolutionarily closely related (Niimura and Nei 2003). This observation suggests that the number of OR genes have increased by repeated tandem gene duplications, and duplication of the entire OR gene cluster appears to be rare. At the same time, however, one OR gene cluster often contains evolutionarily distantly related genes, and OR genes with a close evolutionary relationship reside at different clusters or chromosomes. These observations can be explained by assuming that several rearrangements have occurred at the regions of OR gene clusters and the genes in different clusters were shuffled (Niimura and Nei 2003). In fact, it is known that the chromosome fission event that generated human chromosomes 14 and 15 occurred at a cluster of OR genes in the common ancestor of great apes (Rudd et al. 2009).

The organization of OR gene clusters is generally well conserved between humans and mice, and the numbers of OR gene clusters are similar between the two species (Niimura and Nei 2005b). On average, each mouse cluster contains a larger number of OR genes than a human cluster. It is therefore suggested that the greater OR gene repertoire in mice than in humans has been generated mainly by gene duplications within each cluster.

4.2.5 Classification

Mammalian OR genes can be clearly classified into two groups named Class I and Class II by sequence similarity. The functional difference between the two classes is unclear, but the ligands for Class I genes tend to be more hydrophobic than those for Class II genes (Saito et al. 2009). Interestingly, all Class I genes are located in a single genomic cluster.

Class I and Class II genes are sometimes called "fish like" and "mammal like" because previously it was thought that Class I genes are similar to fish OR genes (Glusman et al. 2000). However, this view was based on inaccurate phylogenetic analyses, and later it was shown that the majority of fish OR genes are close to neither Class I nor Class II genes (Niimura and Nei 2005a, 2006; Niimura 2009a). Therefore, the terminology of "fish-like" and "mammal-like" is misleading and should not be used.

Extensive phylogenetic analyses showed that vertebrate OR genes are classified into seven groups, named α–η (Niimura 2009a). Each group of genes is basically present in either terrestrial organisms (mammals, birds, and reptiles) or aquatic organisms (fishes) in an exclusive manner, such that groups α and γ are in the former whereas groups δ, ε, ζ, and η are in the latter; on the other hand, amphibians have both types. This observation strongly suggests that the former groups of OR genes are for detecting airborne odorants and the latter groups are for water-soluble odorants (Niimura and Nei 2005a; Niimura 2009a, b).

4.2.6 Gene Repertoires and Evolution

4.2.6.1 Identification of OR Genes

Because of the simple structure of OR genes, it is not a particularly difficult task to identify OR genes from the genome sequence. By conducting extensive homology searches and phylogenetic analyses, nearly complete OR gene repertories have been identified from a variety of species for which the whole genome sequences are available, revealing enormous variation in number of OR genes and fraction of pseudogenes (Fig. 4.2b). In this figure, the genes from each species are classified into three categories: intact, truncated, and disrupted genes. An intact gene encodes a complete coding sequence with an initiation codon and a stop codon at proper positions, whereas a disrupted gene includes nonsense or frameshift mutations or long deletions. We should note that a low-quality genome sequence tends to give an underestimate of the number of intact genes. Especially, when the genome data contains many short contigs because of incomplete assembly, a considerable number of genes are truncated at the end of a contig. In Fig. 4.2b, a truncated gene is defined as a partial intact sequence that is located at the contig end, which may become an intact gene when genome sequencing is completed. Here I assume that both intact and truncated genes are functional.

In fact, there is no guarantee that an intact gene is *functional*, because it will take some time for a given gene to accumulate disruptive mutations after it becomes a pseudogene. Menashe et al. (2006) developed a computer program to distinguish functional and nonfunctional OR genes by examining several conserved motifs. They predicted that about 35% of human intact genes might be nonfunctional, whereas the fraction of predicted pseudogenes among mouse intact genes is only about 5%. Therefore, the number of functional genes in humans may be much smaller than that in Fig. 4.2.

We should also note, however, that it is not easy to determine experimentally whether a given gene is functional. Zhang et al. (2007) examined the expression of nearly all predicted human OR genes using a DNA microarray. They confirmed that approximately 80% of putatively functional genes are expressed in the MOE, but interestingly, a considerable fraction of putative OR pseudogenes (~67%) are also expressed in the MOE. Therefore, apparently it is necessary to see the translation of a given gene for the verification of its functionality. Nevertheless, the results in Fig. 4.2 obtained by bioinformatic analyses would give us much information on the evolution of OR genes, as will be seen in the following.

4.2.6.2 Human

From the latest version of the human genome, about 820 OR genes were identified, but about 52% of them are pseudogenes (Fig. 4.2b; Niimura and Nei 2003; Matsui et al. 2010). The number of functional OR genes is approximately 400; however, the actual number might be much smaller, as already mentioned.

Humans have a group of pseudogenes named 7E (Newman and Trask 2003) or H* pseudogenes (Niimura and Nei 2005a), which seem to have been generated by gene duplications after they were pseudogenized. In the human genome, there are about 90 H* pseudogenes, and most of them are present as a singleton rather than forming a cluster. Hominoids have much larger numbers of H* pseudogenes than Old World monkeys (OWMs) or New World monkeys (NWMs), suggesting that the duplication events of H* pseudogenes were activated in the hominoid lineage (Matsui et al. 2010).

Human OR genes are known to be highly polymorphic among individuals, which is arguably one of the most pronounced case of genetic variation in humans (Hasin-Brumshtein et al. 2009). Is has been reported that more than 60 OR loci are segregating pseudogenes, in which both an intact and a pseudogenized allele exist in the human population (Menashe et al. 2003), but this number is expected to increase by recent efforts of individual genome sequencing. Recently it has been indicated that copy number variations (CNVs), that is, structural variants affecting the copy number of DNA segments with a length of 1 kb or more, rather than SNPs, are responsible for most of the genomic variation in humans. Several studies showed that OR genes are significantly enriched in the CNV regions (Nozawa et al. 2007; Young et al. 2008; Hasin et al. 2008).

4.2.6.3 Chimpanzee

Previous studies claimed that the fraction of OR pseudogenes is significantly higher in humans than in chimpanzees and that the loss of OR genes have accelerated in the human lineage after the divergence from chimpanzees about 6 million years ago (Gilad et al. 2003, 2004, 2005, 2007). However, our analysis using the deep-coverage (6×) chimpanzee genome sequences has revealed that the number of functional OR genes and the fractions of pseudogenes are very similar between the two species (Fig. 4.2b; Go and Niimura 2008). Moreover, the rates of pseudogenization, the numbers of predicted pseudogenes among intact genes (see foregoing), and the numbers of genes under positive selection are also similar.

The extent of functional constraint to a gene can be evaluated by the ratio (ω) of the rate of nonsynonymous substitutions to the rate of synonymous substitutions: $\omega = 1$ implies neutrality, and a smaller value of ω indicates stronger purifying selection (Nei and Kumar 2000). The comparison of orthologous OR genes between humans and chimpanzees showed that the mean ω value is 0.94 (Go and Niimura 2008), whereas the mean ω value for mouse–rat OR orthologues is 0.19 (Nei et al. 2008). Gimelbrant et al. (2004) indicated that the effect of the positive selection to human and chimpanzee OR genes is minor. It is therefore suggested that the functional constraint is relaxed in the human and chimpanzee lineages, which is consistent with a diminished importance of olfaction in higher primates (see below).

Regardless of the similarity in number of OR genes in humans and chimpanzees, their OR gene repertoires are considerably diversified. Only about 75% of their functional OR genes are orthologous to each other, and about 25% of the repertoires are species specific (Go and Niimura 2008). This observation suggests that the most recent common ancestor (MRCA) between the two species had a larger repertoire of functional OR genes, but multiple gene loss events have occurred in a lineage-specific manner after the divergence of the two species. In fact, it was estimated that the MRCA had more than 500 functional OR genes and a smaller fraction (~41%) of pseudogenes, suggesting that OR gene repertoires are in the process of deterioration in both human and chimpanzee lineages (Go and Niimura 2008).

The large difference in OR gene repertoire between humans and chimpanzees imply that the spectrum of the detectable odorants might be quite different between the two species, which would be responsible for a species-specific ability of odor perception. It was reported that wild chimpanzees use several olfactory cues in social and sexual contexts (Nishida 1997; Boesch and Boesch-Achermann 2000). However, the phenotypic difference in olfactory perception between humans and chimpanzees is still largely unknown.

4.2.6.4 Other Primates

Primates are generally thought to be vision oriented, and their olfactory ability has been retrogressed. In fact, the relative size of the main olfactory bulb (the forebrain center receiving olfactory inputs) and that of the region in the nasal cavity covered with the MOE is smaller in primates than in other mammals (Smith and Bhatnagar 2004). Among primates, haplorhines (tarsiers, NWMs, OWMs, and hominoids) show even smaller relative sizes of the MOEs than strepsirrhines (lemurs and lorises; Barton 2006). Morphologically, haplorhines and strepsirrhines are distinguished from each other by the absence and presence, respectively, of the rhinarium, the moist and naked surface around the tip of the nose. The rhinarium is very sensitive to olfaction and is able to detect the direction of odors, being also present in many other mammals.

The visual specializations of primates include convergent orbits associated with stereoscopic vision, a distinctive pattern of objections between the eye and the brain, and the enlarged visual cortices (Barton 2006). These specializations are more marked in haplorhines than in strepsirrhines and are correlated with the evolution of large brains. Therefore, the reliance on olfaction is reduced in haplorhines compared with strepsirrhines.

Trichromatic color vision is well developed only in haplorhines, whereas strepsirrhines are mostly dichromatic. All catarrhine species (OWMs and hominoids) have trichromatic vision, which is mediated by three opsins that are activated by different wavelengths, whereas tarsiers are dichromats. In NWMs, typically red–green color vision is controlled by multiple alleles at a single opsin gene locus on the X chromosome; therefore, heterozygous females are trichromats, while homozygous females and all males are dichromats. Gilad et al. (2004, 2007) examined a possible link between olfaction and color vision in primate evolution. They sequenced 100 OR genes that were randomly chosen from each of 19 primate species including strepsirrhines, NWMs, OWMs, and hominoids, and reported that the fractions of OR pseudogenes in catarrhines are significantly higher than NWMs and strepsirrhines. From this observation, they hypothesized that primates have lost the function of OR genes consequent to the acquisition of full trichromatic vision. This hypothesis is called the "color vision priority hypothesis" (Nei et al. 2008; Matsui et al. 2010).

Figure 4.2b shows the numbers of OR genes from seven primate species, including hominoids (human, chimpanzee, and orangutan), OWMs (macaque), NWMs (marmoset), and strepsirrhines (mouse lemur and bushbaby), as well as the tree shrew belonging to the order Scandentia, which is a close relative of primates. The results indicate that the estimated numbers of functional OR genes (intact + truncated genes) are similar (320–400) among simians (hominoids, OWMs, and NWMs). No significant difference was observed between catarrhines and NWMs, which does not support the color vision priority hypothesis. Rather, interestingly, orangutans and macaques showed significantly smaller numbers than the others (Matsui et al. 2010).

The numbers of truncated genes are large for the mouse lemur, bushbaby, and tree shrew, because the genome sequences of these species are at low coverage (<2x; see foregoing). Therefore, the estimation of the number of functional genes by the total of intact and truncated genes may be inaccurate. With this caveat in mind, the results in Fig 4.2b indicate that the numbers of functional OR genes in simians are considerably smaller than those in strepsirrhines or the tree shrew, which is consistent with the reduced olfactory ability in haplorhines compared to strepsirrhines.

Matsui et al. (2010) identified orthologous gene sets among the five simian species in Fig. 4.2b and estimated that the MRCA among simians had approximately 550

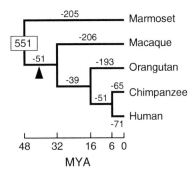

Fig. 4.3 Loss of OR genes in primate evolution. The most recent common ancestor (MRCA) among the five primate species was estimated to have had 551 functional OR genes. A *number at each branch* represents the number of genes that were present in the MRCA but were lost at the branch. The *arrowhead* indicates the acquisition of full trichromatic vision caused by the duplication of red- and green-opsin genes. (From Matsui et al. 2010)

functional OR genes. Figure 4.3 indicates the loss of OR genes in each branch of their phylogeny. The results show a gradual loss of OR genes in every branch from the MRCA to the human rather than a sudden loss at the branch of the catarrhine ancestor, at which the acquisition of full trichromatic vision occurred. Therefore, this analysis again does not support the color vision priority hypothesis.

Hiramatsu et al. (2008) carried out field observations of a group of wild spider monkeys in natural habitats and did not find any significant differences in the foraging efficiency between dichromats and trichromats. Considering these observations together, it is highly probable that the degeneration of OR genes in primates is associated with well-developed vision system, but it cannot be explained by a sole factor of the trichromatic vision (see also Chap. 7).

4.2.6.5 Other Mammals

As already explained, the number of OR genes can change rapidly during evolution depending on each species living environment. It appears that the reduction of OR gene repertoires has occurred many times in the evolution of mammals. As shown in Fig. 4.2b, the platypus also showed a small number of functional genes and a large fraction of pseudogenes (Niimura and Nei 2007). This characteristic is probably associated with the platypus being semiaquatic, because mammalian OR genes for detecting airborne odorants are apparently useless in water (see foregoing). In addition, the platypus has a bill sense, which is a combination of electroreception and mechanoreception and is used for finding prey in the mud at the bottom of a stream.

A more drastic example is seen in dolphins. Dolphins are fully adapted to the aquatic lifestyle. They completely lack the olfactory apparatus and instead have developed the echolocation system. In fact, the analysis of the low-coverage dolphin genome suggests that the number of functional OR genes is about ten, and the fraction of pseudogenes is greater than 90% (Niimura, unpublished data). It has been reported that in baleen whales, which have a significantly reduced olfactory apparatus but an intact olfactory system, the extent of OR gene degeneration is less remarkable than in toothed whales, including dolphins (Kishida et al. 2007; McGowen et al. 2008).

In contrast, animals heavily dependent on olfaction for their survival, such as mice or rats, have gained many new OR genes by repeated gene duplications (Niimura and Nei 2007). In the case of dogs, however, although they are thought to have a very acute olfactory sense, the number of OR genes is not particularly large (Fig. 4.2b). The actual reason is unclear, but this might be because they are carnivorous and thus have less need to distinguish variable odors than do omnivorous animals.

In general, the turnover of OR gene repertoires is extremely rapid during mammalian evolution. The number of functional OR genes in the MRCA of placental mammals was estimated to be approximately 800, but hundreds of gene gains and losses have occurred in the lineage-specific manner for every different order (Niimura and Nei 2007). Therefore, the OR gene family has undergone an extreme form of the birth-and-death evolution (Nei and Rooney 2005).

4.3 Vomeronasal Receptor Genes

4.3.1 Expression, Ligands, and Gene Structure

In most mammals, the olfactory systems comprise at least two major, functionally distinct systems: the main olfactory system (MOS), which was already discussed, and the vomeronasal system (VNS). The second olfactory organ, called the VNO, is located proximal to the vomer bone in the nasal cavity. Previously, the VNO was regarded to be specialized for the detection of pheromones – chemical cues that are released and detected by individuals of the same species and evoke physiological and behavioral effects. However, the MOS and the VNS are now thought to have some overlapping functions (Baxi et al. 2006).

Mammalian vomeronasal receptor neurons express one of the two families of chemosensory receptors, V1Rs (Dulac and Axel 1995) and V2Rs (Herrada and Dulac 1997; Matsunami and Buck 1997; Ryba and Tirindelli 1997). One of the mouse V1Rs responds to 2-heptanone, a mouse pheromone (Boschat et al. 2002), and the deletion of a mouse genomic cluster containing 16 V1R genes results in impaired social behaviors and loss of vomeronasal neuron responses to specific pheromones (Del Punta et al. 2002). Mouse V2Rs recognize several pheromone candidates such as a major histocompatibility complex ligand peptide (Leinders-Zufall et al. 2004) or exocrine gland-secreting peptide 1 (Kimoto et al. 2005). Therefore, at least in mice, V1Rs and V2Rs function as pheromone receptors.

As shown in Fig. 4.1, V1R genes are intronless, as are OR genes, but they do not show any significant sequence similarities to OR genes. On the other hand, V2R genes are interrupted by introns and are characterized by a long extracellular N-terminal tail. V2R genes are dissimilar in amino acid sequence to both OR and V1R genes.

4.3.2 Primates

In humans, the VNO develops in the fetus, but it is only vestigial in adults. There is substantial anatomical and genomic evidence that the human VNO is nonfunctional (Wysocki and Preti 2004); for example, no axonal connections of vomeronasal sensory neurons to the brain were found (Meredith 2001), and the gene encoding TRPC2, a calcium channel that is essential to the signal transduction pathway in the mouse VNO, is a pseudogene in humans (Liman and Innan 2003; Zhang and Webb 2003). The existence of pheromonal communication in humans is still greatly controversial. For example, it is known that menstrual cycles tend to synchronize among women living together (Stern and McClintock 1998). However, its molecular mechanism is unclear, and it is likely not to be mediated by VNOs.

Bioinformatic analyses showed that there are five intact V1R genes and more than 100 V1R pseudogenes in the human genome (Fig. 4.2a; Rodriguez and Mombaerts 2002). At least one of the five genes is expressed in the human olfactory mucosa (Rodriguez et al. 2000), and these five genes can activate an OR-like signal transduction pathway, in contrast to mouse V1Rs (Shirokova et al. 2008). Therefore, it is possible that major functional characteristics of these human V1Rs might be similar to those of ORs. As indicated in Fig. 4.2a, V2R gene families are completely degenerated in humans, chimpanzees, and macaques (Young and Trask 2007).

Catarrhines lack the accessory olfactory bulb (the brain region excited by the VNS), and their VNOs are significantly reduced in adults, but NWMs clearly have VNOs (Barton 2006). It was reported that the TRPC2 genes in NWMs are functional whereas those in catarrhines are pseudogenes, suggesting that the pseudogenization of the TRPC2 gene occurred in the common ancestor of catarrhines (Liman and Innan 2003; Zhang and Webb 2003). Zhang and Webb (2003) proposed that the degeneration of the VNS would be associated with the acquisition of trichromatic vision in males. Females in many catarrhine species develop a prominent reddening and swelling of the sexual skin surrounding the perineum around the time of ovulation, although NWMs do not have true sexual skins. In catarrhines, the chemical-based system in social activities may have been in part replaced by the vision-based system.

The evolutionary pattern of the VNS is apparently different from that of the MOS. It is likely that the evolution of the VNS correlates with social and mating behaviors, whereas that of the MOS predominantly correlates with ecological factors (Barton 2006).

4.3.3 Other Mammals

The numbers of V1R and V2R genes show dramatic variation among different mammalian species, and these numbers are roughly correlated to each other (Fig. 4.2a) (Grus et al. 2005; Young et al. 2005). Mice, rats, and opossums have about 100 or more functional V1R genes while dogs have only 8. Dogs possess a functional VNO, but the dog VNO is relatively thin and the accessory olfactory bulb is small. In general, the size of a V1R gene repertoire is correlated with the morphological complexity of the VNO (Grus et al. 2005). Interestingly, the platypus has the largest repertoire of V1R genes ever examined, and the VNO complexity of the platypus is ranked the highest among all vertebrates (Grus et al. 2007). Mice, rats, and opossums have large repertoires of V2R genes as well as V1R genes, whereas dogs and cows lack functional V2R genes (Fig. 4.2a) (Young and Trask 2007).

For both V1R and V2R genes, the mouse and rat repertoires have been highly diversified since the divergence of the two species, and very few one-to-one orthologues were detected (Grus and Zhang 2008). It is therefore suggested that V1R and V2R genes contribute to species-specific recognition and have played certain roles in specification of rodents. However, in the case of ruminants, V1R gene repertoires are well conserved among cows, sheep, and goats (Ohara et al. 2009). Moreover, all goat V1Rs examined are expressed in both VNO and MOE, raising a possibility that ruminant and rodent V1Rs may have distinct functions.

4.4 Trace Amine-Associated Receptor Genes

TAARs were originally identified as receptors for "trace amines" in the brain (Borowsky et al. 2001). Trace amines designate a collection of amines that are present in the central nervous system at very low concentrations. Trace amines were suspected to be involved in psychiatric disorders, and TAARs have been postulated to play a role in depression and schizophrenia (Lindemann and Hoener 2005). Later, it was demonstrated that TAARs are expressed in the MOE in mice and function as a second class of ORs (Liberles and Buck 2006). It seems that the expression of TAARs and ORs is mutually exclusive. An olfactory neuron expresses either a TAAR gene or an OR gene.

The numbers of TAAR genes in mammalian genomes are much smaller than those of ORs (Fig. 4.2a). Some mouse TAARs respond to volatile amines that are present in urine, one of which is reportedly a pheromone, suggesting a function associated with the detection of social cues (Liberles and Buck 2006). This result underscores the importance of the MOE in receiving pheromonal cues. In contrast to vomeronasal receptors, however, TAARs are relatively well conserved among various organisms, implying a common olfactory role across diverse species (Grus and Zhang 2008; Niimura 2009a).

TAARs are intronless and belong to the rhodopsin-like GPCR superfamily as do ORs, although TAARs are more closely related to receptors for neurotransmitters such as dopamine or serotonin.

4.5 Formyl Peptide Receptor Genes

FPRs were first identified two decades ago (Boulay et al. 1990). They are involved in host defense against pathogens and are expressed in the immune system (Migeotte et al. 2006). FPR genes are activated by formylated peptides that are released by bacteria and other peptides associated with disease or inflammation. Recently, it was reported that five of the seven members of the FPR genes in mice are predominantly expressed in the VNO (Liberles et al. 2009; Rivière et al. 2009). Each gene is expressed in a different subset of neurons that do not express other known vomeronasal receptors. These FPRs may have a role for avoiding potential mates that are infected with pathogens.

FPR genes are present in most mammals, but it seems that cows, sheep, and pigs lack this gene family (Liberles et al. 2009). Humans have three FPR genes, which are expressed in the immune system. A phylogenetic analysis showed that the five mouse FPR genes expressed in the VNO form a rodent-specific clade; therefore, the FPR function in the VNO may have been evolved only in rodents (Liberles et al. 2009).

FPRs are also intronless and belong to the rhodopsin-like GPCR superfamily. However, FPR genes are more closely related to chemokine receptors than OR or TAAR genes (Fredriksson et al. 2003).

4.6 Taste Receptor Genes

Two types of taste receptors, T1Rs and T2Rs, are expressed in taste buds of the tongue. There are five modalities of tastes: sweet, sour, bitter, salty, and umami. Umami means "deliciousness" in Japanese and is a taste of L-glutamate (Ikeda 2002). ("Hot" is not a taste; capsaicin, an ingredient of chili peppers, is recognized by a receptor for high temperature.) Of these five modalities, salty and sour tastes are detected by ion channels. On the other hand, sweet and umami tastes are perceived by T1Rs (Nelson et al. 2001, 2002), and a bitter taste is recognized by T2Rs (Adler et al. 2000; Matsunami et al. 2000). T1R and T2R genes have significant sequence similarities to V2R and V1R genes, respectively (Fig. 4.1).

Most mammals have three T1R genes, named *T1R1*, *T1R2*, and *T1R3*. T1R3 combines with T1R2 to form a sweet taste receptor responding to a variety of sweet tastants, whereas a T1R1 + T1R3 heterodimer functions as an umami receptor. In domestic cats, tigers, and cheetahs, however, the *T1R2* gene is pseudogenized, accounting for cats being indifferent to and apparently unable to detect the sweetness of sugars (Li et al. 2005).

The numbers of T2R genes are larger than those of T1R genes (Fig. 4.2a), reflecting the importance of bitter taste perception, which enables organisms to avoid ingesting potentially toxic and harmful substances. In contrast to T1R genes, T2R genes have undergone lineage-specific expansions and contractions during mammalian evolution, similar to OR genes (Shi and Zhang 2006). Although the fraction of pseudogenes is similar between humans and apes, several studies showed that the functional constraint of T2R genes has been relaxed in the human lineage compared with other primate species (Wang et al. 2004; Go et al. 2005). This observation may be explained by the change in diet and use of fire that emerged in human evolution. Among mammals, the numbers of T2R genes are relatively small in cows and dogs, possibly because cows are ruminants and have high detoxification ability, and therefore the detection of poisons in diet may be relatively unimportant. On the other hand, dogs, which are exclusive carnivores, may less frequently encounter toxic chemicals, because toxic compounds are more common in plant than in animal tissues (Shi and Zhang 2006) (see also Chap. 6).

4.7 Concluding Remarks

As shown in Fig. 4.2a, in general, the repertoires of chemosensory receptor genes have shrunk in higher primates, which has likely occurred because of a well-developed visual sense. It should be noted that, however, the repertories of chemosensory receptor genes in humans is not small compared with other higher primates, implying that our olfactory ability may not be particularly poorer (Matsui et al. 2010). It appears that the reduction of gene repertoires in different families did not happen concomitantly; the color vision priority hypothesis was not supported for OR genes, but the hypothesis may partially account for the degeneration of V1R and V2R genes (see foregoing). Obviously, more detailed analyses using larger number of species would be necessary.

The repertoire of each chemosensory receptor gene family is roughly determined by the living environment of each species. However, extensive gene gains and losses observed in OR, V1R, V2R, and T2R gene families suggest that the effect of the change in gene number to the fitness of an organism would be minor. Nei (2007) proposed a concept of "genomic drift" in which the importance of randomness in the change of gene number is emphasized. However, the relative contribution of adaptive and neutral changes to the evolution of chemosensory receptor genes still remains elusive, and further studies should be conducted for a variety of gene families.

In this chapter, we reviewed the evolution of chemosensory receptor genes from the perspective of bioinformatics. Of course, experimental verification is always necessary, including biochemical, physiological, and behavioral studies. However, we should note that experimental studies are usually performed using model organisms such as mice or rats, and this may cause some biases. As we have seen, large repertoires of V1R and V2R genes or function of FPRs in the VNO might be rodent specific and rather exceptional in mammals. On the other hand, bioinformatic analyses can be performed for any species with the same criteria. Because of the accumulation of overwhelming amounts of genomic data, bioinformatic analyses are now becoming more and more important.

References

Adler E, Hoon MA, Mueller KL et al (2000) A novel family of mammalian taste receptors. Cell 100:693–702

Amoore JE (1967) Specific anosmia: a clue to the olfactory code. Nature (Lond) 214:1095–1098

Barton RA (2006) Olfactory evolution and behavioral ecology in primates. Am J Primatol 68:545–558

Baxi KN, Dorries KM, Eisthen HL (2006) Is the vomeronasal system really specialized for detecting pheromones? Trends Neurosci 29:1–7

Boesch C, Boesch-Achermann H (2000) The chimpanzees of the Taï Forest: behavioural ecology and evolution. Oxford University Press, New York

Borowsky B, Adham N, Jones KA et al (2001) Trace amines: identification of a family of mammalian G protein-coupled receptors. Proc Natl Acad Sci USA 98:8966–8971

Boschat C, Pélofi C, Randin O et al (2002) Pheromone detection mediated by a V1r vomeronasal receptor. Nat Neurosci 5:1261–1262

Boulay F, Tardif M, Brouchon L et al (1990) The human N-formylpeptide receptor. Characterization of two cDNA isolates and evidence for a new subfamily of G-protein-coupled receptors. Biochemistry 29:11123–11133

Bryson-Richardson RJ, Logan DW, Currie PD et al (2004) Large-scale analysis of gene structure in rhodopsin-like GPCRs: evidence for widespread loss of an ancient intron. Gene (Amst) 338:15–23

Buck L, Axel R (1991) A novel multigene family may encode odorant receptors: a molecular basis for odor recognition. Cell 65:175–187

De la Cruz O, Blekhman R, Zhang X et al (2009) A signature of evolutionary constraint on a subset of ectopically expressed olfactory receptor genes. Mol Biol Evol 26:491–494

Del Punta K, Leinders-Zufall T, Rodriguez I et al (2002) Deficient pheromone responses in mice lacking a cluster of vomeronasal receptor genes. Nature (Lond) 419:70–74

Dong D, Jones G, Zhang S (2009) Dynamic evolution of bitter taste receptor genes in vertebrates. BMC Evol Biol 9:12

Dulac C, Axel R (1995) A novel family of genes encoding putative pheromone receptors in mammals. Cell 83:195–206

Feldmesser E, Olender T, Khen M et al (2006) Widespread ectopic expression of olfactory receptor genes. BMC Genomics 7:121

Fredriksson R, Lagerström MC, Lundin LG et al (2003) The G-protein-coupled receptors in the human genome form five main families. Phylogenetic analysis, paralogon groups, and fingerprints. Mol Pharmacol 63:1256–1272

Gilad Y, Man O, Pääbo S et al (2003) Human specific loss of olfactory receptor genes. Proc Natl Acad Sci USA 100:3324–3327

Gilad Y, Przeworski M, Lancet D (2004) Loss of olfactory receptor genes coincides with the acquisition of full trichromatic vision in primates. PLoS Biol 2:e5

Gilad Y, Man O, Glusman G (2005) A comparison of the human and chimpanzee olfactory receptor gene repertoires. Genome Res 15:224–230

Gilad Y, Wiebe V, Przeworski M et al (2007) Correction: loss of olfactory receptor genes coincides with the acquisition of full trichromatic vision in primates. PLoS Biol 5:e148

Gimelbrant AA, Skaletsky H, Chess A (2004) Selective pressures on the olfactory receptor repertoire since the human–chimpanzee divergence. Proc Natl Acad Sci USA 101:9019–9022

Glusman G, Bahar A, Sharon D et al (2000) The olfactory receptor gene superfamily: data mining, classification, and nomenclature. Mamm Genome 11:1016–1023

Glusman G, Yanai I, Rubin I et al (2001) The complete human olfactory subgenome. Genome Res 11:685–702

Go Y, Niimura Y (2008) Similar numbers but different repertoires of olfactory receptor genes in humans and chimpanzees. Mol Biol Evol 25:1897–1907

Go Y, Satta Y, Takenaka O et al (2005) Lineage-specific loss of function of bitter taste receptor genes in humans and nonhuman primates. Genetics 170:313–326

Grus WE, Zhang J (2008) Distinct evolutionary patterns between chemoreceptors of 2 vertebrate olfactory systems and the differential tuning hypothesis. Mol Biol Evol 25:1593–1601

Grus WE, Shi P, Zhang YP et al (2005) Dramatic variation of the vomeronasal pheromone receptor gene repertoire among five orders of placental and marsupial mammals. Proc Natl Acad Sci USA 102:5767–5772

Grus WE, Shi P, Zhang J (2007) Largest vertebrate vomeronasal type 1 receptor gene repertoire in the semiaquatic platypus. Mol Biol Evol 24:2153–2157

Hasin Y, Olender T, Khen M et al (2008) High-resolution copy-number variation map reflects human olfactory receptor diversity and evolution. PLoS Genet 4:e1000249

Hasin-Brumshtein Y, Lancet D, Olender T (2009) Human olfaction: from genomic variation to phenotypic diversity. Trends Genet 25:178–184

Herrada G, Dulac C (1997) A novel family of putative pheromone receptors in mammals with a topographically organized and sexually dimorphic distribution. Cell 90:763–773

Hiramatsu C, Melin AD, Aureli F et al (2008) Importance of achromatic contrast in short-range fruit foraging of primates. PLoS One 3:e3356

Ikeda K (2002) New seasonings. Chem Senses 27:847–849

Keller A, Zhuang H, Chi Q et al (2007) Genetic variation in a human odorant receptor alters odour perception. Nature (Lond) 449:468–472

Kimoto H, Haga S, Sato K et al (2005) Sex-specific peptides from exocrine glands stimulate mouse vomeronasal sensory neurons. Nature (Lond) 437:898–901

Kishida T, Kubota S, Shirayama Y et al (2007) The olfactory receptor gene repertoires in secondary-adapted marine vertebrates: evidence for reduction of the functional proportions in cetaceans. Biol Lett 3:428–430

Lévai O, Feistel T, Breer H et al (2006) Cells in the vomeronasal organ express odorant receptors but project to the accessory olfactory bulb. J Comp Neurol 498:476–490

Leinders-Zufall T, Brennan P, Widmayer P et al (2004) MHC class I peptides as chemosensory signals in the vomeronasal organ. Science 5:1033–1037

Li X, Li W, Wang H et al (2005) Pseudogenization of a sweet-receptor gene accounts for cats' indifference toward sugar. PLoS Genet 1:27–35

Liberles SD, Buck LB (2006) A second class of chemosensory receptors in the olfactory epithelium. Nature (Lond) 442:645–650

Liberles SD, Horowitz LF, Kuang D et al (2009) Formyl peptide receptors are candidate chemosensory receptors in the vomeronasal organ. Proc Natl Acad Sci USA 106:9842–9847

Liman ER, Innan H (2003) Relaxed selective pressure on an essential component of pheromone transduction in primate evolution. Proc Natl Acad Sci USA 100:3328–3332

Lindemann L, Hoener MC (2005) A renaissance in trace amines inspired by a novel GPCR family. Trends Pharmacol Sci 26:274–281

Malnic B, Hirono J, Sato T et al (1999) Combinatorial receptor codes for odors. Cell 96:713–723

Matsui A, Go M, Niimura Y (2010) Comparative evolutionary analyses of olfactory receptor gene repertoires in primates do not support a trade-off between olfaction and color vision. Mol Biol Evol 27:1192–1200

Matsunami H, Buck LB (1997) A multigene family encoding a diverse array of putative pheromone receptors in mammals. Cell 90:775–784

Matsunami H, Montmayeur JP, Buck LB (2000) A family of candidate taste receptors in human and mouse. Nature (Lond) 404:601–614

McGowen MR, Clark C, Gatesy J (2008) The vestigial olfactory receptor subgenome of odontocete whales: phylogenetic congruence between gene-tree reconciliation and supermatrix methods. Syst Biol 57:574–590

Menashe I, Man O, Lancet D et al (2003) Different noses for different people. Nat Genet 34:143–144

Menashe I, Aloni R, Lancet D (2006) A probabilistic classifier for olfactory receptor pseudogenes. BMC Bioinformatics 7:393

4 Evolution of Chemosensory Receptor Genes in Primates and Other Mammals

Menashe I, Abaffy T, Hasin Y et al (2007) Genetic elucidation of human hyperosmia to isovaleric acid. PLoS Biol 5:e284

Meredith M (2001) Human vomeronasal organ function: a critical review of best and worst cases. Chem Senses 26:433–445

Migeotte I, Communi D, Parmentier M (2006) Formyl peptide receptors: a promiscuous subfamily of G protein-coupled receptors controlling immune responses. Cytokine Growth Factor Rev 17:501–519

Nei M (2007) The new mutation theory of phenotypic evolution. Proc Natl Acad Sci USA 104:12235–12242

Nei M, Kumar S (2000) Molecular evolution and phylogenetics. Oxford University Press, New York

Nei M, Rooney AP (2005) Concerted and birth-and-death evolution of multigene families. Annu Rev Genet 39:121–152

Nei M, Niimura Y, Nozawa M (2008) The evolution of animal chemosensory receptor gene repertoires: roles of chance and necessity. Nat Rev Genet 9:951–963

Nelson G, Hoon MA, Chandrashekar J et al (2001) Mammalian sweet taste receptors. Cell 106:381–390

Nelson G, Chandrashekar J, Hoon MA et al (2002) An amino-acid taste receptor. Nature (Lond) 416:199–202

Newman T, Trask BJ (2003) Complex evolution of 7E olfactory receptor genes in segmental duplications. Genome Res 13:781–793

Niimura Y (2009a) On the origin and evolution of vertebrate olfactory receptor genes: comparative genome analysis among 23 chordate species. Genome Biol Evol 1:34–44

Niimura Y (2009b) Evolutionary dynamics of olfactory receptor genes in chordates: interaction between environments and genomic contents. Hum Genomics 4(2):107–118

Niimura Y, Nei M (2003) Evolution of olfactory receptor genes in the human genome. Proc Natl Acad Sci USA 100:12235–12240

Niimura Y, Nei M (2005a) Evolutionary dynamics of olfactory receptor genes in fishes and tetrapods. Proc Natl Acad Sci USA 102:6039–6044

Niimura Y, Nei M (2005b) Comparative evolutionary analysis of olfactory receptor gene clusters between humans and mice. Gene (Amst) 346:13–21

Niimura Y, Nei M (2006) Evolutionary dynamics of olfactory and other chemosensory receptor genes in vertebrates. J Hum Genet 51:505–517

Niimura Y, Nei M (2007) Extensive gains and losses of olfactory receptor genes in mammalian evolution. PLoS One 2:e708

Nishida T (1997) Sexual behavior of adult male chimpanzees of the Mahale Mountains National Park, Tanzania. Primates 38:379–398

Nozawa M, Kawahara Y, Nei M (2007) Genomic drift and copy number variation of sensory receptor genes in humans. Proc Natl Acad Sci USA 104:20421–20426

Ohara H, Nikaido M, Date-Ito A et al (2009) Conserved repertoire of orthologous vomeronasal type 1 receptor genes in ruminant species. BMC Evol Biol 9:233

Rivière S, Challet L, Fluegge D et al (2009) Formyl peptide receptor-like proteins are a novel family of vomeronasal chemosensors. Nature (Lond) 459:574–577

Rodriguez I, Mombaerts P (2002) Novel human vomeronasal receptor-like genes reveal species-specific families. Curr Biol 12:R409–R411

Rodriguez I, Greer CA, Mok MY et al (2000) A putative pheromone receptor gene expressed in human olfactory mucosa. Nat Genet 26:18–19

Rudd MK, Endicott RM, Friedman C et al (2009) Comparative sequence analysis of primate subtelomeres originating from a chromosome fission event. Genome Res 19:33–41

Ryba NJ, Tirindelli R (1997) A new multigene family of putative pheromone receptors. Neuron 19:371–379

Saito H, Chi Q, Zhuang H et al (2009) Odor coding by a mammalian receptor repertoire. Sci Signal 2:ra9

Serizawa S, Miyamichi K, Nakatani H et al (2003) Negative feedback regulation ensures the one receptor-one olfactory neuron rule in mouse. Science 302:2088–2094

Shi P, Zhang J (2006) Contrasting modes of evolution between vertebrate sweet/umami receptor genes and bitter receptor genes. Mol Biol Evol 23:292–300

Shirokova E, Raguse JD, Meyerhof W et al (2008) The human vomeronasal type-1 receptor family—detection of volatiles and cAMP signaling in HeLa/Olf cells. FASEB J 22:1416–1425

Smith TD, Bhatnagar KP (2004) Microsmatic primates: reconsidering how and when size matters. Anat Rec B New Anat 279:24–31

Spehr M, Gisselmann G, Poplawski A et al (2003) Identification of a testicular odorant receptor mediating human sperm chemotaxis. Science 299:2054–2058

Stern K, McClintock MK (1998) Regulation of ovulation by human pheromones. Nature (Lond) 392:177–179

Wang X, Thomas SD, Zhang J (2004) Relaxation of selective constraint and loss of function in the evolution of human bitter taste receptor genes. Hum Mol Genet 13:2671–2678

Wysocki CJ, Preti G (2004) Facts, fallacies, fears, and frustrations with human pheromones. Anat Rec A Discov Mol Cell Evol Biol 281:1201–1211

Young JM, Trask BJ (2007) V2R gene families degenerated in primates, dog and cow, but expanded in opossum. Trends Genet 23:212–215

Young JM, Shykind BM, Lane RP et al (2003) Odorant receptor expressed sequence tags demonstrate olfactory expression of over 400 genes, extensive alternate splicing and unequal expression levels. Genome Biol 4:R71

Young JM, Kambere M, Trask BJ et al (2005) Divergent V1R repertoires in five species: amplification in rodents, decimation in primates, and a surprisingly small repertoire in dogs. Genome Res 15:231–240

Young JM, Endicott RM, Parghi SS et al (2008) Extensive copy-number variation of the human olfactory receptor gene family. Am J Hum Genet 83:228–242

Zhang J, Webb DM (2003) Evolutionary deterioration of the vomeronasal pheromone transduction pathway in catarrhine primates. Proc Natl Acad Sci USA 100:8337–8341

Zhang X, De la Cruz O, Pinto JM et al (2007) Characterizing the expression of the human olfactory receptor gene family using a novel DNA microarray. Genome Biol 8:R86

Zozulya S, Echeverri F, Nguyen T (2001) The human olfactory receptor repertoire. Genome Biol 2:research0018.1–0018.12

Chapter 5
Functional Evolution of Primate Odorant Receptors

Kaylin A. Adipietro, Hiroaki Matsunami, and Hanyi Zhuang

Abbreviations

AC	Adenylyl cyclase
ATP	Adenosine-5′-triphosphate
cAMP	Cyclic adenosine monophosphate
CNG	Cyclic nucleotide-gated
dN	Number of nonsynonymous nucleotide substitutions per nonsynonymous site
dS	Number of synonymous nucleotide substitutions per synonymous site
FPR	Formyl peptide receptor
GC-D	Guanylyl cyclase type D
GPCR	Heterotrimeric G-protein-coupled receptor
$G_{\alpha olf}$	Olfactory heterotrimeric G-protein alpha-subunit
GTP	Guanosine-5′-triphosphate
NWM	New World monkey

K.A. Adipietro
Department of Molecular Genetics and Microbiology, Duke University
Medical Center, Durham, NC, USA

H. Matsunami (✉)
Department of Molecular Genetics and Microbiology, Duke University
Medical Center, Durham, NC, USA

Department of Neurobiology, Duke University Medical Center, Durham, NC, USA
e-mail: matsu004@mc.duke.edu

H. Zhuang
Department of Pathophysiology, Key Laboratory of Cell Differentiation
and Apoptosis of National Ministry of Education, Shanghai Jiaotong
University School of Medicine, Shanghai, China

Institute of Health Sciences, Shanghai Institutes for Biological
Sciences of Chinese Academy of Sciences and Shanghai Jiaotong
University School of Medicine, Shanghai, China

H. Hirai et al. (eds.), *Post-Genome Biology of Primates*, Primatology Monographs,
DOI 10.1007/978-4-431-54011-3_5, © Springer 2012

OR	Olfactory receptor
OWM	Old World monkey
RTP	Receptor transport protein
SNP	Single nucleotide polymorphism
TM	Transmembrane
TAAR	Trace amine-associated receptor
V1R	Vomeronasal type I receptor
V2R	Vomeronasal type II receptor

5.1 Introduction

The ability to detect and discriminate volatile chemical cues in the environment is crucial for the fitness of an individual and a species because odorants can signify favorable or toxic food sources, mating preferences, predators, and habitats (Mombaerts 2004; Keller and Vosshall 2008). This capacity to detect a vast variety of odorous chemicals is accomplished through a repertoire of hundreds of odorant receptors (ORs), which are expressed at the cell surface of olfactory sensory neurons in the main olfactory epithelium. It is believed that each odorous chemical activates a specific combination of multiple ORs, leading to odor perception. In addition, biologically relevant chemicals such as urine or food odor are usually complex mixtures. These factors make it challenging to address questions regarding mammalian olfaction. Nevertheless, an important step toward understanding olfaction in the context of primate evolution is to link together changes in anatomy, sequence, function, and behavior.

5.2 Anatomy of the Primate Olfactory System

The mammalian olfactory system can be divided into two major subsystems based on broad functional distinctions: The main olfactory system comprises the olfactory epithelium and the main olfactory bulb and is traditionally viewed as being involved in detecting odorous chemicals in the animal's environment; the accessory olfactory system includes the vomeronasal organ and the accessory olfactory bulb and has been implicated in interindividual pheromone detection controlling innate behaviors and physiological changes. Recent studies show that the mammalian olfactory system is more complicated than the dichotomy of the odor and pheromone detections, respectively, assigned to the main and accessory olfactory systems because many functional interactions exist between the two subsystems (Dorries et al. 1995; Sam et al. 2001; Swann et al. 2001; Trinh and Storm 2003).

Evolution played a key in role in determining the morphology of the primate nose. Based upon physical features of the nose, primates can be classified into two suborders, strepsirrhines and haplorrhines. Strepsirrhines are characterized by the presence of a sensitive, moist naked surface around the tip of their nose, called a rhinarium. This suborder consists of lemurs, lorsies, and galagos. Haplorrhines lack a rhinarium and encompass tarsiers, New World monkeys (NWMs), Old World monkeys (OWMs), and the great apes. In addition to the absence of a rhinarium, the size of the main olfactory bulb is reduced in haplorrhines compared to strepsirrhines, and OWMs and great apes lack a functional vomeronasal organ and the accessory olfactory bulb (Barton 2006). The size of the main olfactory bulb is also greater in nocturnal animals (mostly strepsirrhines) and frugivores/insectivores in comparison to diurnal animals (most haplorrhines) and folivores, suggesting the evolution of the main olfactory system is correlated with activity period and diet (Barton 2006).

On a histological level, the olfactory epithelium is located in the dorsal and dorso-posterior aspect of the nasal cavity, lining the turbinates. Compared to rodents, primates have relatively simple turbinates and thus a significantly reduced percentage of surface area in the nasal cavity that is lined with olfactory epithelium (Harkema et al. 2006). It was shown that in rodents, rabbits, and dogs approximately 50% of the nasal cavity surface area is covered with olfactory epithelium compared to the mere 3% in primate species (Gross et al. 1982); furthermore, the size of the olfactory epithelium in dogs is approximately 100 cm^2 compared to 10 cm^2 in humans (Issel-Tarver and Rine 1997).

5.3 The Odorant Receptors and Signal Transduction

Since their initial discovery by Buck and Axel in 1991, it is now known that ORs are the largest gene superfamily in the mammalian genome, composed of 800–1,500 receptors (including pseudogenes), depending upon the species. As members of family I heterotrimeric G-protein-coupled receptors (GPCRs), ORs are rhodopsin-like receptors with a characteristic seven-transmembrane (TM) domain structure. According to one sequence-based classification, mammalian ORs are classified into two classes, 17 families, and 250 subfamilies (Mombaerts 2001; Zozulya et al. 2001; Gilad et al. 2005; Man et al. 2007). ORs that are members of the same sub-family are assumed to have similar functional properties (Man et al. 2007), but despite the remarkable progress made in the field, our knowledge of specific receptor–ligand interactions is limited and lacks functional classification of ORs based on odor selectivity.

Extracellular ligands, or odor molecules, are predicted to interact with certain residues in a pocket formed by the combination of transmembrane domains (Man et al. 2004, 2007; Katada et al. 2005) (Fig. 5.1). Once an odor is bound to the receptor, the guanosine 5'-triphosphate (GTP)-bound form of the G_α subunit, $G_{\alpha olf}$, stimulates adenylyl cyclase III to catalyze the conversion of adenosine triphosphate (ATP) into cyclic adenosine monophosphate (cAMP), which causes the opening of cyclic nucleotide-gated cation channels and depolarization of the olfactory sensory neuron (Mombaerts 2001). Until recently, it was not easy to determine the ligands of a given OR because of difficulty in efficiently expressing these

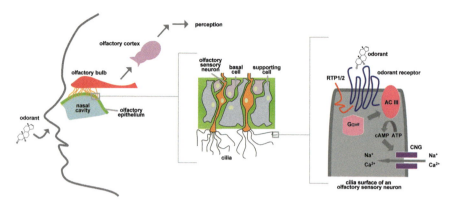

Fig. 5.1 Olfactory signal transduction in primates. Olfactory receptors are trafficked to the cell surface of the cilia of olfactory sensory neurons in the olfactory epithelium by receptor transport protein (RTP) *RTP1/2*. When an odorant passes through the nasal cavity, the odor binds to its cognate odorant receptor, causing the guanosine triphosphate guanosine triphosphate (GTP)-bound form of G$_{\alpha olf}$ to stimulate adenylyl cyclase III (*AC III*) to catalyze the conversion of ATP into *cAMP*. This increase in cAMP opens cyclic nucleotide-gated (*CNG*) cation channels, causing a depolarization of the neuron and transmission of the signal to the olfactory bulb and higher brain regions

receptors in a heterologous system. With the identification of OR accessory proteins that enhance the functional expression of ORs and development of a cell-based expression system and functional assay, ligands for a large number of ORs from different species have been gradually uncovered (Saito et al. 2004; Matsunami 2005; Zhuang and Matsunami 2007, 2008; Saito et al. 2009).

The current assumption on ligand–OR interaction is that an OR is functional if the open reading frame is intact. Similarities between coding sequences are used to predict functionality in lieu of real experimental data (Malnic et al. 1999; Glusman et al. 2000; Kajiya et al. 2001; Zhang and Firestein 2002; Godfrey et al. 2004; Man et al. 2007). This reasoning has also led to the idea that OR orthologues from different species will maintain similar binding properties (Godfrey et al. 2004; Man et al. 2007). As a more reliable predictor of functionality, Man et al. used coding region alignments to identify 22 amino acid residues that are conserved among orthologues but variable among paralogues as candidate residues important for odor binding (Man et al. 2004). More recently, Saito et al. functionally characterized 52 mouse ORs and 10 human ORs and used these data with sequence alignments to examine how well descriptors of amino acid residues could predict their functional data. They found 16 amino acid sites with descriptors of volume, polarity, and composition that could explain approximately 50% of the variance in their data set (Saito et al. 2009). Intriguingly, the set of sites in these two studies did not overlap. One potential explanation is that OR orthologues do not have similar responses against the same odorant, as suggested by a recent study (Zhuang et al. 2009), in which the orthologues of an OR, OR7D4, from several primate species were cloned and characterized using a heterologous expression assay and were shown to have a

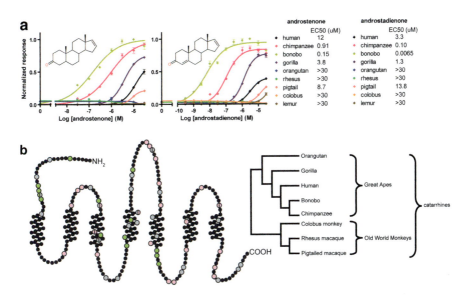

Fig. 5.2 Primate odorant receptor (OR)7D4 orthologues exhibit functional divergence in response to sex steroid-derived odors. (**a**) Dose–response curves of all intact primate OR7D4 orthologues to androstenone (*left*) and androstadienone (*right*) are assessed in a heterologous odorant receptor expression system using a luciferase reporter gene assay. y-axis denotes normalized response ± SEM. The EC_{50} value of each dose–response curve is shown. (**b**) Nonsynonymous mutations in OR7D4 contribute to its functional divergence. OR7D4 snake plot with amino acid changes in the catarrhine (Old World monkeys and great apes) lineages indicated by *non-black circles*. Residues are colored in *pink* for decreased function, *gray* for no significant change, and *green* for increased function. The phylogenetic relationships of the catarrhine species with an intact OR7D4 gene are shown in the tree. (Modified from Zhuang et al. 2009)

very diverse response profile, supporting this idea (Fig. 5.2a). Future studies involving additional ORs are necessary to determine if the functional differences among orthologues are restricted to specific ORs or are more general phenomena across all families.

5.4 Odorant Receptor Repertoire in Primates

With the availability of newly sequenced genomes, it is now possible to appreciate how members of the OR gene family have evolved, presumably to fulfill the needs of different species (Kambere and Lane 2007; Niimura and Nei 2007). The size of the OR repertoire varies dramatically between species: mouse and rat genomes encode more than 1,000 intact ORs, although the chimpanzee and the human contain fewer than 400 intact genes (Spehr and Munger 2009). In addition to the size of the gene family, the OR repertoire has been subject to extensive gains and losses of receptors over time, likely reflecting species- and niche-specific adaptations and innovations (Rouquier et al. 2000; Kambere and Lane 2007; Niimura and Nei 2007). The rate of pseudogenization caused by coding region disruptions in the mouse genome is approximately 20%, whereas this rate appears to increase to 30% in NWMs and to greater than 50% in humans and chimpanzees (Gilad et al. 2003, 2005; Keller and Vosshall 2008). The result is a significant decline in the number of intact ORs in the primate lineage and the categorization of a small subset of putative orthologues across species (Rouquier et al. 2000; Go and Niimura 2008). Some bioinformatic studies have suggested this decline in the number of functional OR genes was gradual in all branches of primates, while others suggest this loss occurred suddenly when OWM and great apes diverged from a common ancestor and is most pronounced in humans (Rouquier et al. 2000; Gilad et al. 2003, 2004, 2005; Go and Niimura 2008). Gilad et al. speculated that this loss of functional ORs and significant increase in pseudogenes in the primate genome reflects a decreased reliance on smell and an increased dependence on visual cues for survival (Gilad et al. 2004).

5.5 Olfaction and the Color Vision Hypothesis

The evolution and elaboration of the primate visual system has led to the belief that primates are in general exceptionally vision-oriented species; consequently, primates are regarded as animals which have decreased olfactory capabilities because of the lessened requirement of this sensory system for survival (Jacobs 1996; Gilad et al. 2004; Barton 2006; Liman 2006; Niimura and Nei 2007; Go and Niimura 2008). Although there are anatomical and sequence data that indicate deterioration of both the accessory and main olfactory systems in primates, there are behavioral data suggesting that olfaction may play an underestimated role in primate behavior (Laska et al. 2000; Liman 2006; Barton 2006).

The evolution of the primate visual system has been investigated in great detail. Visual specialization in primates is more discernible in haplorrhines than strepsirrhines, including larger visual cortices, greater acuity, and the development of full trichromatic vision (Barton 2006). Trichromatic vision is mediated by three cone opsins that are activated by different wavelengths (Jacobs 1996, 2008; Jacobs et al. 1996). All OWMs, great apes, and one NWM, the howler monkey, express three opsins, yielding trichromatic vision. Many NWMs are highly polymorphic dichromats, and females heterozygous at the single locus for one of the opsins are trichromatic. Among strepsirrhines, lorises, bushbabies, and some types of lemurs have a single, functional opsin, rendering these species monochromats (Jacobs 2008). Other species such as the mouse lemur and ring-tailed lemur possess two functional opsins, allowing for dichromatic vision (Perry et al. 2007; Leonhardt et al. 2008).

In an attempt to correlate the acquisition of trichromatic vision with deterioration in the main olfactory system, Gilad et al. conducted a sequence-based analysis of random OR gene sequences across 19 primate species from both strepsirrhine and haplorrhine suborders (Gilad et al. 2004, 2007). They found that the fraction of pseudogenes in OWM and great apes was significantly higher than that of NWMs and prosimians, with the exception of the howler monkey, a trichromatic NWM whose genome contained a similar number of pseudogenes to that of OWMs and the great apes; intriguingly, the second phenotype shared by the great apes, OWMs, and the howler monkey is that of full trichromatic vision. Based on this observation, the authors concluded that the acquisition of full trichromatic vision occurred concomitantly with an increase in OR pseudogenization, indicative of a decline in the sense of smell; this is termed the "color vision priority hypothesis" (Gilad et al. 2004, 2007; Matsui et al. 2010). While the authors argue their observations suggest an exchange of importance between these two sensory systems, they have not behaviorally or functionally tested the olfactory abilities in these primates.

To further investigate the color vision priority hypothesis, Matsui et al. (2010) compared the OR repertoire of three great apes (humans, chimpanzees, orangutans), OWMs (rhesus macaque), NWMs (marmoset), bushbaby, mouse lemur, and tree shrew. Their results indicate that the size of the putatively functional OR repertoire between primates is not significantly different and that the degeneration of the repertoire occurred gradually across every lineage, not after the divergence from NWMs. These results suggest that the evolution of the OR repertoire in primates cannot be explained solely by the color vision priority hypothesis.

5.6 Other Chemosensory Receptors Expressed in the Olfactory System

The complexity of the primate olfactory system is also marked by a plethora of different types of sensory cells. Several non-OR chemosensory receptor families are expressed in the olfactory system in mammals. These receptors are likely to play important roles in detecting chemicals relevant for the behavior and fitness of an animal. Some of these receptors could be expressed in primate olfactory systems and are thus of significance to the evolution of species in this order.

A small family of trace amine-associated receptors (TAARs) are shown to be expressed in a subset of mouse olfactory epithelium and activated by amines (Liberles and Buck 2006). Although humans have fewer numbers of TAARs in the genome, some of the members might be expressed and function as sensory receptors.

A fraction of olfactory sensory neurons express guanylyl cyclase type D (GC-D) in mice. GC-D is implicated in detecting carbon dioxide as well as peptides (Hu et al. 2007; Leinders-Zufall et al. 2007; Sun et al. 2009). GC-D is a pseudogene in primates except for lemurs (Young et al. 2007).

The vomeronasal organ does not express ORs but instead it expresses different GPCR families of chemosensory receptors, the vomeronasal type I receptors (V1Rs), the vomeronasal type II receptors (V2Rs), and the formyl peptide receptors (FPRs). Although OWMs and great apes do not have a functional vomeronasal organ, the human genome contains a few intact V1R genes that are shown to be expressed in the main olfactory epithelium (Rodriguez et al. 2000). Even though NWMs possess the vomeronasal organ, the marmoset has only 8 intact V1R genes in the genome whereas lemurs have as many as 200 V1Rs (Young et al. 2009). No intact V2Rs are found in the primate genome (Young and Trask 2007). The FPRs are a class of newly identified vomeronasal receptors in mice (Liberles et al. 2009; Riviere et al. 2009). It is not yet clear whether primates have the vomeronasal FPRs.

5.7 Pheromones in Primates

Pheromones function as chemical signals that influence behavior or physiology within the same species. The traditional view of primate pheromone-sensing was based on the assumption that most species belonging to this order rely mostly on visual and acoustic cues rather than chemical ones. However, there is evidence pointing to the preponderance of olfactory abilities in some primate species (Laska et al. 2005a, 2006a).

The prosimians possess the most highly developed olfactory system among primates and probably rely heavily on chemical communications. The ring-tailed lemur is the first non-human primate in which olfactory individual recognition has been demonstrated on the basis of the specialized scent glands on their wrists and around their anus and genitals that deposit secretions via a variety of scent-marking display behaviors (Jolly 1966). Gas chromatography studies also showed that the male brachial gland secretions of individual lemurs have a particular signature that is maintained through time for enhanced distinctiveness in individual recognition (Dapporto 2008; Boulet et al. 2009).

Similarly to the ring-railed lemurs, common marmosets, a NWM, communicate with secretions from anogenital and brachial glands to convey a variety of information, especially reproductive status (Lazaro-Perea et al. 1999; Ziegler et al. 2005). In a given marmoset social group, only one to two dominant females can ovulate and breed because the presence of a dominant female prevents the ovulation of unrelated female marmosets so that they are unable to breed (Abbott 1984; Digby and Barreto 1993; Nievergelt et al. 2000; Ziegler and Sousa 2002). It has been suggested that the mechanism by which this occurs involves contact of subordinate females with olfactory stimuli from dominant females. When subordinate females are physically separated from the dominant female but presented with scents transferred from the dominant female, there is a delay in the onset of ovulation, indicating possible involvement of personal cues in the maintenance of the suppression of ovulation in the subordinate females (Barrett et al. 1990). Although it is likely that either the main olfactory epithelium or the vomeronasal organ is involved in the reception of these olfactory stimuli, it is not known which receptors are functioning in these behaviors.

A few sex hormone-related steroidal odors have been proposed to function as pheromones in humans. In the domestic pig, a testosterone-related compound, androstenone, has been identified as a male pheromone that activates the neurons in the main olfactory epithelium and induces a receptive mating stance in estrous females (Dorries et al. 1997). Interestingly, androstenone is found in human saliva, sweat, and urine (Brooksbank and Haslewood 1961; Brooksbank et al. 1974; Claus and Alsing 1976; Bird and Gower 1983), and a structurally related compound, androstadienone, has been shown to influence human behaviors in aspects such as brain activity, endocrine levels, and sexual arousal (Lundstrom et al. 2003; Preti et al. 2003; Bensafi et al. 2004; Berglund et al. 2006; Wyart et al. 2007). Furthermore, a few reports have proposed that female urine contains the estrogen-like steroid, estratetraenol, which could also be a candidate human pheromone (Jacob et al. 2001; Berglund et al. 2006). However, the effects of these compounds on human reproductive activities remain controversial (Pause 2004).

It is not known so far whether non-human primates exhibit physiological responses to these steroidal compounds. Squirrel monkeys and pigtailed macaques can detect androstenone at micromolar concentrations and estratetraenol at millimolar concentrations (Laska et al. 2005b, 2006b). More interestingly, it was reported that there is a sex-specific olfactory sensitivity in spider monkeys, with males not responding to the highest concentrations of the androstenone tested and females not responding to estratetraenol.

The recent report showing dramatic functional differences of primate OR7D4, a receptor that is activated by androstenone and androstadienone, made it tempting to speculate that sensitivity to these two chemicals, which is at least partly determined by OR7D4 in humans, could play a role in the reproductive behavior in some primate species. For example, bonobo OR7D4 is by far the most sensitive receptor to both compounds among all the orthologues tested (Zhuang et al. 2009) (Fig. 5.2a). Intriguingly, the bonobos are distinguished by the prominent role of sexual activities in their society (de Waal 1995), although the role of olfaction in their behavior is unexplored.

5.8 Sequence-Based and Function-Based OR Evolutionary Analyses

Because olfaction is closely linked to how animals evaluate their conspecifics and perceive the environment, genes involved in olfaction are likely subject to strong selective pressures. On an evolutionary scale, positive selection may be acting on certain residues of the OR sequence to select for advantageous genetic variants. However, it is also possible that olfaction has become increasingly less important in the primate lineage, thus relaxing the selective pressures acting on this gene family. Evidence for positive selection can be determined using well-established statistical methods comparing coding regions of orthologous genes across species and analyzing the rates of nonsynonymous (those that change the amino acid) to synonymous (those that do not change the amino acid) substitutions relative to the number of possible changes for a particular sequence (Gilad et al. 2003; Gimelbrant et al. 2004). In most comparisons of related genes, the number of synonymous substitutions per synonymous site (dS) exceeds the number of nonsynonymous substitutions per nonsynonymous site (dN); this is because most nonsynonymous substitutions are detrimental to the protein whereas synonymous mutations are selectively neutral, or nearly so (Hughes 2008). Purifying selection acts to eliminate the majority of nonsynonymous mutations lowering the dN relative to dS, thus, a dN/dS < 1 indicates purifying selection acting on a gene. In a number of cases, natural selection may favor a nonsynonymous change at certain residues, leading to an excess of dN relative to dS, causing dN/dS > 1, indicative of positive selection (Hughes 2008).

Using this type of analysis, several studies using a genome-wide approach have indicated many OR genes are under positive selection in mammals, including chimpanzees and humans (Gilad et al. 2000, 2003; Clark et al. 2003; Nielsen et al. 2005; Sabeti et al. 2007; Moreno-Estrada et al. 2008; Zhuang et al. 2009). In a comparison of 13,731 human–chimpanzee orthologues, Nielsen et al. calculated dN/dS ratios to uncover evidence of positive selection acting on certain genes. This analysis returned genes involved in the immune response, followed by genes related to olfactory perception (Nielsen et al. 2005). By comparing trios of human–chimpanzee–mouse OR orthologues, one group found evidence of positive selection acting on ORs in both humans and chimpanzees (Gilad et al. 2005). Although these studies show evidence for positive selective pressure shaping ORs, others have found no evidence for positive selection in the human–chimpanzee lineage (Gimelbrant et al. 2004; Go and Niimura 2008). A comparison of 186 OR orthologous pairs between human and chimpanzee demonstrated no evidence of positive selection acting on the genes, consistent with relaxation of selective constraints on OR genes (Gimelbrant et al. 2004). These discrepancies are likely caused by the differences in sequences and genes used in the analysis, as well as the methodology behind each analysis.

The computational methods used to identify evidence for positive selection were recently challenged by several studies (Hughes 2008; Yokoyama et al. 2008; Nozawa et al. 2009; Hurst 2009a). Nozawa et al. showed that site-prediction methods often falsely identify sites where amino acid substitutions are unlikely to be of functional importance (Nozawa et al. 2009). Hurst argued that sequences showing accelerated rates of change and a dN/dS ratio exceeding 1 may not be under positive selection, but rather were subject to a biased DNA repair process (Hurst 2009b). Each of these studies stresses that a convincing demonstration of positive selection requires more than just computational analyses.

In the case of visual opsins, Yokoyama et al. reconstructed ancestral opsin genes from vertebrates and experimentally tested their maximal absorption in an attempt to correlate evolutionary changes associated with habitat to changes in function (Yokoyama et al. 2008). Having experimentally identified important amino acid replacements at 12 sites, they compared their data to those generated using parsimony and Bayesian methods to predict positive selection. They found the computational predictions were unsuccessful at identifying sites implicated in function using experimental methods. Importantly, the authors and others stress the need for experimental conformation to better understand adaptive evolution (Yokoyama et al. 2008; Nozawa et al. 2009).

Recently, a similar type of analysis was conducted using OR7D4, a human OR shown to be selectively activated by the steroidal ligands androstenone and androstadienone (Zhuang et al. 2009). The authors reconstructed and analyzed ligand-mediated activation of putative ancestral ORs to investigate the functional evolution of this receptor (Fig. 5.2b). Similar to the opsin study, they experimentally identified functionally important residues but found these were not predicted by computational methods detecting positive selection. Future analysis on additional ORs is required to determine whether this finding is specific to a particular OR or if it is a shared characteristic among different OR types.

5.9 Functional Variance of Odorant Receptors Within Species

Individuals with higher sensitivity to odors related to food, mates, or predators could have a fitness advantage over their conspecifics. On the other hand, lack of aversive response to beneficial or medicinal materials could also increase the fitness among the population. Individual differences in the perception of various odorant compounds have been noted in many behavioral studies in humans (Amoore 1967; Brown et al. 1968; Wysocki and Gilbert 1989; Jafek et al. 1990). Although culture and upbringing can provide sensory experiences that influence olfaction perception, twin studies point to the genetics underlying sensitivity to odors (Wysocki and Beauchamp 1984). Furthermore, segregating pseudogenes and copy number variations in the human OR repertoire further validate a genetic basis underlying perceptual variability (Menashe et al. 2003; Nozawa et al. 2007; Young et al. 2008; Hasin et al. 2008).

For example, about 30% of our population is reported to be unable to smell androstenone (Amoore 1977). Keller et al. showed that a human OR, OR7D4, which selectively responds to androstenone and androstadienone, harbors two linked single nucleotide polymorphisms (SNPs), R88W and T133M, that severely impair function in vitro compared to the reference OR7D4 (Keller et al. 2007). This functional variation is well correlated with variability in human perception of these steroidal odors (Keller et al. 2007). Two rarer variants, P79L and S84N, which have severely impaired or dramatically increased function, respectively, also showed a correlation between subjects who possess these variants and respective perception to the odors (Keller et al. 2007).

Similarly, a genetic basis for high sensitivity to isovaleric acid has been recently elucidated through the combination of a genotype–phenotype association study and an in vitro assay validation (Menashe et al. 2007). Menashe et al. tested the association between the SNPs in 43 OR segregating loci and human olfactory thresholds to four odorants (Menashe et al. 2007). They found association between a nonsense SNP in OR11H7, which is activated by isovaleric acid in vitro, and the detection threshold of the ligand (Menashe et al. 2007). Although these examples are likely to represent a tiny fraction of functional variation of ORs in a given species, these studies provide evidence that variation in OR sequences could affect odor perception. Future studies will investigate if other primate species also have polymorphic ORs that cause differences in function and show interindividual differences in odor perception. It would be interesting to ask whether variations in OR function are correlated with the social status of an individual. In addition, it would be important to delineate whether OR variants resulted from either relaxation of selective constraints or positive selection in the population. Future studies involving sequence analysis of ORs from a large number of geologically diverse populations may help distinguish the two possibilities.

5.10 Conclusion

By combining functional analysis of OR orthologues with sequence-based computational analysis of selective pressures, we can gain deeper insights into the functional evolution of the ORs. We can compare the histories of a variety of ORs across primate species in hopes of identifying residues that are critical for the regulation of OR ligand selectivity and other functional properties. Comparing results in species that differ in their visual abilities will gain insight into the idea of a visual–olfactory trade-off. Relying on sequence analysis alone is likely to be insufficient for predicting receptor function, and therefore experimentally matching receptors with ligands is indispensable for a thorough understanding of the evolutionary mechanisms in the olfactory system in primates and other animals.

Acknowledgments We are indebted to Dr. Hirohisa Hirai, Dr. Hiroo Imai, and Dr. Yasuhiro Go for their invitation to contribute to the book. Our work described here is supported by National Institutes of Health, Defense Advanced Research Projects Agency, Chinese National Natural Science Foundation, Shanghai Municipal Education Commission, Shanghai Education Development Foundation, and the Science and Technology Commission of Shanghai.

References

Abbott DH (1984) Behavioral and physiological suppression of fertility in subordinate marmoset monkeys. Am J Primatol 6:169–186

Amoore JE (1967) Specific anosmia: a clue to the olfactory code. Nature (Lond) 214:1095–1098

Amoore JE (1977) Specific anosmia and the concept of primary odors. Chem Senses Flav 2:267–281

Barrett J, Abbott DH, George LM (1990) Extension of reproductive suppression by pheromonal cues in subordinate female marmoset monkeys, *Callithrix jacchus*. J Reprod Fertil 90:411–418

Barton RA (2006) Olfactory evolution and behavioral ecology in primates. Am J Primatol 68:545–558

Bensafi M, Brown WM, Khan R et al (2004) Sniffing human sex-steroid derived compounds modulates mood, memory and autonomic nervous system function in specific behavioral contexts. Behav Brain Res 152:11–22

Berglund H, Lindstrom P, Savic I (2006) Brain response to putative pheromones in lesbian women. Proc Natl Acad Sci USA 103:8269–8274

Bird S, Gower DB (1983) Estimation of the odorous steroid, 5 alpha-androst-16-en-3-one, in human saliva. Experientia (Basel) 39:790–792

Boulet M, Charpentier MJ, Drea CM (2009) Decoding an olfactory mechanism of kin recognition and inbreeding avoidance in a primate. BMC Evol Biol 9:281

5 Functional Evolution of Primate Odorant Receptors

Brooksbank BW, Haslewood GA (1961) The estimation of androst-16-en-3-alpha-ol in human urine. Partial synthesis of androstenol and of its beta-glucosiduronic acid. Biochem J 80:488–496

Brooksbank BW, Brown R, Gustafsson JA (1974) The detection of 5-alpha-androst-16-en-3-alpha-ol in human male axillary sweat. Experientia (Basel) 30:864–865

Brown KS, Maclean CM, Robinette RR (1968) The distribution of the sensitivity to chemical odors in man. Hum Biol 40:456–472

Clark AG, Glanowski S, Nielsen R et al (2003) Inferring nonneutral evolution from human-chimp-mouse orthologous gene trios. Science 302:1960–1963

Claus R, Alsing W (1976) Occurrence of 5-alpha-androst-16-en-3-one, a boar pheromone, in man and its relationship to testosterone. J Endocrinol 68:483–484

Dapporto L (2008) The asymmetric scent: ringtailed lemurs (*Lemur catta*) have distinct chemical signatures in left and right brachial glands. Naturwissenschaften 95:987–991

de Waal FB (1995) Bonobo sex and society. Sci Am 272:82–88

Digby LJ, Barreto CE (1993) Social organization in a wild population of *Callithrix jacchus*. I. Group composition and dynamics. Folia Primatol (Basel) 61:123–134

Dorries K, Adkins-Regan E, Halpern B (1995) Olfactory sensitivity to the pheromone, androstenone, is sexually dimorphic in the pig. Physiol Behav 57:255–259

Dorries K, Adkins-Regan E, Halpern B (1997) Sensitivity and behavioral responses to the pheromone androstenone are not mediated by the vomeronasal organ in domestic pigs. Brain Behav Evol 49:53–62

Gilad Y, Segre D, Skorecki K et al (2000) Dichotomy of single-nucleotide polymorphism haplotypes in olfactory receptor genes and pseudogenes. Nat Genet 26:221–224

Gilad Y, Man O, Paabo S et al (2003) Human specific loss of olfactory receptor genes. Proc Natl Acad Sci USA 100:3324–3327

Gilad Y, Przeworski M, Lancet D (2004) Loss of olfactory receptor genes coincides with the acquisition of full trichromatic vision in primates. PLoS Biol 2:E5

Gilad Y, Man O, Glusman G (2005) A comparison of the human and chimpanzee olfactory receptor gene repertoires. Genome Res 15:224–230

Gilad Y, Przeworski M, Lancet D (2007) Loss of olfactory receptor genes coincides with the acquisition of full trichromatic vision in primates. PLoS Biol 5:e148

Gimelbrant AA, Skaletsky H, Chess A (2004) Selective pressures on the olfactory receptor repertoire since the human-chimpanzee divergence. Proc Natl Acad Sci USA 101:9019–9022

Glusman G, Bahar A, Sharon D et al (2000) The olfactory receptor gene superfamily: data mining, classification, and nomenclature. Mamm Genome 11:1016–1023

Go Y, Niimura Y (2008) Similar numbers but different repertoires of olfactory receptor genes in humans and chimpanzees. Mol Biol Evol 25:1897–1907

Godfrey PA, Malnic B, Buck LB (2004) The mouse olfactory receptor gene family. Proc Natl Acad Sci USA 101:2156–2161

Gross EA, Swenberg JA, Fields S et al (1982) Comparative morphometry of the nasal cavity in rats and mice. J Anat 135:83–88

Harkema JR, Carey SA, Wagner JG (2006) The nose revisited: a brief review of the comparative structure, function, and toxicologic pathology of the nasal epithelium. Toxicol Pathol 34:252–269

Hasin Y, Olender T, Khen M et al (2008) High-resolution copy-number variation map reflects human olfactory receptor diversity and evolution. PLoS Genet 4:e1000249

Hu J, Zhong C, Ding C et al (2007) Detection of near-atmospheric concentrations of CO_2 by an olfactory subsystem in the mouse. Science 317:953–957

Hughes AL (2008) The origin of adaptive phenotypes. Proc Natl Acad Sci USA 105:13193–13194

Hurst LD (2009a) Evolutionary genomics and the reach of selection. J Biol 8:12

Hurst LD (2009b) Evolutionary genomics: a positive becomes a negative. Nature (Lond) 457:543–544

Issel-Tarver L, Rine J (1997) The evolution of mammalian olfactory receptor genes. Genetics 145:185–195

Jacob S, Kinnunen LH, Metz J et al (2001) Sustained human chemosignal unconsciously alters brain function. Neuroreport 12:2391–2394

Jacobs GH (1996) Primate photopigments and primate color vision. Proc Natl Acad Sci USA 93:577–581

Jacobs GH (2008) Primate color vision: a comparative perspective. Vis Neurosci 25:619–633

Jacobs GH, Neitz M, Deegan JF et al (1996) Trichromatic colour vision in new world monkeys. Nature (Lond) 382:156–158

Jafek BW, Gordon AS, Moran DT et al (1990) Congenital anosmia. Ear Nose Throat J 69:331–337

Jolly A (1966) Lemur social behavior and primate intelligence. Science 153:501–506

Kajiya K, Inaki K, Tanaka M et al (2001) Molecular bases of odor discrimination: reconstitution of olfactory receptors that recognize overlapping sets of odorants. J Neurosci 21:6018–6025

Kambere MB, Lane RP (2007) Co-regulation of a large and rapidly evolving repertoire of odorant receptor genes. BMC Neurosci 8(suppl 3):S2

Katada S, Hirokawa T, Oka Y et al (2005) Structural basis for a broad but selective ligand spectrum of a mouse olfactory receptor: mapping the odorant-binding site. J Neurosci 25:1806–1815

Keller A, Vosshall LB (2008) Better smelling through genetics: mammalian odor perception. Curr Opin Neurobiol 18:364–369

Keller A, Zhuang H, Chi Q et al (2007) Genetic variation in a human odorant receptor alters odour perception. Nature (Lond) 449:468–472

Laska M, Seibt A, Weber A (2000) 'Microsmatic' primates revisited: olfactory sensitivity in the squirrel monkey. Chem Senses 25:47–53

Laska M, Miethe V, Rieck C et al (2005a) Olfactory sensitivity for aliphatic ketones in squirrel monkeys and pigtail macaques. Exp Brain Res 160:302–311

Laska M, Wieser A, Hernandez Salazar LT (2005b) Olfactory responsiveness to two odorous steroids in three species of nonhuman primates. Chem Senses 30:505–511

Laska M, Rivas Bautista RM, Hernandez Salazar LT (2006a) Olfactory sensitivity for aliphatic alcohols and aldehydes in spider monkeys (*Ateles geoffroyi*). Am J Phys Anthropol 129:112–120

Laska M, Wieser A, Salazar LT (2006b) Sex-specific differences in olfactory sensitivity for putative human pheromones in nonhuman primates. J Comp Psychol 120:106–112

Lazaro-Perea C, Snowdon CT, Arruda MD (1999) Scent-marking behavior in wild groups of common marmosets (*Callithrix jacchus*). Behav Ecol Sociobiol 46:313–324

Leinders-Zufall T, Cockerham RE, Michalakis S et al (2007) Contribution of the receptor guanylyl cyclase GC-D to chemosensory function in the olfactory epithelium. Proc Natl Acad Sci USA 104:14507–14512

Leonhardt SD, Tung J, Camden JB et al (2008) Seeing red: behavioral evidence of trichromatic color vision in strepsirrhine primates. Behav Ecol 20:1–12

Liberles SD, Buck LB (2006) A second class of chemosensory receptors in the olfactory epithelium. Nature (Lond) 442:645–650

Liberles SD, Horowitz LF, Kuang D et al (2009) Formyl peptide receptors are candidate chemosensory receptors in the vomeronasal organ. Proc Natl Acad Sci USA 106:9842–9847

Liman E (2006) Use it or lose it: molecular evolution of sensory signaling in primates. Pflügers Arch Eur J Physiol 453:125–131

Lundstrom JN, Goncalves M, Esteves F et al (2003) Psychological effects of subthreshold exposure to the putative human pheromone 4,16-androstadien-3-one. Horm Behav 44:395–401

Malnic B, Hirono J, Sato T et al (1999) Combinatorial receptor codes for odors. Cell 96:713–723

Man O, Gilad Y, Lancet D (2004) Prediction of the odorant binding site of olfactory receptor proteins by human-mouse comparisons. Protein Sci 13:240–254

Man O, Willhite DC, Crasto CJ et al (2007) A framework for exploring functional variability in olfactory receptor genes. PLoS One 2:e682

Matsui A, Go Y, Niimura Y (2010) Degeneration of olfactory receptor gene repertories in primates: no direct link to full trichromatic vision. Mol Biol Evol 27(5):1192–1200

Matsunami H (2005) Functional expression of mammalian odorant receptors. Chem Senses 30(suppl 1):i95–i96

Menashe I, Man O, Lancet D et al (2003) Different noses for different people. Nat Genet 34:143–144

5 Functional Evolution of Primate Odorant Receptors

Menashe I, Abaffy T, Hasin Y et al (2007) Genetic elucidation of human hyperosmia to isovaleric acid. PLoS Biol 5:e284

Mombaerts P (2001) How smell develops. Nat Neurosci 4(suppl 1):1192–1198

Mombaerts P (2004) Genes and ligands for odorant, vomeronasal and taste receptors. Nat Rev Neurosci 5:263–278

Moreno-Estrada A, Casals F, Ramirez-Soriano A et al (2008) Signatures of selection in the human olfactory receptor OR5I1 gene. Mol Biol Evol 25:144–154

Nielsen R, Bustamante C, Clark AG et al (2005) A scan for positively selected genes in the genomes of humans and chimpanzees. PLoS Biol 3:e170

Nievergelt CM, Digby LJ, Ramakrishnan U et al (2000) Genetic analysis of group composition and breeding system in a wild common marmoset (*Callithrix jacchus*) population. Int J Primatol 21:1–20

Niimura Y, Nei M (2007) Extensive gains and losses of olfactory receptor genes in mammalian evolution. PLoS One 2:e708

Nozawa M, Kawahara Y, Nei M (2007) Genomic drift and copy number variation of sensory receptor genes in humans. Proc Natl Acad Sci USA 104:20421–20426

Nozawa M, Suzuki Y, Nei M (2009) Reliabilities of identifying positive selection by the branch-site and the site-prediction methods. Proc Natl Acad Sci USA 106:6700–6705

Pause BM (2004) Are androgen steroids acting as pheromones in humans? Physiol Behav 83:21–29

Perry GH, Martin RD, Verrelli BC (2007) Signatures of functional constraint at aye-aye opsin genes: the potential of adaptive color vision in a nocturnal primate. Mol Biol Evol 24:1963–1970

Preti G, Wysocki CJ, Barnhart KT et al (2003) Male axillary extracts contain pheromones that affect pulsatile secretion of luteinizing hormone and mood in women recipients. Biol Reprod 68:2107–2113

Riviere S, Challet L, Fluegge D et al (2009) Formyl peptide receptor-like proteins are a novel family of vomeronasal chemosensors. Nature (Lond) 459:574–577

Rodriguez I, Greer CA, Mok MY et al (2000) A putative pheromone receptor gene expressed in human olfactory mucosa. Nat Genet 26:18–19

Rouquier S, Blancher A, Giorgi D (2000) The olfactory receptor gene repertoire in primates and mouse: evidence for reduction of the functional fraction in primates. Proc Natl Acad Sci USA 97:2870–2874

Sabeti PC, Varilly P, Fry B et al (2007) Genome-wide detection and characterization of positive selection in human populations. Nature (Lond) 449:913–918

Saito H, Kubota M, Roberts RW et al (2004) Rtp family members induce functional expression of mammalian odorant receptors. Cell 119:679–691

Saito H, Chi Q, Zhuang H et al (2009) Odor coding by a mammalian receptor repertoire. Sci Signal 2:ra9

Sam M, Vora S, Malnic B et al (2001) Neuropharmacology. Odorants may arouse instinctive behaviours. Nature (Lond) 412:142

Spehr M, Munger SD (2009) Olfactory receptors: G protein-coupled receptors and beyond. J Neurochem 109:1570–1583

Sun L, Wang H, Hu J et al (2009) Guanylyl cyclase-D in the olfactory CO_2 neurons is activated by bicarbonate. Proc Natl Acad Sci USA 106:2041–2046

Swann J, Rahaman F, Bijak T et al (2001) The main olfactory system mediates pheromone-induced fos expression in the extended amygdala and preoptic area of the male Syrian hamster. Neuroscience 105:695–706

Trinh K, Storm DR (2003) Vomeronasal organ detects odorants in absence of signaling through main olfactory epithelium. Nat Neurosci 6:519–525

Wyart C, Webster WW, Chen JH et al (2007) Smelling a single component of male sweat alters levels of cortisol in women. J Neurosci 27:1261–1265

Wysocki CJ, Beauchamp GK (1984) Ability to smell androstenone is genetically determined. Proc Natl Acad Sci USA 81:4899–4902

Wysocki CJ, Gilbert AN (1989) National Geographic smell survey. Effects of age are heterogeneous. Ann NY Acad Sci 561:12–28

Yokoyama S, Tada T, Zhang H et al (2008) Elucidation of phenotypic adaptations: molecular analyses of dim-light vision proteins in vertebrates. Proc Natl Acad Sci USA 105:13480–13485

Young JM, Trask BJ (2007) V2R gene families degenerated in primates, dog and cow, but expanded in opossum. Trends Genet 23:212–215

Young JM, Waters H, Dong C et al (2007) Degeneration of the olfactory guanylyl cyclase D gene during primate evolution. PLoS One 2:e884

Young JM, Endicott RM, Parghi SS et al (2008) Extensive copy-number variation of the human olfactory receptor gene family. Am J Hum Genet 83:228–242

Young JM, Massa HF, Hsu L et al (2009) Extreme variability among mammalian V1R gene families. Genome Res

Zhang X, Firestein S (2002) The olfactory receptor gene superfamily of the mouse. Nat Neurosci 5:124–133

Zhuang H, Matsunami H (2007) Synergism of accessory factors in functional expression of mammalian odorant receptors. J Biol Chem 282:15284–15293

Zhuang H, Matsunami H (2008) Evaluating cell-surface expression and measuring activation of mammalian odorant receptors in heterologous cells. Nat Protoc 3:1402–1413

Zhuang H, Chien MS, Matsunami H (2009) Dynamic functional evolution of an odorant receptor for sex-steroid-derived odors in primates. Proc Natl Acad Sci USA 106:21247–21251

Ziegler TE, Sousa MBC (2002) Parent-daughter relationships and social controls on fertility in female common marmosets, *Callithrix jacchus*. Horm Behav 42:356–367

Ziegler TE, Schultz-Darken NJ, Scott JJ et al (2005) Neuroendocrine response to female ovulatory odors depends upon social condition in male common marmosets, *Callithrix jacchus*. Horm Behav 47:56–64

Zozulya S, Echeverri F, Nguyen T (2001) The human olfactory receptor repertoire. Genome Biol 2:RESEARCH0018

Chapter 6
Post-Genome Biology of Primates Focusing on Taste Perception

Tohru Sugawara and Hiroo Imai

Abbreviations

AceK   Acesulfame K
bp   Base pair
CL   Cytoplasmic loop
EL   Extracellular loop
GMP   Guanosine-5′-monophosphate
IMP   Inosine-5′-monophosphate
indels Insertions/deletions
MSG   Monosodium glutamate
ORFs   Open reading frames
PTC   Phenylthiocarbamide
SNPs   Single nucleotide polymorphisms
TM   Transmembrane
TRCs   Taste receptor cells

T. Sugawara
Primate Research Institute, Kyoto University, 41-2 Kanrin,
Inuyama, Aichi 484-8506, Japan

Department of Reproductive Biology, National Research Institute for Child Health
and Development, 2-10-1 Okura, Setagaya-ku, Tokyo 157-8535, Japan

H. Imai (✉)
Primate Research Institute, Kyoto University, 41-2 Kanrin,
Inuyama, Aichi 484-8506, Japan
e-mail: imai@pri.kyoto-u.ac.jp

H. Hirai et al. (eds.), *Post-Genome Biology of Primates*, Primatology Monographs,
DOI 10.1007/978-4-431-54011-3_6, © Springer 2012

6.1 Introduction

The sense of taste allows mammals to evaluate their food and to decide which foods they will eat. Tastants taken into the oral cavity are first detected by the taste buds, which are small structures distributed on the upper surface of the tongue. Each taste bud forms an onion-like shape and is composed of tens of taste receptor cells (TRCs). Taste receptors are expressed in the apical end of TRCs (Chandrashekar et al. 2006). Tastant–receptor interaction triggers intracellular signal transduction in TRCs, resulting in depolarization of TRCs and subsequent neurotransmitter release from TRCs to the gustatory nerves innervating them. Taste is finally perceived in the gustatory area in the brain via the secondary and higher-order neurons.

Mammals can perceive and distinguish five basic taste qualities, namely, sweet, bitter, sour, salty, and umami, the taste of glutamate. Of these five modalities, sweet and umami tastes are mediated by taste receptors type 1, T1Rs (TAS1Rs) and bitter taste is mediated by taste receptors type 2, T2Rs (TAS2Rs), both of which belong to the large family of seven-transmembrane (TM) G-protein-coupled receptors (Adler et al. 2000; Chandrashekar et al. 2000; Matsunami et al. 2000; Conte et al. 2003; Shi et al. 2003; Mueller et al. 2005, Fig. 6.1). The cytoplasmic loops (CLs) and their adjacent TM segments are the predicted sites of G-protein interaction, whereas distinctive extracellular loops (ELs) for T1R and TM domains for T2Rs are the predicted regions of ligand binding (Adler et al. 2000; Margolskee 2002). Despite this structural similarity, these two subfamilies of molecules exhibit no obvious amino acid similarities. Because taste receptors are directly involved in the interaction between mammals and their dietary sources, it is likely that these genes evolved to reflect species-specific dietary and ecological changes during mammalian evolution for maximizing the intake of nutrients and minimizing the ingestion of potentially harmful substances. An understanding of the correlation between the variation of taste perception among species and their environments may provide us the chance to understand how ecological changes influence their genomes and adaptation to the environment as well as other sensory systems (see also Chaps. 4, 5, and 7).

6.1.1 Recent Studies of Taste Receptor Genes

Recent genomic analyses using comparison with reference sequences have revealed the evolution of taste receptor genes in various vertebrates (Shi and Zhang 2006; Go 2006) (Table 6.1). In mammals, the T1R family includes three members: *T1R1*, *T1R2*, and *T1R3* (Hoon et al. 1999; Kitagawa et al. 2001; Max et al. 2001; Montmayeur et al. 2001; Nelson et al. 2001; Sainz et al. 2001; Zhao et al. 2003). The coding regions of *T1R* genes are about 2,000 base pairs (bp) long and contain multiple introns. The proteins of *T1R* genes are characterized by a long extracellular N-terminal domain, often referred to as the Venus flytrap domain, that may be involved in ligand binding. Sweet compounds are detected by T1R2/T1R3 heteromers, whereas umami compounds are detected by T1R1/T1R3 heteromers (Fig. 6.1). Orthologous *T1R* sequences are relatively conserved in evolution. Moreover,

6 Post-Genome Biology of Primates Focusing on Taste Perception

Table 6.1 Number of genes and pseudogenes in vertebrate *T1R* and *T2R* gene families

	Taste receptors type 1 (*T1R*s)		Taste receptors type 2 (*T2R*s)	
	Intact genes	Total	Intact genes	Pseudogenes
Human	3	36	25	11
Mouse	3	41	35	6
Rat	3	42	37	5
Dog	3	21	16	5
Opossum	3	34	29	5
Chicken	2	3	3	0
Frog	0	64	52	12
Zebrafish	1	4	4	0

Source: Modified from Shi and Zhang (2006)

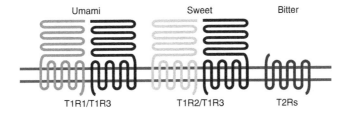

Fig. 6.1 Schematic drawing of taste receptors for *umami*, *sweet*, and *bitter*. Taste buds are composed of 50–100 taste receptor cells. Tastants are received by taste receptors, which are expressed in the taste pores at the apical end of taste buds. Taste signals are finally transmitted to afferent sensory nerve fibers. Taste receptor type 1 (*T1R*s) and taste receptor type 2 (*T2R*s) belong to the large family of seven transmembrane G-protein-coupled receptors. T1Rs are characterized by a long extracellular N-terminal domain

the *T1R* gene repertoire remains virtually constant in size across most vertebrates, with some exceptions, such as cats and chickens. The cat *T1R2* gene shows a 247-bp microdeletion in exon 3 and stop codons in exons 4 and 6. It was reported that there was no evidence of detectable mRNA from cat *T1R2* by reverse transcription-polymerase chain reaction (RT-PCR) or in situ hybridization, and no evidence of protein expression by immunohistochemistry. *T1R2* in tigers and cheetahs and in six healthy adult domestic cats all showed similar deletion and stop codons (Li et al. 2005). In chicken, *T1R1* and *T1R3* are present, but *T1R2* is missing. Interestingly, electrophysiological studies showed that chicken taste buds do not generate impulses when stimulated with sucrose or saccharine (Duncan 1960; Halpern 1962). Moreover, chickens show no preference for sweet stimuli in behavioral reaction tests (Ganchrow et al. 1990). Thus, the loss of *T1R2* provides a genetic explanation of the chicken's indifference to sweet tastants.

In contrast with *T1R*s, the *T2R* family contains approximately 30 genes (Conte et al. 2003; Shi et al. 2003). The coding regions of *T2R* genes are about 900 bp long and are not interrupted by introns. Orthologous *T2R* sequences are variable, and the *T2R* gene repertoire diverges among species and within species. The number of *T2R* genes and the proportion of pseudogenes range from 21 in dog to 42 in rat and from 12% in rat to 44% in cow, respectively (Go et al. 2005;

Go 2006; Shi and Zhang 2006). Here, pseudogenes mean genes whose open reading frames (ORFs) are disrupted by insertions/deletions (indels) or nonsense mutations. All remaining *T2R* genes that have intact ORFs are counted as functional genes. These variations in *T2R* genes are likely to respond to and reflect dietary changes during mammalian evolution. Substantial modification of the *T2R* gene repertoire is likely to reflect different responses to changes in the environment and to result from species-specific food preferences during vertebrate evolution.

6.1.2 Tastes in Humans and Chimpanzees

Taste perception varies considerably across different species. Here we review recent studies about the evolution of *T1R* and *T2R* genes in humans and chimpanzees. Chimpanzees, close relatives of humans, rely on plants for a considerable amount of their natural food (Chivers 1994) and do not have the benefits of fire. They are estimated to take some 87–98% of their food from plant sources (Milton 2003). They routinely and systematically hunt colobus monkeys and other smaller mammals, but the amounts consumed are generally minor (Stanford 1998), and the typical chimpanzee diet is composed largely of ripe fruits (Stanford 1998). In contrast, humans can control and make fire, which has led to cooking and had a direct and profound effect on human diet by expanding the range of foods available. Namely, humans also have the ability to detoxify poisonous foods by using fire, and can avoid ingesting poisonous substances based on experience and culture, not based on taste alone. Accordingly, diets and feeding behaviors seem considerably different between humans and chimpanzees, and therefore taste receptor genes, especially *T2R* genes, are good candidate genes that may contribute to phenotypic differences humans and chimpanzees.

6.2 T1Rs, Umami and Sweet Taste Receptors

Umami and sweet tastes are said to be caloric detectors, sensing protein-rich or carbohydrate-rich food. The ability to identify sweet-tasting foodstuffs provides us and other vertebrates with a means to seek out needed carbohydrates with high nutritive value. Umami allows the recognition of amino acids that act as biosynthetic precursors of various molecules, metabolic fuels, and neurotransmitters.

6.2.1 Ligands

Ligand screenings of T1Rs using calcium imaging analysis have revealed that human T1Rs function as sweet and umami receptors. In most studies, human embryonic kidney (HEK) 293 cells were used; they were transfected with candidate taste receptor genes and loaded with calcium indicator, and then stimulated with various taste substances. Li et al. (2002) showed that human T1R2/T1R3 responded to many sweet taste stimuli: the sugars sucrose, fructose, galactose, glucose, lactose, and maltose; the amino acids glycine and D-tryptophan (but not its bitter enantiomer); the sweet proteins monellin and thaumatin; and the synthetic sweeteners acesulfame K (AceK), aspartame, cyclamate, dulcin, neotame, saccharin, and sucralose. Although T1R3 combines with T1R2 to form a sweet taste receptor that responds to all classes of sweet tastants, including natural sugars, artificial sweeteners, D-amino acids, and intensely sweet proteins, the receptor exhibits stereo-selectivity for certain molecules. For example, it responds to D-tryptophan but not L-tryptophan, which correlates with the behavioral data. They also showed that human T1R1/T1R3 responded to L-glutamate, but did not respond to the weak umami taste stimuli L-aspartate and L-2-amino-4-phosphonobutanoic acid. In the presence of inosine-5′-monophosphate (IMP) and guanosine-5′-monophosphate (GMP), however, these compounds activated T1R1/T1R3 at physiologically relevant concentrations. The response of T1R1/T1R3 was selective for umami taste stimuli; T1R1/T1R3 did not respond to sweet taste stimuli, amino acids such as D-glutamate and D-aspartate, or binary mixtures of these compounds with IMP.

6.2.2 Functional Domains of T1Rs

Xu et al. (2004) analyzed functional domains of human T1Rs such as ligand-binding domains and G-protein-coupled domains using mutated and chimeric human and rat *T1R* genes. The amino acid alterations containing Ser144Ala and Glu302Ala in human T1R2, which change amino acids that correspond to those which are crucial in ligand binding in mGluR1, selectively affected the response to aspartame and neotame but not cyclamate. Thus, the N-terminal domain of human T1R2 is required for recognizing aspartame and neotame. Cyclamate binds directly to the TM domain of human T1R3 and activates the receptor. Replacing EL2 or EL3 of human T1R3 with rat sequences abolished the cyclamate response without affecting the sucrose or aspartame responses. In contrast, replacing EL1 had no obvious effect on the response to cyclamate, suggesting important roles for EL2 and EL3 in recognizing cyclamate. Similar to cyclamate, lactisole, an aralkyl carboxylic acid, which is a human-specific sweet taste inhibitor, also requires the human T1R3 C-terminal domain, in this case to inhibit the receptor's response to sucrose and AceK. In addition, the C-terminal half of human T1R2 is required for the response to sucrose and AceK, suggesting that T1R2 is the subunit responsible for G-protein coupling in taste cells. Thus, T1R2 is required for both ligand recognition and G-protein coupling. In summary, functional studies have demonstrated that both T1R2 and T1R3 are required in a functional sweet taste receptor, that aspartame and neotame require the N-terminal extracellular domain of T1R2, that G-protein coupling requires the C-terminal half of T1R2, and that cyclamate and lactisole require the TM domain of T1R3.

6.2.3 Polymorphisms in Humans and Chimpanzees

Genetic variation of human *T1R1* and *T1R3* has been reported. Shigemura et al. (2009) sequenced the entire coding regions of human *T1R1* (2,526 bp) and *T1R3* (2,559 bp) from 254 individuals and performed functional expression analysis using HEK293 cells. They found three single nucleotide polymorphisms (SNPs; Gln12His, Ala372Thr in T1R1, and Arg757Cys in T1R3) that they considered to be common SNPs among the samples. They demonstrated by functional analysis that human T1R1–372Thr creates a more sensitive umami receptor than –372Ala, while T1R3–757Cys creates a less sensitive receptor than –757Arg for monosodium glutamate (MSG) and MSG plus IMP, and they found a strong correlation between the recognition thresholds and in vitro dose–response relationships. They argued that amino acid position 372 in human T1R1 may modulate the function of the binding domain for MSG but not IMP, and that the binding region for MSG may differ from that for IMP. The amino acid at position 372 in T1R1 is present in the N-terminal domain, which contains the predicted ligand-binding domain, suggesting that the human T1R1–Ala372Thr amino acid substitution may lead to a conformational change of the ligand-binding site for glutamate, thereby affecting umami taste sensitivity. In addition, they argued that the human T1R3–Arg757Cys amino acid substitution may affect general functions of T1R1/T1R3, such as cell-surface expression, dimerization of T1R1 and T1R3, or the dynamic equilibrium between the active and resting conformations modulated by the presence/absence of ligand.

Furthermore, we sequenced *T1R1–T1R3* in eight individual chimpanzees and compared the sequences between human and chimpanzees. The entire coding regions of chimpanzee *T1R1*, *T1R2* and *T1R3* are 1,764, 2,520, and 2,559 bp long, respectively. We found several synonymous and nonsynonymous polymorphic sites in these genes. For example, there are two synonymous and three nonsynonymous polymorphic sites in chimpanzee *T1R2* among these eight individuals. In future studies, we will attempt to understand the meanings of these variations in humans and chimpanzees.

6.3 T2Rs, Bitter Taste Receptors

Bitter sensitivity has a particularly important role: many naturally poisonous substances taste bitter to humans, and virtually all animals show an aversive response to such tastants, suggesting that bitter transduction evolved as a key defense mechanism against the ingestion of harmful substances. The perception of bitter is essential for its protective value, enabling humans to avoid potentially harmful plant alkaloids and other environmental toxins. However, adult humans are not always averse to bitter taste, and some degree of bitterness is preferred in certain foodstuffs and beverages, including chocolate, coffee, and beer. In these instances, some bitter taste works as a component of flavor and is indispensable to improve the palatability of the food or beverage. It is not necessarily in an animal's best interest to trust the bitter rejection response as an accurate measure of toxicity (Glendinning 1994). High bitter sensitivity may occasionally cause animals to taste-reject nontoxic and nutritious foods or increase the risk of disease by lowering the intake of such beneficial compounds.

6.3.1 Ligands

As described for T1R, ligand screening of T2Rs using calcium imaging analysis has revealed human T2R ligands. There is a great excess of numerous, structurally diverse bitter compounds relative to the approximately 30 T2Rs. Thus, T2Rs are broadly tuned to detect multiple substances (Behrens et al. 2004, 2009; Brockhoff et al. 2007). For example, Brockhoff et al. (2007) identified the human bitter taste receptor T2R46 as a sensor for bitter sesquiterpene lactones, or related clerodane diterpenoids and labdane diterpenes, and for strychnine, strychnine-related bitter substances, denatonium, chloramphenicol, and sucrose octaacetate. A complete understanding of the physiology of bitter taste will require exact knowledge about the interaction of T2R with bitter substances. There have been some reports of the identification of bitter tastant–receptor combinations. Recently, Meyerhof and his colleagues deorphanized 25 human taste 2 receptors (hTAS2Rs) with 104 natural or synthetic bitter chemicals in a heterologous expression system (Meyerhof et al. 2010). This study provides deep insight into the functional properties of human T2Rs.

6.3.2 Polymorphisms in Humans and Chimpanzees

A database search of the human genome sequence identified 25 putatively functional h*T2R*s (with intact ORFs) and 11 pseudogenes (with ORFs disrupted by indels or nonsense mutations) and mapped these genes to chromosomes 5, 7, and 12 (Shi and Zhang 2006; Go et al. 2005). In the chimpanzee, 25 apparently functional c*T2R*s and 10 pseudogenes were identified (Go et al. 2005; Go 2006). It is well known that there is intraspecies copy number variation of sensory receptor genes as a result of DNA polymorphisms (Nozawa et al. 2007). Therefore, it is not sufficient to compare a few reference sequences of the species to investigate the changes in DNA responsible for human- or chimpanzee-specific sensors of their environments. Two studies have analyzed intraspecies variations of the 25 putatively functional genes of the h*T2R*s (Wang et al. 2004; Kim et al. 2005). However, as described previously, humans can cook and modify natural food to eat, and therefore, the taste receptors of humans cannot reflect adaptation to natural food. Thus, it is important to examine T2Rs of wild animals. Now we have analyzed polymorphisms in the entire repertoire of *T2R* genes in wild chimpanzees, the species most closely related to humans.

6.3.3 Human T2R Genes

Wang et al. (2004) sequenced all 25 functional h*T2R* genes in 22 humans of diverse geographic origin, and identified 72 nonsynonymous and 33 synonymous polymorphic sites from 21,408 nucleotide sites, including 15,242 nonsynonymous and 6,166 synonymous sites, respectively. In addition to the many nonsynonymous polymorphisms, two nonsense polymorphisms were observed in human *T2R*s. The first was a C-to-T mutation at position 640 of *T2R7*, which changed an arginine residue to a premature stop codon and resulted in a receptor that contains only five TM domains. The second nonsense polymorphism (G to A at position 749) changed a tryptophan to a premature stop codon in T2R46, resulting in a truncated receptor with six TM domains. In addition to these seemingly clear nonsense polymorphisms, there are some defective amino acid changes in apparently functional genes. For example, the amino acid changes in three sites (Pro49Ala, Ala262Val, Val296Ile) in hT2R38 strongly reduce the activity of the receptor, providing the molecular basis of "nontaster" individuals for a specific bitter taste, phenylthiocarbamide (PTC) (Kim et al. 2003; Bufe et al. 2005). In addition, two amino acid changes in T2R43 reduce the activity of the receptor, resulting in the difference in the sensitivity for specific bitter taste, aloin (Pronin et al. 2007).

6.3.4 Chimpanzee T2Rs

We examined the *T2R* genes in the chimpanzee genome sequence (NCBI; Build 2.1) using BLASTN and TBLASTN with all annotated hT2Rs as queries. We identified 24 apparently functional genes and 13 pseudogenes in the chimpanzee genome sequence and mapped them to chromosomes 5, 6, 7, and 12 (Sugawara et al. 2011). These numbers of genes were almost the same as those reported previously (Go et al. 2005; Go 2006; Fischer et al. 2005).

To examine the possible adaptive changes of the c*T2R*s, we sequenced 28 *T2R* genes in 46 apparently unrelated West African chimpanzees (*Pan troglodytes verus*) and identified 57 nonsynonymous (including 9 singletons) and 18 synonymous polymorphic sites from 26,172 nucleotide sites of West African chimpanzee *T2R*s (Sugawara et al. 2011). The SNP data are also available in Primate Genome Database (http://gcoe.biol.sci.kyoto-u.ac.jp/pgdb/) for the individuals in the Primate Research Institute. In addition to the many nonsynonymous polymorphisms, we found novel nonsense polymorphisms for six c*T2R*s besides c*T2R38*, in which a T-to-G mutation at position 2, which changes a start codon, was found as the genetic basis of nontasters, as reported previously (Wooding et al. 2006). The fraction of pseudogenes of c*T2R38* in this study (70/92) was close to that found in a previous study (30/58; see Wooding et al. 2006). All these mutations appeared to make c*T2R*s defective. The number of defective c*T2R*s should vary among individuals. However, few individuals have these pseudogenes as homozygotes. One individual examined here had 25 functional c*T2R*s, and 3 individuals had 26 functional c*T2R*s. All the remaining individuals had 27 or 28 functional c*T2R*s. These results indicate that chimpanzees have two or three more *T2R* genes than humans. It is possible that chimpanzees have the ability to distinguish more kinds of tastants than humans. To confirm this at the behavioral level, we are constructing behavioral test systems. Using PTC as test tastants, the correlation between genetic and behavioral results, including cognitional information, has been confirmed (Morimura et al., unpublished data).

6.3.5 Analysis of Polymorphisms in Humans and Chimpanzees: Behavioral Experiments

To obtain an overview of the evolution of *T2R*s in humans and chimpanzees, we considered three aspects of the data to assess the support for different models of natural selection. First, we calculated the nucleotide diversity, π, the mean pairwise difference between sequences per nucleotide. Second, we estimated the synonymous and nonsynonymous nucleotide substitution and calculated per site nonsynonymous substitutions over per site synonymous substitutions (*Ka/Ks*). An average *Ka/Ks* of 1 is expected if both amino acid replacement and silent mutations are neutral with respect to selection. Lower values are consistent with selection against amino acid replacements (i.e., purifying selection). Third, we considered the summary of the allelic frequency spectrum, Tajima's *D* (Tajima 1989), the mean of which is expected to be approximately 0 under the standard neutral model. Negative values for this statistic reflect an excess of low-frequency variants in the population, consistent with positive directional selection or a population expansion. Positive values reflect an excess of intermediate-frequency variants in the population, consistent with balancing selection or a population contraction.

It has been reported that these parameters for human *T2R*s are close to those of noncoding regions of the human genome, suggesting that human *T2R*s are evolutionally relaxed from selective constraint (Wang et al. 2004; Meyerhof 2005). In the case of humans, cultural factors, such as the use of fire, affect the simple biological selection, and therefore it is of interest to examine the parameters for chimpanzees as wild animals and compare them to those for human *T2R*s and noncoding regions. Our results showed that Tajima's *D* values of some c*T2R*s were significantly positive, different from those of chimpanzee noncoding sequences and h*T2R*s (Sugawara et al. 2011). Because this fact suggests the existence of balancing selection for some c*T2R*s, it could be responsible for the great variety of bitter taste reception in wild animals. Thus, it will be of interest to determine whether the differences in haplotypes are reflected in functional differences at the protein and higher levels. To examine these things, functional expression of each protein and behavioral tests, respectively, would provide additional information.

6.3.6 Functional Importance of Polymorphisms in T2Rs

How is variation related to adaptation among Western chimpanzees, and does the variation in cT2Rs have any functional consequences? CLs of T2Rs are the domains that contain critical sites for proper coupling with G proteins on the intracellular side of the plasma membrane. Amino acid variations in the CLs in human T2R38 and T2R43 alter the sensitivity to their ligands (Bufe et al. 2005; Pronin et al. 2007). Variations of amino acids in the CLs of cT2Rs were thus expected to affect receptor sensitivity. A previous study demonstrated that hT2R43 controls human taste sensitivity to aloin, saccharin, and aristolochic acid (Pronin et al. 2007). Aristolochic acid is a component of *Aristolochia* and *Asarum* (which both belong to the Aristolochiaceae family of plants) that has been found to be toxic to the kidney in humans (Cosyns et al. 1999). Retaining the sensitivity to cT2R43 thus seemed likely to be important for the survival of chimpanzees. Functional assays of the receptors encoded by the cT2R43 alleles are now in progress. If the observed replacements in the CL region affect cT2R43 receptor sensitivity, this gene could be the second case of parallel evolution of bitter-taste sensitivity in humans and chimpanzees (Wooding et al. 2006), resembling the case in vision (Sugawara et al. 2005, 2010).

The TM domain and EL of the receptor are very important for the activation of the receptor by ligands. Mutational analysis of tryptophan 88 in EL1 of hT2R44 revealed that this residue contributes to ligand selectivity (Pronin et al. 2004). Therefore, polymorphism of the amino acid in this region is also predicted to alter the ligand specificity. The fact that we found a polymorphism in the corresponding position of cT2R46 (Sugawara et al. 2011) implies that this gene could receive more kinds of substances than if there were invariant amino acids in the same receptor as a result of the presence of polymorphisms in the group and individual animals. It has been reported that hT2R46 can respond to various types of sesquiterpene lactones (Brockhoff et al. 2007), which are a major class of bitter compounds occurring naturally in vegetables and culinary herbs as well as in aromatic and medicinal plants. In some studies, when wild chimpanzees were diagnosed as being sick or wounded, they were observed to eat bitter-tasting plants such as *Vernoia amygdalina*, which contains specific sesquiterpenes (Huffman and Seifu 1989; Koshimizu et al. 1994). The ability to perceive bitterness may have evolved not only to provide protection against ingested toxic compounds but also for utilizing bitter food as medicine.

6.4 Taste Receptors of Other Primates

In this review, we mainly focused on the taste receptors of humans and chimpanzees as a model case for post-genome studies because of the early open access of the genome sequences of these two primates. There are also many reports on the feeding behavior of other primates, including Old World monkeys, New World monkeys, prosimians, and other apes. Because more than ten primate genomes have been opened now, it is possible to use the genomic information and to analyze polymorphism as related to protein functional assays and behavioral tests. For example, how mammals can detect numerous bitter compounds with a limited set of receptors is an important question. T2Rs are broadly tuned to detect multiple substances (Brockhoff et al. 2007; Behrens et al. 2004, 2009), which answers this question in part. We performed, for the first time, a population genetic analysis of *T2R* genes of wild-born primates other than humans. As a result, we found a clue suggesting that balancing selection acts to maintain polymorphisms in several c*T2R*s. Heterozygous individuals might have the ability to taste a broader range of substances, also partly answering the foregoing question. It will be necessary to further analyze other wild primates and share information widely among primate researchers such as molecular biologists, behavioral psychologists, and ecologists. For this purpose, a primate genome database was constructed and opened by the Primate Research Institute of Kyoto University and the Kyoto University global COE program for Formation of Strategic Base for Biodiversity Research: http://gcoe.biol.sci.kyoto-u.ac.jp/pgdb/index.en.html.

We hope it will be helpful for furthering explorations of the interactions between genomes and ecosystems.

Acknowledgments We thank Dr. E. Nakajima for English correction, and the Great Ape Information Network (GAIN) for the information of individual chimpanzees. Our work described here was financially supported by global COE program A06 and by Grants-in-Aid from the Ministry of Education, Culture, Sports, Science, and Technology of Japan (20657044, 21370109, and 22247036), Environment Research and Technology Development Fund (D-1007) of the Ministry of the Environment, Japan, and grants from the Takeda Foundation for Science and the Suzuken Memorial Foundation to H.I.

References

Adler E, Hoon MA, Mueller KL et al (2000) A novel family of mammalian taste receptors. Cell 100:693–702

Behrens M, Brockhoff A, Kuhn C et al (2004) The human taste receptor hTAS2R14 responds to a variety of different bitter compounds. Biochem Biophys Res Commun 319:479–485

Behrens M, Brockhoff A, Batram C et al (2009) The human bitter taste receptor hTAS2R50 is activated by the two natural bitter terpenoids andrographolide and amarogentin. J Agric Food Chem 57:9860–9866

Brockhoff A, Behrens M, Massarotti A et al (2007) Broad tuning of the human bitter taste receptor hTAS2R46 to various sesquiterpene lactones, clerodane and labdane diterpenoids, strychnine, and denatonium. J Agric Food Chem 55:6236–6243

Bufe B, Breslin PA, Kuhn C et al (2005) The molecular basis of individual differences in phenylthiocarbamide and propylthiouracil bitterness perception. Curr Biol 15:322–327

Chandrashekar J, Mueller KL, Hoon MA et al (2000) T2Rs function as bitter taste receptors. Cell 100:703–711

Chandrashekar J, Hoon MA, Ryba NJ et al (2006) The receptors and cells for mammalian taste. Nature (Lond) 444:288–294

Chivers DJ (1994) Diets and guts. In: Martin RD (ed) The Cambridge encyclopedia of human evolution. Cambridge University Press, New York

Conte C, Ebeling M, Marcuz A et al (2003) Evolutionary relationships of the Tas2r receptor gene families in mouse and human. Physiol Genomics 14:73–82

Cosyns JP, Jadoul M, Squifflet JP et al (1999) Urothelial lesions in Chinese-herb nephropathy. Am J Kidney Dis 33:1011–1017

Duncan CJ (1960) The sense of taste in birds. Ann Appl Biol 48:409–414

Fischer A, Gilad Y, Man O et al (2005) Evolution of bitter taste receptors in humans and apes. Mol Biol Evol 22:432–436

Ganchrow JR, Steiner JE, Bartana A (1990) Behavioral reactions to gustatory stimuli in young chicks (*Gallus gallus domesticus*). Dev Psychobiol 23:103–117

Glendinning JI (1994) Is the bitter rejection response always adaptive? Physiol Behav 56:1217–1227

Go Y (2006) Lineage-specific expansions and contractions of the bitter taste receptor gene repertoire in vertebrates. Mol Biol Evol 23:964–972

Go Y, Satta Y, Takenaka O et al (2005) Lineage-specific loss of function of bitter taste receptor genes in humans and nonhuman primates. Genetics 170:313–326

Halpern BP (1962) Gustatory nerve impulses in the chicken. Am J Physiol 203:541–544

Hoon MA, Adler E, Lindemeier J et al (1999) Putative mammalian taste receptors: a class of taste-specific GPCRs with distinct topographic selectivity. Cell 96:541–551

Huffman MA, Seifu M (1989) Observations on the illness and consumption of a possibly medicinal plant *Vernonia amygdalina* (Del.) by a wild chimpanzee in the Mahale Mountains National Park, Tanzania. Primates 30:51–63

Kim UK, Jorgenson E, Coon H et al (2003) Positional cloning of the human quantitative trait locus underlying taste sensitivity to phenylthiocarbamide. Science 299:1221–1225

Kim U, Wooding S, Ricci D et al (2005) Worldwide haplotype diversity and coding sequence variation at human bitter taste receptor loci. Hum Mutat 26:199–204

Kitagawa M, Kusakabe Y, Miura H et al (2001) Molecular genetic identification of a candidate receptor gene for sweet taste. Biochem Biophys Res Commun 283:236–242

Koshimizu K, Ohigashi H, Huffman MA (1994) Use of *Vernonia amygdalina* by wild chimpanzee: possible roles of its bitter and related constituents. Physiol Behav 56:1209–1216

Li X, Staszewski L, Xu H et al (2002) Human receptors for sweet and umami taste. Proc Natl Acad Sci USA 99:4692–4696

Li X, Li W, Wang H et al (2005) Pseudogenization of a sweet-receptor gene accounts for cats' indifference toward sugar. PLoS Genet 1:27–35

Margolskee RF (2002) Molecular mechanisms of bitter and sweet taste transduction. J Biol Chem 277:1–4

Matsunami H, Montmayeur JP, Buck LB (2000) A family of candidate taste receptors in human and mouse. Nature (Lond) 404:601–604

Max M, Shanker YG, Huang L et al (2001) Tas1r3, encoding a new candidate taste receptor, is allelic to the sweet responsiveness locus Sac. Nat Genet 28:58–63

Meyerhof W (2005) Elucidation of mammalian bitter taste. Rev Physiol Biochem Pharmacol 154:37–72

Meyerhof W, Batram C, Kuhn C et al (2010) The molecular receptive ranges of human TAS2R bitter taste receptors. Chem Senses 35:157–170

Milton K (2003) The critical role played by animal source foods in human (Homo) evolution. J Nutr 133:3886S–3892S

Montmayeur JP, Liberles SD, Matsunami H et al (2001) A candidate taste receptor gene near a sweet taste locus. Nat Neurosci 4:492–498

Mueller KL, Hoon MA, Erlenbach I et al (2005) The receptors and coding logic for bitter taste. Nature (Lond) 434:225–229

Nelson G, Hoon MA, Chandrashekar J et al (2001) Mammalian sweet taste receptors. Cell 106:381–390

Nozawa M, Kawahara Y, Nei M (2007) Genomic drift and copy number variation of sensory receptor genes in humans. Proc Natl Acad Sci USA 104:20421–20426

Pronin AN, Tang H, Connor J et al (2004) Identification of ligands for two human bitter T2R receptors. Chem Senses 29:583–593

Pronin AN, Xu H, Tang H et al (2007) Specific alleles of bitter receptor genes influence human sensitivity to the bitterness of aloin and saccharin. Curr Biol 17:1403–1408

Sainz E, Korley JN, Battey JF et al (2001) Identification of a novel member of the T1R family of putative taste receptors. J Neurochem 77:896–903

Shi P, Zhang J (2006) Contrasting modes of evolution between vertebrate sweet/umami receptor genes and bitter receptor genes. Mol Biol Evol 23:292–300

Shi P, Zhang J, Yang H et al (2003) Adaptive diversification of bitter taste receptor genes in mammalian evolution. Mol Biol Evol 20:805–814

Shigemura N, Shirosaki S, Sanematsu K et al (2009) Genetic and molecular basis of individual differences in human umami taste perception. PLoS One 21:e6717

Stanford CB (1998) Chimpanzee and red colobus: the ecology of predator and prey. Harvard University Press, Cambridge

Sugawara T, Terai Y, Imai H et al (2005) Parallelism of amino acid changes at the RH1 affecting spectral sensitivity among deep-water cichlids from Lakes Tanganyika and Malawi. Proc Natl Acad Sci USA 102:5548–5553

Sugawara T, Imai H, Nikaido M et al (2010) Vertebrate rhodopsin adaptation to dim light via rapid meta-II intermediate formation. Mol Biol Evol 27:506–516

Sugawara T, Go Y, Udono T et al (2011) Diversification of bitter taste receptor gene family in western chimpanzees. Mol Biol Evol 28:921–931

Tajima F (1989) Statistical method for testing the neutral mutation hypothesis by DNA polymorphism. Genetics 123:585–595

Wang X, Thomas SD, Zhang J (2004) Relaxation of selective constraint and loss of function in the evolution of human bitter taste receptor genes. Hum Mol Genet 13:2671–2678

Wooding S, Bufe B, Grassi C et al (2006) Independent evolution of bitter-taste sensitivity in humans and chimpanzees. Nature (Lond) 440:930–934

Xu H, Staszewski L, Tang H et al (2004) Different functional roles of T1R subunits in the heteromeric taste receptors. Proc Natl Acad Sci USA 101:14258–14263

Zhao GQ, Zhang Y, Hoon MA et al (2003) The receptors for mammalian sweet and umami taste. Cell 115:255–266

Chapter 7
Polymorphic Color Vision in Primates: Evolutionary Considerations

Shoji Kawamura, Chihiro Hiramatsu, Amanda D. Melin, Colleen M. Schaffner, Filippo Aureli, and Linda M. Fedigan

Abbreviations

λ_{max}	Wavelength of maximal absorbance
cDNA	Complementary DNA
ERG	Electroretinogram
LCR	Locus control region
M/LWS	Middle to long wavelength-sensitive
MSP	Microspectrophotometry
PCR	Polymerase chain reaction
RH1	Rhodopsin

S. Kawamura (✉)
Department of Integrated Biosciences, Graduate School of Frontier Sciences,
The University of Tokyo, Kashiwa, Chiba, Japan
e-mail: kawamura@k.u-tokyo.ac.jp

C. Hiramatsu
Department of Integrated Biosciences, Graduate School of Frontier Sciences,
The University of Tokyo, Kashiwa, Chiba, Japan

Department of Psychology, Graduate School of Letters, Kyoto University, Kyoto, Japan

A.D. Melin
Department of Anthropology, University of Calgary, Calgary, AB, Canada

Department of Anthropology, Dartmouth College, Hanover, NH, USA

C.M. Schaffner
Psychology Department, University of Chester, Chester, UK

Instituto de Neuroetologia, Universidad Veracruzana, Xalapa, Mexico

F. Aureli
Research Centre in Evolutionary Anthropology and Palaeoecology,
Liverpool John Moores University, Liverpool, UK

Instituto de Neuroetologia, Universidad Veracruzana, Xalapa, Mexico

L.M. Fedigan
Department of Anthropology, University of Calgary, Calgary, AB, Canada

H. Hirai et al. (eds.), *Post-Genome Biology of Primates*, Primatology Monographs,
DOI 10.1007/978-4-431-54011-3_7, © Springer 2012

RH2   Rhodopsin-like
SWS1   Short wavelength-sensitive type 1
SWS2   Short wavelength-sensitive type 2

7.1 Color Vision and Opsins

Color vision is based on the ability to discriminate light by differences in wavelength (or hue). At least two different spectral types of photoreceptors in the retina are necessary to compare signals from these wavelengths. Generally speaking, the number of discriminable colors increases as the number of spectrally distinct photoreceptors increases and as the spectral overlap among them is reduced (Vorobyev 2004). However, this does not always hold true under dim light conditions. It should also be noted that some colors may be more important than others and that it may not even be necessary to perceive certain colors, depending on the ecological requirements for a given animal (Vorobyev 2004).

Vertebrate retinas contain two types of visual photoreceptor cells: rods and cones. Rods allow dim-light vision and cones allow daylight and color vision. Photosensitive molecules, called visual pigments, are located in the outer segments of these cells. A visual pigment consists of a protein moiety, opsin, and a chromophore, either 11-*cis* retinal or 11-*cis* 3,4-dehydroretinal (vitamin A_1 or A_2 aldehyde, respectively) (Nathans 1987). Opsins in general are retinal-mediated light-sensing proteins found in a variety of organisms from bacteria to vertebrates. These proteins have various visual and nonvisual functions. The vertebrate and invertebrate opsins belong to a multigene superfamily of the G-protein-coupled receptors (GPCR), which commonly form a characteristic seven-transmembrane structure. The chemosensory receptors, such as odor, taste, and pheromone receptors, also belong to the GPCR gene families (see also Chaps. 4–6). Amino acid sequences of the opsins modulate absorption spectra of the chromophore. Absorption spectra of visual pigments are bell shaped when plotted against wavelengths. The wavelength where peak absorbance occurs is called the lambda max, "λ_{max}." Given the uniformity of the absorption curve, the λ_{max} is commonly used to represent the whole absorption spectra of a visual pigment.

In diurnal birds, reptiles, and lungfishes, colored oil droplets, located in cones, further modulate absorption spectra of the cones as colored filters (Walls 1942; Robinson 1994). These retinal filters reduce the overlap in sensitivity between spectrally adjacent cones and hence increase the number of discriminable colors (Govardovskii 1983). Thus, opsins, chromophores, and oil droplets together shape the spectral properties of cones. However, the A2 type chromophore is not used in most terrestrial vertebrates, and oil droplets are absent in many groups of animals (Lythgoe 1979; Goldsmith 1990). Thus, opsins play a universal and pivotal role in evolution of color vision. In addition, a major advantage to molecular evolution studies that focus on opsins lies in the feasibility of functional assays of these proteins, coupled with site-directed mutagenesis, by transfection of opsin cDNAs to cultured cells, reconstitution of functional photopigments in vitro with a chromophore, and spectral measurement of the purified pigments (Yokoyama 2000b).

For the evolution of color vision, animals need (1) a set of spectrally differentiated cone opsins, (2) the genetic mechanisms that allow the opsins to be expressed in different cones, (3) the neural mechanisms that enable the comparison of signals from the different spectral classes of cones and the extraction of color information, and (4) natural selection to promote and maintain a particular color vision status. Perhaps the most commonly endorsed interpretation of human trichromacy would still be the simple one that trichromacy is universally superior to dichromacy for any visual task, and that the dichromatic and vision-polymorphic animals may eventually acquire routine trichromacy, if the necessary events were to occur, that is, duplication and divergence of opsin genes leading to fixation of two spectrally different opsin genes. However, recent studies indicate a paradigm shift toward the view that the adaptive value of trichromacy is conditional rather than universal, depending on the specific ecological demands on animals in their environments. The question does remain about what exactly these conditions are. Because of the wide variation of color vision both within and between species, New World monkeys are excellent subjects for study of the utility of color vision in natural environments, which will help us to elucidate the selective advantage of being a trichromat or dichromat.

In the following section, we first introduce the current state of knowledge of variation in vertebrate visual opsins, and then shift our main focus to primates, covering all four requisites for the evolution of color vision just listed.

7.2 Visual Opsins in Vertebrates

The visual opsins in vertebrates are classified into five phylogenetic types: RH1 (rhodopsin or rod opsin), and four cone opsins [RH2 (rhodopsin-like, or green), SWS1 (short wavelength-sensitive type 1, or ultraviolet-blue), SWS2 (short wavelength-sensitive type 2, or blue), and M/LWS (middle to long wavelength-sensitive, or red–green)] (Yokoyama 2000a). It is well established that these five types were present in the common ancestor of all vertebrates, including jawless fish (Yokoyama 2000a; Collin et al. 2003, 2009; Davies et al. 2009). Thus, early vertebrates could already have had four-dimensional color vision (tetrachromacy). Many fish are known to possess a rich repertoire of visual opsins, including two or more opsin subtypes within the five types: for example, zebrafish (*Danio rerio*) have nine visual opsin genes including spectrally distinct two M/LWS and four RH2 opsin subtypes (Chinen et al. 2003). Many species of birds and reptiles retain one each of the four cone opsin types (and a rod opsin) and are tetrachromatic in color vision (Ebrey and Koutalos 2001). In contrast, mammals are considered to have lost RH2 and either the SWS2 (placental mammals and marsupials) or SWS1 (monotremes) opsin gene in a nocturnal ancestor that lived during the Mesozoic Period (Ahnelt and Kolb 2000; Davies et al. 2007). Extant placental mammals are basically dichromatic with only SWS1 and M/LWS as cone opsins besides a rod opsin (Jacobs 1993). Placental mammals use only 11-*cis* retinal as a chromophore, and all spectral variations of their visual pigments are caused by the opsins. Primates are the only exception among placental mammals in attaining trichromatic vision by diversifying the M/LWS opsin gene through either gene duplication or allelic diversification (Jacobs 1999).

Shadows generally yield strong variation in the intensity of illumination, but comparison of chromatic signals from different spectral types of cones provides a value that remains constant across different levels of illumination intensity (Foster and Nascimento 1994). Thus, color vision is especially useful for object detection in conditions of patchy and changing illumination or against a dappled background. In addition, color facilitates the identification of an object by its surface reflectance irrespective of spectral distribution of the illuminant, the phenomenon known as color constancy (Pokorny et al. 1991). Patchy and spectrally varying illumination is common in shallow water and in forests, and hence these are the places where color vision would be strongly selected for and be diversified most dramatically (Vorobyev 2004). Appearance of the four cone opsin types in early vertebrates before the divergence of jawed and jawless forms (Yokoyama 2000a; Collin et al. 2003, 2009; Davies et al. 2009) is intriguing in this context. Early vertebrates could already have had tetrachromatic color vision in their shallow aquatic habitat in the early Cambrian, approximately 540 million years ago (Maximov 2000). This idea also could explain why current fish are so varied and rich in the visual opsin repertoire, presumably reflecting their evolutionary adaptation to diverse aquatic light environments (Levine and MacNichol 1982). Similarly, in terms of the evolution of primate color vision, forest light may have been a key factor that prompted reclamation of the third opsin in primates, enabling their trichromacy (Mollon 1989).

7.3 Variation in Color Vision and Visual Opsin Repertoire Among Primates

7.3.1 Nomenclature

Conventionally, the primate SWS1 opsin is called the "blue" or "S" opsin, with λ_{max} at around 420–430 nm. The longer-wave subtype of the M/LWS opsin (λ_{max} at around 560 nm) is often called the "red" or "L" opsin, and the shorter-wave subtype (λ_{max} at around 530 nm) is called the "green" or "M" opsin. In the case of New World monkeys, in which three M/LWS opsin alleles are often found, alleles are called by their λ_{max} values or by conventional color names, such as red, yellow (or orange), and green, or by the abbreviations "L," "A" ("A" stands for "anomalous" in human sense) (or "I"; "I" stands for "intermediate"), and "M." It should be noted that the color name does not necessarily match the corresponding color of the pure light of the λ_{max}: for example, the 560-nm light, a typical λ_{max} value of "red" opsins, appears yellowish to trichromatic human observers.

7.3.2 Variation of Color Vision in Primates

Among primates, variation in number and spectral properties of the M/LWS and SWS1 opsins results in marked variation in color vision (Fig. 7.1). Routine trichromacy

7 Polymorphic Color Vision in Primates: Evolutionary Considerations

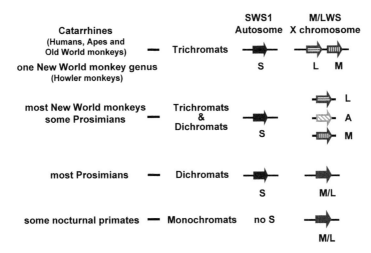

Fig. 7.1 Variation in color vision and visual opsin repertoire among primates. Catarrhines and howler monkeys have L and M opsin genes (as the M/LWS, middle to long wavelength-sensitive or red-green class of opsin gene) on the same X chromosome and the S opsin gene (as the SWS1, short wavelength-sensitive type 1 or ultraviolet-blue class of opsin gene) on an autosome and are therefore uniformly trichromatic in color vision. Many New World monkeys and some prosimians have two or more M/LWS opsin alleles of a single X-chromosomal locus in addition to the autosomal S opsin gene and are polymorphic in color vision, consisting of both trichromats and dichromats. Three alleles are shown and labeled as L, A, and M in this figure. Most prosimians have a single M/LWS opsin locus (labeled M/L) with no allelic variation in addition to the autosomal S opsin gene and are uniformly dichromatic. Some nocturnal primates are monochromatic as a result of the loss of the functional S opsin gene

is attained by a gene duplication of the M/LWS opsin on the X chromosome and by spectral differentiation of the resulting L and M opsins in catarrhines (humans, apes, and Old World monkeys) (Nathans et al. 1986; Jacobs 1996). A similar gene duplication is reported for howler monkeys (*Alouatta*), a genus of New World monkeys (Jacobs et al. 1996a), although this was attained by an independent evolutionary event.

Color vision polymorphism, that is, a mixed population of dichromats and trichromats resulting from allelic variation of the X-linked M/LWS opsins (Fig. 7.2), has been documented in many species of New World monkeys (Fig. 7.3) (Jacobs 1998). In this system, females are either trichromatic or dichromatic, whereas males are all dichromatic. A wide variation of L–M opsin allelic composition occurs among the vision-polymorphic New World monkeys, ranging from diallelic, seen typically in *Ateles* (spider monkeys) and *Lagothrix* (woolly monkeys), up to pentallelic, reported for *Callicebus moloch* (dusky titi monkeys) (Jacobs and Deegan 2001, 2005; Talebi et al. 2006; Jacobs 2007). Similar allelic variation is reported for a few species of diurnal or cathemeral prosimians (Tan and Li 1999; Heesy and Ross 2001; Jacobs et al. 2002; Veilleux and Bolnick 2009).

Uniform dichromacy is considered a norm in many prosimians with a monomorphic M/LWS opsin and an SWS1 opsin, as is found in most placental mammals. Finally, monochromacy resulting from the loss of a functional SWS1 opsin has been documented in some nocturnal species, notably lorisiform prosimians and owl monkeys (*Aotus*), a genus of New World monkeys that is the only nocturnal simian

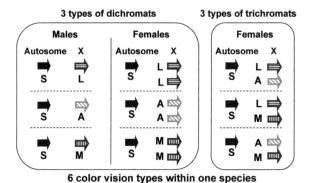

Fig. 7.2 Color vision polymorphism of New World monkeys. Typically, three alleles are found in the X-linked M/LWS opsin gene, here labeled L, A, and M. Males have only one X chromosome and are therefore obligate dichromats, having a single M/LWS opsin allele on the X chromosome and the single autosomal S opsin gene. However, there are three types of dichromatic males within the same species, each having different M/LWS allele types. If a female has the same M/LWS opsin allele on both X chromosomes, she is also dichromatic, as are the males. If a female has two different M/LWS alleles, she is a trichromat. There are three types of trichromatic females in the species because three heterozygote combinations are possible in a triallelic M/LWS system. In total, six different color vision phenotypes can exist in one species if there are three M/LWS alleles

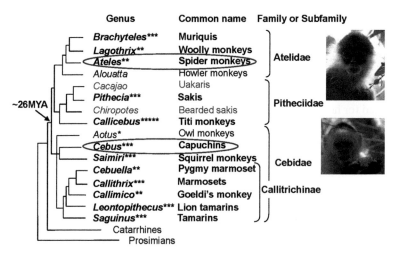

Fig. 7.3 Phylogenetic distribution of color vision polymorphism among New World monkeys. Polymorphism has been reported for the genera indicated with *boldfaced letters*. The number of *asterisks* indicates the number of alleles reported for the genus (Jacobs 1998; Jacobs and Deegan II 2001; Surridge and Mundy 2002; Jacobs and Deegan 2003; Jacobs and Deegan 2005; Talebi et al. 2006). Spider monkeys and capuchins have been targeted in our field study and are *circled* with associated photographs. Howler monkeys (*Alouatta*) are regarded as uniformly trichromatic and owl monkeys (*Aotus*) as monochromatic. Color vision of monkeys indicated with *gray letters* (*Cacajao* and *Chiropotes*) is not known. (The phylogenetic tree is after Schneider 2000)

(catarrhines and New World monkeys) (Jacobs et al. 1996b; Kawamura and Kubotera 2004; Tan et al. 2005). An intact copy or multiple pseudogene copies of M/LWS opsin gene are present on the Y chromosome of owl monkeys in addition to the original X-linked one, but their functional significance remains unknown (Kawamura et al. 2002; Nagao et al. 2005).

Among routinely trichromatic catarrhines, humans constitute a notable exception. Approximately 3–8% of males have "color vision defects," mainly because of unequal meiotic recombination between L and M opsin genes (Deeb 2006). The L and M opsin genes are highly similar in nucleotide sequence (~96% identity) and are closely juxtaposed (Nathans et al. 1986). The recombination results in an L–M hybrid gene if the crossover point lies within a gene region. If the crossover is in an intergenic region, it results in gene deletion in one chromosome and duplication in the other. Many humans have multiple copies of the M opsin gene in the M/LWS opsin gene array, wherein the most upstream gene is typically L and the others are M. Only the upper two genes are expressed, and when a hybrid gene occupies either position, it causes anomalous trichromacy (Hayashi et al. 1999). When there is only one M/LWS opsin gene on an X chromosome or when the two positions are occupied by the same genes, this causes dichromacy (red–green colorblindness: more specifically, protanope when L is lost, and deuteranope when M is lost). Color vision defects caused by these mutations in the M/LWS opsin gene include both anomalous trichromacy and dichromacy. These phenotypes are typically found in men because women have two X chromosomes and thus are more likely to have a "normal" gene array in either one. There are rare cases of individuals, irrespective of sex, who have no functional blue cones (tritanopes, <1:10,000) because of mutations in the S opsin gene on chromosome 7 (Sharpe et al. 1999).

Compared to humans, the incidence of color vision defects in non-human catarrhines is low (Onishi et al. 1999; Jacobs and Williams 2001; Terao et al. 2005). Among 744 male long-tailed macaques (*Macaca fascicularis*) examined, only 3 were found to have a single hybrid M/LWS opsin gene and to be dichromats (Onishi et al. 1999; Hanazawa et al. 2001). Among 58 male chimpanzees (*Pan troglodytes*), 1 was found to have a hybrid gene in addition to one normal M opsin gene on the X chromosome and to be an anomalous (protanomalous) trichromat (Saito et al. 2003; Terao et al. 2005). Thus, frequencies of color vision deficiencies in male long-tailed macaques and male chimpanzees can be calculated to be about 0.4% and about 1.7%, respectively. These frequencies could be overestimated because no defects were found in 455 male monkeys from other macaque species (Onishi et al. 1999) and the chimpanzees examined were from limited numbers of breeding colonies (Terao et al. 2005). Other researchers have reported an absence of color vision defects in Old World monkeys and apes (Jacobs and Williams 2001). Multiple copies of M opsin genes are likely to increase the frequency of unequal recombination events. Although multiple M copies are found in 66% of human (Caucasian) males (Drummond-Borg et al. 1989), they were found in only 5% of 130 male long-tailed macaques (Onishi et al. 1999, 2002) and 6% of the 58 male chimpanzees (Terao et al. 2005). The low incidence of the multiple M copies may partly explain the low incidence of color vision defects in these primates.

However, other studies report that multiple copies are common among Old World monkeys and apes (Ibbotson et al. 1992; Dulai et al. 1994).

7.3.3 Molecular Evolution of the L–M Opsin Gene in Primates

Mutagenesis studies indicate that the λ_{max} of the vertebrate M/LWS opsins can be predicted from the amino acid composition at the three sites, 180, 277, and 285 (the residue numbers hereafter follow those in the human L opsin), together with two additional sites, 197 and 308 ("five-site rule") (Yokoyama and Radlwimmer 1998, 1999; Yokoyama et al. 2008). The residue 180 is encoded in exon 3, 197 in exon 4, and 277, 285, and 308 in exon 5. Among primate M/LWS opsins, however, the residues 197 and 308 are not varied and are irrelevant to spectral differences among them. Therefore, the five-site rule can be reduced to the "three-site rule" in primates, wherein amino acid changes from Ser to Ala at site 180 (denoted Ser180Ala), Tyr277Phe, and Thr285Ala shift the λ_{max} values by −7, −8, and −15 nm, respectively, and the reverse amino acid changes cause opposite spectral shifts to the same extent in a nearly additive manner (Neitz et al. 1991; Kawamura et al. 2001; Yokoyama and Radlwimmer 2001; Kawamura and Kubotera 2003; Hiramatsu et al. 2004).

Molecular phylogenies reconstructed from the M/LWS opsin gene sequences show that alleles often cluster by species rather than by allele types and that catarrhine L and M genes form a separate cluster from the New World monkey alleles, implying that alleles were formed in many species independently and that spectral differentiation of L and M opsins in catarrhines occurred after the gene duplication (Shyue et al. 1995; Hunt et al. 1998). However, gene conversion is known to occur frequently between loci and between alleles of M/LWS opsin genes, homogenizing and masking sequence variations among them (Zhou and Li 1996; Boissinot et al. 1998). The spectral differentiation among M/LWS alleles in New World monkeys and prosimians, as well as that between duplicated L and M opsins in catarrhines and howler monkeys, all result from the three-site combinations of the alternative amino acids at each site in common. Thus, it is likely that color vision polymorphism occurred in the common primate ancestor of simians and prosimians. It is also likely that the allelic variation was incorporated and fixed in the same chromosome by a gene duplication in the catarrhine ancestor and independently in howler monkeys (Boissinot et al. 1998; Surridge and Mundy 2002; Surridge et al. 2003; Tan et al. 2005).

The possible antiquity of the M/LWS polymorphism in the primate ancestors, together with the presence of functional SWS1 in many non-lorisiform nocturnal prosimians, implies that trichromacy occurred in the ancestral primates in a polymorphic manner. It is thus likely that the primate ancestor had already shifted to a diurnal or cathemeral pattern from a nocturnal pattern at an early stage. The current nocturnality in many prosimians and in owl monkeys could be a derived state that recurred in several lineages (Tan and Li 1999; Tan et al. 2005).

7.4 The Genetic Mechanism of Differential Expression of M/LWS Opsin Subtypes in Primates

Given the spectral differentiation among opsin types and subtypes, a genetic mechanism is required to achieve primate trichromacy that enables differential expression of spectrally distinct opsin genes in different cones. Distinction of expression between SWS1 and M/LWS opsin genes is ancient, dating back to the early vertebrates. The relevant regulatory regions and transcription factors, however, largely remain unsolved. In contrast, the mechanism regulating the expression between M/LWS subtypes in different cones, enabling primate trichromacy, is well understood.

The X-chromosomal locality of the M/LWS type opsin gene in placental mammals is extremely important for the mutually exclusive expression of its subtypes. In females, an allele from only one X chromosome is expressed by virtue of random X-chromosomal inactivation (lyonization), and in males there is simply only one X chromosome. Trichromatic color vision in female New World monkeys is attained by this X-chromosomal locality of the M/LWS opsin gene.

In catarrhines, an additional mechanism is required to selectively express only one gene from the M/LWS gene array on the X chromosome. This expression is achieved through a process of stochastic interaction of a locus control region (LCR), situated upstream of the gene array, with the promoter of only one gene from the array in one cone (Fig. 7.4a) (Wang et al. 1999; Smallwood et al. 2002). The LCR is originally an enhancer element for the single-copy M/LWS opsin gene of mammals (Wang et al. 1992). The gene duplication excluding the LCR from the duplication unit enabled the catarrhines to achieve mutually exclusive expressions of the duplicated M/LWS opsin genes. Hence, primate trichromacy was achieved through the preexisting gene regulatory systems including the X-chromosome lyonization.

In howler monkeys, a gene duplication of the M/LWS opsin has been documented (Jacobs et al. 1996a). Absorption spectra of the resulting L and M opsins were measured indirectly by electroretinogram (ERG) (Jacobs et al. 1996a), which measures gross electric potential of the retina. The estimated absorption spectra of the two photopigments are in good agreement with those of catarrhine L and M, respectively. Mutually exclusive expression of the L and M opsins has been supported by ganglion recording and microspectrophotometry (MSP) (Saito et al. 2004). A recent behavioral experiment also supported the existence of uniform trichromacy in male and female howler monkeys (Araujo et al. 2008). In contrast to catarrhines, however, the howler LCR is duplicated together with the opsin gene (Fig. 7.4b) (Dulai et al. 1999). This finding implies that howler monkeys have evolved extra machinery to attain the mutually exclusive expression of the L and M opsin genes. Studies of this mechanism have not yet been reported. It is also important to study whether trichromatic color vision is really "uniform" in howler monkey populations with no "defective" variants.

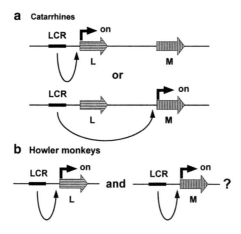

Fig. 7.4 Mutually exclusive expression of L and M opsin genes. (**a**) In catarrhines, the locus control region (LCR) interacts with either the uppermost (typically L) or the second (typically M) opsin gene in an M/LWS gene array on an X chromosome and turns on the expression of the associated opsin gene. (**b**) In howler monkeys, LCR is duplicated together with the opsin gene and is adjacent to both L and M opsin genes. Without some regulatory mechanism, both L and M opsin genes would be expressed from the same chromosome by the interaction with their LCRs in a given M/LWS cone. The mechanism is currently not known, however

7.5 Neuronal Requirement for Evolution of Primate Color Vision

The last mechanistic requirement to achieve primate trichromacy is to establish a neuronal network to convey the color-opponent L–M signals. As in the genetic mechanism for the exclusive gene expression, the necessary neural circuitry was provided serendipitously from a preexisting system for spatial vision.

Most primate features seemingly evolved for a predominantly arboreal life. The visual system of primates, especially that of diurnal simians, is characterized not only by the presence of trichromacy. It is also characterized by forward-facing eyes, the postorbital plate (a bony cup surrounding the eye), and the fovea (a major central peak in cone density in the retina). The forward-facing eyes are seen in all primates and enable stereoscopic vision. The postorbital plate is found in simians and prevents the chewing muscles from disrupting eye position, which could serve to improve visual acuity (Fleagle 1999; Heesy et al. 2007). The simian fovea also allows for very high visual acuity. These features are essential for agile movement and saltatory locomotion from branch to branch.

High spatial acuity is also realized via the evolution of a one-to-one midget ganglion pathway, which receives input from only one M/LWS cone cell in the center of the receptive field of a midget ganglion cell (Martin 1998). This pathway compares the center input to inputs from surrounding M/LWS cones. A more primitive midget ganglion pathway is seen even in nocturnal prosimians (bushbabies) (Yamada et al. 1998). In bushbabies, the cone–ganglion convergence ratio in the central retina is higher than one (5 cones per ganglion) but is still much lower than that, for example, in cats (~30 cones per ganglion).

The midget ganglion pathway is present only in primates among mammals and conveys information about fine spatial details of the image. When the center-surrounding inputs are from different spectral types of M/LWS cones, the pathway also provides a color opponent mechanism, which is the neural basis of primate trichromacy (Martin 1998).

Recently, knock-in mice that express a human L opsin gene in the form of an X-linked polymorphism have been created (Jacobs et al. 2007). The heterozygous females carry the human L opsin gene and the native mouse M opsin gene. Behavioral tests demonstrated that these mice showed enhanced long-wavelength sensitivity and acquired a new capacity for chromatic discrimination. However, mice lack a midget system. Most retinal ganglion cells in mice have a receptive field center that receives inputs from multiple cone cells. Its antagonistic surround also receives inputs from multiple cone cells. The knock-in mice could extract chromatic information based on the differences in total M versus L input to the two regions. The demonstrated plasticity of the nervous system to accommodate the altered receptor signals is surprising. However, it is questionable whether the new color dimension presented over the coarse spatial image would be useful and adaptive for mice.

Prosimians lack a fovea, and their midget system is relatively unspecialized. Trichromacy by the L–M gene polymorphism is not present in very many species of prosimians. Thus, the selection pressure to maintain trichromacy may not have been strong enough for these mammals with poor spatial resolution of visual images. Without the midget ganglion pathway, evolution of trichromacy would not have been possible in other mammals even if a similar spectral differentiation of opsin subtypes had occurred (Surridge et al. 2003; Vorobyev 2004).

7.6 Unsolved Mystery: What Selects for Primate Trichromacy?

7.6.1 Fruit Theory

Although our understanding of the mechanistic aspects of primate trichromacy has greatly advanced in recent years, its adaptive nature is still controversial. It has long been hypothesized that primate trichromacy was selected for finding ripe fruit against a mature leaf background ("fruit theory") (Allen 1879). In favor of this hypothesis, an analysis of the spectra of fruits eaten by humans showed that the spectral separation of L and M cones is close to optimum for detection of fruit against foliage (Osorio and Vorobyev 1996). This result was later supported by spectral analysis of fruit eaten by primates and collected in the rainforest (Regan et al. 1998, 2001; Sumner and Mollon 2000).

Recently, however, there have been arguments against the fruit theory because many fruits eaten by primates are, in fact, also distinguishable from background leaves by dichromats via the blue–yellow (S vs. L/M) signal and luminance signal (Dominy and Lucas 2001). Furthermore, some fruits do not develop conspicuous colors, yet constitute a significant portion of primate diets, and fruit color seemed to provide no consistent nutritional cue in a study of the 12 plant species most commonly consumed by the primates of Kibale Forest, Uganda (Dominy 2004). Primates are not the sole seed dispersers, and plants also rely on many other frugivores, such as parrots, bats, and peccaries, for successful seed dispersal. In fact, some Old World monkeys are known to often select unripe fruits, which is detrimental for seeds (Dominy and Lucas 2001). In addition, many fruits are highly seasonal and become scarce in dry season.

Figs and palm nuts are not seasonal and can function as keystone resources during the periods of fruit dearth (Terborgh 1986). Cryptic coloration is frequent in figs and palms, and it is suggested that early primates in warm Paleocene-Eocene forests, which were characterized by figs and palms (Morley 2000), relied on them as keystone resources (Dominy et al. 2003b). However, the global cooling and drying during the Eocene–Oligocene interval (~40–30 million years ago) coupled with increasing seasonal fluctuations dramatically reduced densities and availability of figs and palms, especially in Africa (Morley 2000). Africa, where early simians are supposed to have evolved (Fleagle 1999), is still highly seasonal, with a phenology characterized by alternating periods of fruiting and leafing (Dominy et al. 2003b).

7.6.2 Young Leaf Theory

Another theory regarding the evolution of trichromacy is the "young leaf theory." This theory states that, given the seasonality of Africa, young leaves provide a critical fallback resource during periods of fruit shortage (Lucas et al. 1998). Regardless of tree species, young leaves are tender and rich in proteins as well as free amino acids (Dominy and Lucas 2001). Young leaves are often reddish and thus distinct from mature leaves only via the red–green color channel of trichromats. Hence, the ability to discern between young and mature leaves may have been a major selective force for primate trichromacy (Lucas et al. 1998, 2003; Dominy and Lucas 2001). The young leaf theory is strengthened in the context of the historical biogeography of figs and palms; in Africa, where early catarrhines evolved, figs and palms are scarce and routine trichromatic vision was selected for exploiting proteinaceous young leaves as a replacement resource. However, in the Neotropics and Madagascar, where polymorphic color vision is seen in most New World monkeys and some prosimians, figs and palms remained abundant and some New World monkeys, such as marmosets, do not depend on young leaves at all (Dominy et al. 2003b). Thus, the young leaf theory does not seem to explain the evolution and maintenance of trichromacy outside Africa.

7.6.3 Other Theories

Another longstanding hypothesis to explain the evolution of trichromacy has been the detection of social signals or the detection of predators (Allen 1879; Surridge et al. 2003; Vorobyev 2004; Changizi et al. 2006). A recent study showed, however, that primate trichromacy appeared before the evolution of red pelage and red skin, as well as gregarious mating systems, and therefore the social signals could not be a factor in the evolution of trichromacy from dichromacy (Fernandez and Morris 2007).

Another hypothesis recently put forward could be named the "long-distance foliage hypothesis" (Sumner and Mollon 2000). For trichromatic primates, perceived color, that is, chromaticity, can be described as a ratio of the quantum catch among their L, M, and S cones and expressed as a point in color space analogous to the MacLeod–Boynton diagram (MacLeod and Boynton 1979) consisting of $L/(L+M)$ and $S/(L+M)$ axes. The former axis represents a ratio of quantum catch of L cones to that of L and M cones whereas the latter represents that ratio for S cones to L and M cones (Regan et al. 1998). $L/(L+M)$ indicates the redness that is provided by the "red–green" chromatic channel, for which only trichromats are equipped, and subserved by the midget ganglion cells. The $S/(L+M)$ indicates the blueness that is provided by the "blue–yellow" chromatic channel, for which all mammals are equipped. This more ancient system is subserved by the small bistratified ganglion cells. Colorimetric measurements of natural scenes in forests reveal that the chromaticity of mature leaves falls in a very narrow range of $L/(L+M)$ values but spreads widely along the $S/(L+M)$ axis and also in luminance values. Thus, the chromaticity of fruits, young leaves, pelage, and skin often deviates from mature leaves in $L/(L+M)$ value but largely overlaps with them in $S/(L+M)$ and luminance values (Regan et al. 1998, 2001; Sumner and Mollon 2000, 2003), leading to a hypothesis that primate trichromacy could be adaptive for and have evolved for detecting *anything* differing from the background foliage in $L/(L+M)$ value. This trichromat advantage is supposed to be maximized during long-distance viewing because the scene would contain a larger variety of background $S/(L+M)$ and luminance values than would a closer view. In addition, during close viewing, other sensory cues, such as odors, are available and visual cues could be less important (see also Chap. 5).

On the other hand, the "short-distance foliage hypothesis" is also suggested by psychophysical studies. The human visual system shows a relatively greater sensitivity to low spatial frequencies of chromatic spatial modulation than to luminance spatial modulation (Mullen 1985). In addition, a statistical analysis of spatial frequencies of natural images suggests that the spatiochromatic properties of the red–green system of human color vision may be optimized for the encoding of any reddish or yellowish objects against a background of foliage at relatively small viewing distances commensurate with a typical grasping distance (Parraga et al. 2002).

7.6.4 Advantage of Dichromacy

Recent studies have found that dichromatic vision may be advantageous to primates under some conditions, for example, finding cryptic fruits or insects or detecting cryptic predators, such as snakes (Caine et al. 2003; Saito et al. 2005b). The conceptual basis for this hypothesis is that trichromatic vision compromises the acuity of other visual systems. The neural system of trichromatic individuals must combine signals from the L and M photoreceptors to obtain the luminance signal used for achromatic "colorblind" tasks such as spatial vision and the perception of shape, texture, and motion (Morgan et al. 1992; Kelber et al. 2003). The different spectral inputs from the two receptors can cause corruption, resulting in a weaker overall signal. Additionally, color may compete with texture information, or trichromats may learn to rely on color at the expense of information to be gained by texture. Therefore, dichromats may have an advantage over trichromats in achromatic (colorblind) tasks, such as depth perception and breaking camouflage.

7.6.5 Behavioral Experiments

Behavioral experiments that compared feeding efficiency between vision types in laboratory settings with artificial targets have suggested that there is a selective advantage of trichromacy in foraging on colored foods (Caine and Mundy 2000; Caine 2002; Smith et al. 2003b). On the other hand, these studies and others have found that dichromats are better than trichromatic primates at detecting camouflaged stimuli (Caine et al. 2003; Saito et al. 2005b). It should be noted that such comparisons evaluate whether the difference in visual ability is consistent with the difference in color vision phenotypes but do not evaluate whether one phenotype is more advantageous than another (Saito et al. 2005a). These experiments should only be regarded as tests that determine visual phenotypes, although they do provide useful predictions about the potential foraging advantages of dichromatic and trichromatic phenotypes.

7.7 New World Monkeys as a Model to Study the Adaptive Nature of Primate Color Vision

Whatever the theories and laboratory experiments predict, the adaptive value of primate trichromacy (or dichromacy) can only be evaluated in light of behaviors seen in natural environments. Hence, it is important to compare behaviors between free-ranging dichromats and trichromats and to evaluate whether and how the contrast between a visual target and its background are correlated with behavioral differences in these two types of primates. New World monkeys are an excellent model to test the suggested advantage of trichromacy because of the allelic polymorphism of the L–M opsin gene that results in coexistence of dichromatic and trichromatic individuals in the same population (Mollon et al. 1984) (Fig. 7.2). Prerequisites for realizing these studies are that a method must be established to determine vision phenotypes without harming the study animals and that the monkeys at the study site must be well habituated and individually identified by researchers. In the next two subsections, we review the methodology related to the first prerequisite and introduce our study site as an example. Then, we summarize recent findings, including ours, on vision–behavior relationships in wild New World monkeys. We also introduce our population genetic study on natural selection acting to maintain color vision variation in the monkeys at our study site.

7.7.1 Spectral Genotyping of L–M Opsin Alleles for Field Samples

The spectral genotype of the L–M opsin gene can now be estimated noninvasively through polymerase chain reaction (PCR) and DNA sequencing analyses of fecal samples collected in the field (Surridge et al. 2002; Hiramatsu et al. 2005). This estimation can be done by examining exon 3 (containing the residue 180) and exon 5 (containing the residues 277 and 285) of the L–M opsin gene (Hiramatsu et al. 2004) (see Sect. 7.3.3). The estimated absorption spectra can be confirmed experimentally by reconstitution of the opsin photopigments in vitro and direct measurement of their absorption spectra (Yokoyama 2000b). With a single S opsin gene, female monkeys having two different spectral alleles of the L–M opsin gene are considered trichromats, and females having two identical L–M opsin alleles, and males, because of the hemizygosity of the X chromosome, are considered dichromats. Consistency between the genotype and phenotype has been well demonstrated by behavioral experiments (Caine and Mundy 2000; Saito et al. 2005a).

7.7.2 An Example Study

To study vision–behavior relationships in wild monkeys, it is essential that monkeys in the study site be habituated to human observers and individually identified. It is also important that ecological and sociological information be available, and it is desirable that two or more primate species with different ecology, phylogeny, and vision types be present at the site for comparative purposes. We have chosen the Santa Rosa Sector of the Área de Conservación Guanacaste (ACG), northwestern Costa Rica (10°45′ to 11°00′ N and 85°30′ to 85°45′ W), as an ideal study site where prerequisite conditions are satisfied (Chapman 1990; Fedigan and Jack 2001, 2004; Fedigan 2003). The park is composed of tropical dry forest in various successional stages. Rainfall in Guanacaste Province is highly seasonal; mean annual rainfall is approximately 2 m, almost all of which is accumulated between mid-May and mid-December (Janzen 2002).

There are three sympatric primate species at Santa Rosa: white-faced capuchin monkeys (*Cebus capucinus*), black-handed spider monkeys (*Ateles geoffroyi*), and mantled howler monkeys (*Alouatta palliata*). The first two genera of monkeys are known to have color vision polymorphism (Jacobs 1998). Spider and capuchin monkeys differ in diet (frugivores and omnivores, respectively), social structure (male philopatry with high fission–fusion dynamics and female philopatry with rather cohesive groups, respectively) (Fragaszy et al. 2004; Aureli and Schaffner 2008), and phylogeny (Atelidae and Cebidae, respectively) (Schneider 2000). Capuchins are reported to possess three L–M opsin alleles, whereas spider monkeys possess two (Shyue et al. 1998; Jacobs and Deegan 2001). These conditions make capuchins and spider monkeys excellent study subjects, complementary to cal-litrichine species (marmosets and tamarins), whose vision–behavior relationships are the most intensively investigated among New World monkeys (Caine and Mundy 2000; Caine 2002; Caine et al. 2003; Pessoa et al. 2003; Smith et al. 2003a, b).

On the basis of the three-site rule, we identified three L–M opsin alleles in capu-chin monkeys, one having the three-site composition Ser, Tyr, and Thr at sites 180, 277, and 285, respectively (designated Ser/Tyr/Thr), and the other two having Ala/Phe/Thr and Ala/Phe/Ala (Hiramatsu et al. 2005). The λ_{max} values of the three alleles were directly measured by the method of photopigment reconstitution in vitro and were determined to be 561, 543, and 532 nm (designated P561, P543, and P532), respectively (Hiramatsu et al. 2005). Similarly, we identified two L–M opsin alleles in spider monkeys, one having the three-site composition Ser/Tyr/Thr and the other having Ser/Phe/Thr (Hiramatsu et al. 2005). The λ_{max} values of the two alleles were directly measured in vitro and were determined to be 553 and 538 nm, respectively (designated P553 and P538) (Hiramatsu et al. 2008). In a spider monkey group, the two alleles, P553 and P538, are present at 59.6% and 40.4%, respectively, among 52 X chromosomes examined (20 females and 12 males) (Hiwatashi et al. 2010). In a capuchin monkey group, the three alleles, P561, P543 and P532, are present at 51.6%, 35.5% and 12.9%, respectively, among 31 X chromosomes examined (8 females and 15 males) (Hiwatashi et al. 2010). In both species, the genotypes are present at frequencies that do not deviate significantly from Hardy–Weinberg equi-librium (Hiwatashi et al. 2010). These results suggest that there is little effect of color vision phenotype on mate choice, or with the small sample sizes here, effect of color vision phenotype on mate choice, if any, is not strong.

7.7.3 Behavioral Observations on Wild Populations of New World Monkeys

In recent years, many studies, including ours, have been published on behavioral observations of wild populations of New World monkeys for evaluating vision–behavior relationships. Despite the predicted advantage of trichromacy, these studies have provided only limited support. In a study of a wild mixed-species troop of saddleback (*Saguinus fuscicollis*) and mustached (*S. mystax*) tamarins, during vigilance, trichromats are further from their neighbors than their dichromatic conspecifics, which is explained as resulting from the potentially better perception of predation risk in trichromats (Smith et al. 2005). Our study of a population of white-faced capuchin monkeys (*C. capucinus*) found that dichromats sniffed more figs and had longer foraging sequences than trichromats, especially for cryptic figs (Melin et al. 2009). Among six subtypes of dichromats and trichromats, monkeys possessing the trichromat phenotype with the most spectrally separated L–M opsin alleles showed the highest acceptance index for conspicuous figs, although there were no differences in feeding rates among phenotypes (Melin et al. 2009).

Results of other behavioral observations of wild New World monkeys have produced equivocal results or results contradictory to the predictions from the trichromat advantage hypothesis. The study of the wild mixed-species troops of tamarins showed that the color vision types (dichromatic or trichromatic) did not have a consistent effect on the leadership of the troops to feeding trees (Smith et al. 2003a). Another study of tamarins (*S. imperator imperator* and *S. fuscicollis weddelli*) found no significant difference between females (thought to consist of trichromats and dichromats) and males (all dichromats) in their ability to locate or discriminate between feeding sites (Dominy et al. 2003a). In a population of capuchin monkeys (*C. capucinus*), there was no significant difference between trichromats and dichromats in feeding or energy intake rates (Vogel et al. 2007). In another population of the same capuchin monkey species, we showed that there was no difference between dichromats and trichromats in time spent foraging on different food types (Melin et al. 2008).

Some modeling studies based on field observations have found that many fruits eaten by spider monkeys or squirrel monkeys (*Saimiri sciureus*) are similarly discernible or similarly indiscernible from background foliage for both trichromats and dichromats (Riba-Hernández et al. 2004; Stoner et al. 2005; De Araujo et al. 2006). Our field study of free-ranging black-handed spider monkeys measuring their foraging efficiency on fruits and colorimetric properties of fruits and background leaves revealed that dichromats are not inferior to trichromats in frequency, accuracy, and unit-time intake efficiency of detecting fruits (Hiramatsu et al. 2008). We showed that this is because the luminance contrast of fruits to background leaves is the main determinant of fruit detection in both dichromats and trichromats. Another study of ours on the same social group of spider monkeys also showed that, irrespective of color vision phenotypes, the monkeys sniff visually cryptic fruits more often than visually conspicuous fruits (Hiramatsu et al. 2009). This finding indicates that color vision is not the sole determinant for ingestion or rejection of fruits. Our field study of white-faced capuchin monkeys has even demonstrated a dichromat advantage in foraging for surface-dwelling insects (Melin et al. 2007, 2010). These findings of observational studies in natural environments suggest that the superior ability of trichromats to see the red–green color contrast may not translate into a selective advantage because the use of a variety of sensory modalities may compensate for the inferiority of any one sense (Hiramatsu et al. 2009) (see also Chaps. 4–6).

7.7.4 Population Genetic Analysis of L–M Opsin Gene Polymorphism in the Wild Populations

As we have already summarized, field observations of the foraging behaviors of New World monkeys have thus far mostly failed to detect a clear advantage of trichromacy or have even demonstrated a dichromat advantage for insect foraging. This situation leaves a fundamental question unanswered regarding what maintains trichromatic vision in New World monkeys, because trichromacy (i.e., heterozygosity on the L–M opsin alleles) would have disappeared without a selective force acting to maintain allelic variations of the L–M opsin.

Interspecies comparisons of the L–M opsin gene sequences among primates and others have found signatures of positive natural selection for generating trichromatic color vision and have identified relevant amino acid substitutions for the selection effective in spectral differentiation between the L and M opsin genes (Yokoyama and Yokoyama 1990; Neitz et al. 1991; Shyue et al. 1995, 1998; Boissinot et al. 1998). The color vision polymorphism is transspecific and is documented in all three families of the New World monkeys (Atelidae, Pitheciidae, and Cebidae) (Jacobs 2007). The long duration of the polymorphism in these Neotropical primate families is consistent with balancing selection, a form of positive natural selection, operating to maintain variation via heterozygote advantage of trichromatic females (Boissinot et al. 1998; Surridge and Mundy 2002; Surridge et al. 2003).

However, the effective population size N_e, a major determinant of the duration of allelic turnover, remains unknown for New World monkeys. Assuming that the last common ancestor of all New World monkeys originated 26 million years ago (Schneider 2000), it follows that the opsin polymorphism has persisted over this period. In theory, the expected survival time for a neutral X-linked allele is $3N_e$ generations in a stationary population. If N_e of New World monkeys had been large enough (e.g., of the order of 10^6) in the long-isolated South American continent without formidable eutherian predators, then the polymorphism could have persisted for this length of time. Another problem is that it is difficult to estimate N_e in a natural population. The estimated value of N_e depends on the accuracy of estimates of the mutation rate and generation time and is confounded by demographic effects such as the historical dynamism of population size, migration pattern, and population structure. Although demographic effects influence genetic variation of all genes in the genome alike, the pattern and the intensity of natural selection can vary among genomic regions depending on direct or indirect effects of mutations in that region to fitness.

It is thus necessary to apply a method that compares the pattern of intraspecific genetic variation between a focal region (i.e., the L–M opsin gene) and other reference regions in the same genome using the same population samples to cancel out the effects of demographic factors which both regions share (Verrelli and Tishkoff 2004; Perry et al. 2007; Verrelli et al. 2008). We employed this approach for a group of spider monkeys and a group of capuchin monkeys from the Santa Rosa populations described in Sect. 7.7.2 (Hiwatashi et al. 2010).

7 Polymorphic Color Vision in Primates: Evolutionary Considerations

Hiwatashi et al. (2010) evaluated the three basic parameters, π, θ_w, and Tajima's D, to describe the level of nucleotide variation within a population. Nucleotide diversity (π) is the average number of nucleotide differences per nucleotide site between two sequences (and is also the unbiased estimator of the average heterozygosity among nucleotide sites) (Nei and Kumar 2000). The number of polymorphic (segregating) sites among samples (S) derives a nucleotide polymorphism parameter $\theta_w \equiv S/L/\sum_{i=1}^{n-1}\frac{1}{i}$, where L is the length of the sequence and n is the number of samples (Watterson 1975). In theory, when mutations are selectively neutral and population size is constant through generations, both π and θ_w are expected to converge to the population mutation rate θ ($\equiv 4N_e\mu$ for autosomal and $3N_e\mu$ for X-chromosomal genes of diploid organisms, where N_e is the effective population size and μ is the mutation rate per nucleotide site per chromosome per generation). Tajima's D evaluates the difference between π and θ_w, which is given by $\pi-\theta_w$ divided by the estimated standard error of the difference (Tajima 1989).

The Tajima's D value of neutral references can be regarded as a control measure of severity of demographic effects. If Tajima's D of neutral references is positive, it could imply a long-term reduction of the population size or recent admixture of genetically differentiated subpopulations. If it is negative, it could imply a long-term expansion of population size or recent incorporation of genetically differentiated minority. On top of this, if the Tajima's D value of the L–M opsin gene is positive and significantly larger than that of neutral references, this is taken as an evidence of balancing selection operating on the L–M opsin gene. We used a computer simulation ("coalescence simulation") to determine whether the observed values of the three parameters are deviated with statistical significance from expectation under the assumption of neutrality and constant demography.

We also evaluated nucleotide divergence between spider monkey and capuchin monkey populations. The nucleotide divergence between species is defined as the proportion of nucleotide sites where one species is monomorphic (i.e., fixed) with a nucleotide and the other species is fixed with another nucleotide.

We showed that the nucleotide sequence of the L–M opsin gene was significantly more polymorphic than the sequences of the neutral references in terms of π and θ_w in both spider monkeys and capuchin monkeys. The Tajima's D value of the L–M opsin gene also deviated significantly in a positive direction from the neutral expectation in both species. In particular, this deviation from neutrality was evident in the central part of the L–M opsin gene region, including exon 3 and exon 5, which encode the spectrally important amino acid sites. On the other hand, viewed from nucleotide divergence between the two species, L–M opsin gene sequences were not more divergent than the sequences of the neutral references. Within the L–M opsin gene region, the central region was not more divergent between species than the peripheral region. The nucleotide divergence data confirm that the larger within-species variation in L–M opsin gene than neutral references, especially in its central region, is not the result of a difference of mutation rate among these regions.

In addition to the results shown in Hiwatashi et al. (2010), we evaluated whether the ratio of polymorphic to fixed sites was different between genomic regions by

Table 7.1 The number of polymorphic and fixed nucleotide sites in the neutral references and the L–M (red–green) opsin gene regions in spider and capuchin monkeys

Region	Length (bp)	No. of polymorphic sites (%)	No. of fixed sites (%)
Neutral reference	2,045	20 (1.0)	103 (5.0)
L–M opsin			
Exon 1	881	7 (0.8)	35 (4.0)
Exon 3	949	42 (4.4)	26 (2.7)
Exon 5	827	34 (4.1)	13 (1.6)
Exon 6	835	23 (2.8)	39 (4.7)
Total	3,492	106 (3.0)	113 (3.2)

Note: The set of sequences used is the same as in Table 7 of Hiwatashi et al. (2010)

the conventional χ^2 test (a simplified HKA test) (Hudson et al. 1987) (Table 7.1). The ratio of polymorphic to fixed sites was significantly higher in the L–M opsin gene region than in the neutral references ($\chi^2 = 35.0$, $df = 1$, $P < 0.0001$). Within the L–M opsin gene, the ratio is also significantly higher in the central region (exons 3 + 5) than in the peripheral region (exons 1 + 6) ($\chi^2 = 30.3$, $df = 1$, $P < 0.0001$). All these results are explained only by the action of balancing selection to the spectrally important amino acid sites located in the central region of the L–M opsin gene. The study by Hiwatashi et al. (2010) is the first to statistically demonstrate balancing selection acting on the polymorphic L–M opsin gene of New World monkeys.

7.7.5 What Is the Nature of Balancing Selection on the L–M Opsin Gene?

Given the clear indication of balancing selection obtained in our study and the uncertainty about benefits of trichromacy (i.e., for individuals heterozygous for the L–M opsin alleles) resulting from behavioral studies of wild primates (Dominy et al. 2003a; Smith et al. 2003a; Melin et al. 2007; Vogel et al. 2007; Hiramatsu et al. 2008, 2009), how should we interpret the nature of balancing selection? Several advantages of trichromacy have been proposed, such as long-distance detection of reddish objects under dappled foliage (Sumner and Mollon 2000), foraging on red-dish ripe fruits in severe dry seasons when these could be scarce (Dominy and Lucas 2001), and recognition of social signals and predators (Changizi et al. 2006; Fernandez and Morris 2007). Critical behavioral data to demonstrate these advantages have yet to be gathered, however.

Besides the trichromat advantage, three other mechanisms of balancing selection have been hypothesized to explain color vision polymorphism in New World monkeys (Mollon et al. 1984): (1) negative frequency-dependent selection, which predicts the fitness of any given phenotype to be reciprocal to the frequency of that phenotype in the population; (2) niche divergence, which predicts that individuals of each phenotype will specialize in a distinct visual or ecological niche or visual ability; and

(3) mutual benefit of association, which predicts that individuals of each phenotype benefit from being associated with individuals of other phenotypes in a polymorphic group.

Negative frequency-dependent selection is generally invoked for predator–prey interaction or a disassortative mating system, that is, mating with a different type from oneself in terms of the genetic trait in question (Conner and Hartl 2004). This case is hard to envision in the case of color vision and often appears to be interpreted mistakenly in literatures as a consequence of niche divergence (thus the two hypotheses are often confounded). Under the niche-divergence situation the population size of each phenotype can fluctuate independently from each other and irrespectively of its frequency because individuals with different phenotypes exploit different resources or niches and population size of the phenotype changes as the carrying capacity of their niche changes but not as population size of another phenotype changes. Although negative frequency-dependent selection is often referred as an alternative explanation to the heterozygote advantage hypothesis as a mechanism for maintaining color vision polymorphism, it would be the least likely mechanism.

Few studies evaluate the niche divergence hypothesis. In the food foraging behaviors of capuchin monkeys, there is no difference in foraging time spent on different food types between dichromatic and trichromatic individuals (Melin et al. 2008). However, more studies are needed to test this hypothesis.

Although no study has evaluated the mutual benefit hypothesis, balancing selection may represent an advantage of individuals associated with different color vision phenotypes coexisting in the same population. There is a clear advantage of monkey and ape dichromats, as well as human dichromats, in detecting color-camouflaged objects (Morgan et al. 1992; Caine et al. 2003; Saito et al. 2005b), including surface-dwelling insects (Melin et al. 2007, 2010), an important food source for many primates. Humans are polymorphic in color vision and have a long history of a hunting and gathering lifestyle in which the ability to break camouflage may be advantageous. We thus need to ask whether the selective advantages of dichromacy are applicable to evolution of human color vision. Given such potential selective advantages in dichromats, we also need to ask why dichromats are so rare in nonhuman catarrhines.

7.8 Conclusions

Color vision of primates is unique among vertebrates in its evolutionary history. Trichromatic color vision in primates was generated from dichromatic color vision seen in other mammals by allelic differentiation or gene duplication of the M/LWS type opsin gene that had evolved in an early vertebrate ancestor. The necessary mechanism for mutually exclusive expression of the L and M opsin gene in a cone cell was provided by the X-chromosomal locality of the M/LWS type opsin gene in mammals and a preexisting enhancer element for it. The necessary mechanism for neuronal processing was also provided by a preexisting system for acute spatial vision. We showed that the polymorphic color vision of New World monkeys is maintained by natural selection. But there is still controversy over the advantages of trichromatic color vision and of polymorphic color vision. A deeper knowledge of primate color vision will facilitate our understanding of human color vision. Nonhuman primates are a good reference point for comparison. Studies of New World monkeys are particularly important for understanding a condition where color vision can be polymorphic in the population. Since the isolation of cone opsin genes in the mid-1980s (Nathans et al. 1986), our understanding of the evolution of color vision has progressed rapidly, largely because these studies encompass research on genes, physiology, and behavior. Further interdisciplinary studies will provide a wealth of data for increasing our understanding of the evolution of color vision.

Acknowledgments We thank R. Blanco Segura, M.M. Chavarria, and other staff of the Área de Conservación Guanacaste for local support, and we are grateful to the Ministerio de Ambiente y Energía (MINAE) of Costa Rica for giving us permission to conduct our field study in Santa Rosa. We appreciate the help of E. Murillo Chacon, C. Sendall, K.M. Jack, S. Carnegie, L. Rebecchini, A.H. Korstjens, G. McCabe, and H. Young with collection of fecal samples and behavioral data and assistance in the field. Our field study was supported by Grants-in-Aid for Scientific Research A 19207018 and 22247036 from the Japan Society for the Promotion of Science (JSPS) and Grants-in-Aid for Scientific Research on Priority Areas "Comparative Genomics" 20017008 and "Cellular Sensor" 21026007 from the Ministry of Education, Culture, Sports, Science and Technology of Japan to S.K.; a Grant-in-Aid for JSPS Fellows (15-11926) to C.H.; postgraduate scholarships and grants from the Alberta Ingenuity Fund, the Natural Sciences and Engineering Research Council of Canada, the Leakey Foundation, and the Animal Behavior Society to A.D.M.; the Canada Research Chairs Program and a Discovery Grant from the Natural Sciences and Engineering Research Council of Canada to L.M.F.; the Leakey Foundation and the North of England Zoological Society to F.A.; and the British Academy and the University of Chester small grants scheme to C.M.S.

References

Ahnelt PK, Kolb H (2000) The mammalian photoreceptor mosaic-adaptive design. Prog Retin Eye Res 19:711–777

Allen G (1879) The color sense: its origin and development. Trubner, London

Araujo AC, Didonet JJ, Araujo CS et al (2008) Color vision in the black howler monkey (*Alouatta caraya*). Vis Neurosci 25:243–248

Aureli F, Schaffner CM (2008) Spider monkeys: social structure, social relationships and social interactions. In: Campbell C (ed) Spider monkeys: behavior ecology & evolution of the genus *Ateles*. Cambridge University Press, Cambridge, pp 236–265

Boissinot S, Tan Y, Shyue SK et al (1998) Origins and antiquity of X-linked triallelic color vision systems in New World monkeys. Proc Natl Acad Sci USA 95:13749–13754

Caine NG (2002) Seeing red: consequences of individual differences in color vision in callitrichid primates. In: Miller LE (ed) Eat or be eaten. Cambridge University Press, Cambridge, pp 58–73

Caine NG, Mundy NI (2000) Demonstration of a foraging advantage for trichromatic marmosets (*Callithrix geoffroyi*) dependent on food colour. Proc R Soc Lond B 267:439–444

Caine NG, Surridge AK, Mundy NI (2003) Dichromatic and trichromatic *Callithrix geoffroyi* differ in relative foraging ability for red-green color-camouflaged and non-camouflaged food. Int J Primatol 24:1163–1175

Changizi MA, Zhang Q, Shimojo S (2006) Bare skin, blood and the evolution of primate color vision. Biol Lett 2:217–221

Chapman CA (1990) Association patterns of spider monkeys: the influence of ecology and sex on social organization. Behav Ecol Sociobiol 26:409–414

Chinen A, Hamaoka T, Yamada Y et al (2003) Gene duplication and spectral diversification of cone visual pigments of zebrafish. Genetics 163:663–675

Collin SP, Knight MA, Davies WL et al (2003) Ancient colour vision: multiple opsin genes in the ancestral vertebrates. Curr Biol 13:R864–R865

Collin SP, Davies WL, Hart NS et al (2009) The evolution of early vertebrate photoreceptors. Philos Trans R Soc B 364:2925–2940

Conner JK, Hartl DL (2004) A primer of ecological genetics. Sinauer Associates, Sunderland

Davies WL, Carvalho LS, Cowing JA et al (2007) Visual pigments of the platypus: a novel route to mammalian colour vision. Curr Biol 17:R161–R163

Davies WL, Collin SP, Hunt DM (2009) Adaptive gene loss reflects differences in the visual ecology of basal vertebrates. Mol Biol Evol 26:1803–1809

De Araujo MF, Lima EM, Pessoa VF (2006) Modeling dichromatic and trichromatic sensitivity to the color properties of fruits eaten by squirrel monkeys (*Saimiri sciureus*). Am J Primatol 68:1129–1137

Deeb SS (2006) Genetics of variation in human color vision and the retinal cone mosaic. Curr Opin Genet Dev 16:301–307

Dominy NJ (2004) Color as an indicator of food quality to anthropoid primates: ecological evidence and an evolutionary scenario. In: Ross C, Kay RF (eds) Anthropoid origins. Kluwer, New York, pp 599–628

Dominy NJ, Lucas PW (2001) Ecological importance of trichromatic vision to primates. Nature (Lond) 410:363–366

Dominy NJ, Garber PA, Bicca-Marques JC et al (2003a) Do female tamarins use visual cues to detect fruit rewards more successfully than do males? Anim Behav 66:829–837

Dominy NJ, Svenning JC, Li WH (2003b) Historical contingency in the evolution of primate color vision. J Hum Evol 44:25–45

Drummond-Borg M, Deeb SS, Motulsky AG (1989) Molecular patterns of X chromosome-linked color vision genes among 134 men of European ancestry. Proc Natl Acad Sci USA 86:983–987

Dulai KS, Bowmaker JK, Mollon JD et al (1994) Sequence divergence, polymorphism and evolution of the middle-wave and long-wave visual pigment genes of great apes and Old World monkeys. Vision Res 34:2483–2491

Dulai KS, von Dornum M, Mollon JD et al (1999) The evolution of trichromatic color vision by opsin gene duplication in New World and Old World primates. Genome Res 9:629–638

Ebrey T, Koutalos Y (2001) Vertebrate photoreceptors. Prog Retin Eye Res 20:49–94

Fedigan LM (2003) Impact of male takeovers on infant deaths, births and conceptions in *Cebus capucinus* at Santa Rosa, Costa Rica. Int J Primatol 24:723–741

Fedigan LM, Jack K (2001) Neotropical primates in a regenerating Costa Rican dry forest: a comparison of howler and capuchin population patterns. Int J Primatol 22:689–713

Fedigan LM, Jack K (2004) The demographic and reproductive context of male replacements in *Cebus capucinus*. Behaviour 141:755–775

Fernandez AA, Morris MR (2007) Sexual selection and trichromatic color vision in primates: statistical support for the preexisting-bias hypothesis. Am Nat 170:10–20

Fleagle JG (1999) Primate adaptation and evolution, 2dth edn. Academic, San Diego

Foster DH, Nascimento SM (1994) Relational colour constancy from invariant cone-excitation ratios. Proc R Soc Lond B 257:115–121

Fragaszy DM, Visalberghi E, Fedigan LM (2004) The complete capuchin: the biology of the genus *Cebus*. Cambridge University Press, Cambridge

Goldsmith TH (1990) Optimization, constraint, and history in the evolution of eyes. Q Rev Biol 65:281–322

Govardovskii VI (1983) On the role of oil drops in colour vision. Vision Res 23:1739–1740

Hanazawa A, Mikami A, Sulistyo Angelika P et al (2001) Electroretinogram analysis of relative spectral sensitivity in genetically identified dichromatic macaques. Proc Natl Acad Sci USA 98:8124–8127

Hayashi T, Motulsky AG, Deeb SS (1999) Position of a 'green-red' hybrid gene in the visual pigment array determines colour-vision phenotype. Nat Genet 22:90–93

Heesy CP, Ross CF (2001) Evolution of activity patterns and chromatic vision in primates: morphometrics, genetics and cladistics. J Hum Evol 40:111–149

Heesy CP, Ross CF, Demes B (2007) Oculomotor stability and the functions of the postorbital bar and septum. In: Ravosa MJ, Dagosto M (eds) Primate origins: adaptations and evolution. Springer, New York, pp 257–283

Hiramatsu C, Radlwimmer FB, Yokoyama S et al (2004) Mutagenesis and reconstitution of middle-to-long-wave-sensitive visual pigments of New World monkeys for testing the tuning effect of residues at sites 229 and 233. Vision Res 44:2225–2231

Hiramatsu C, Tsutsui T, Matsumoto Y et al (2005) Color-vision polymorphism in wild capuchins (*Cebus capucinus*) and spider monkeys (*Ateles geoffroyi*) in Costa Rica. Am J Primatol 67:447–461

Hiramatsu C, Melin AD, Aureli F et al (2008) Importance of achromatic contrast in short-range fruit foraging of primates. PLoS One 3:e3356

Hiramatsu C, Melin AD, Aureli F et al (2009) Interplay of olfaction and vision in fruit foraging of spider monkeys. Anim Behav 77:1421–1426

Hiwatashi T, Okabe Y, Tsutsui T et al (2010) An explicit signature of balancing selection for color-vision variation in new world monkeys. Mol Biol Evol 27:453–464

Hudson RR, Kreitman M, Aguade M (1987) A test of neutral molecular evolution based on nucleotide data. Genetics 116:153–159

Hunt DM, Dulai KS, Cowing JA et al (1998) Molecular evolution of trichromacy in primates. Vision Res 38:3299–3306

Ibbotson RE, Hunt DM, Bowmaker JK et al (1992) Sequence divergence and copy number of the middle- and long-wave photopigment genes in Old World monkeys. Proc R Soc Lond B 247:145–154

Jacobs GH (1993) The distribution and nature of colour vision among the mammals. Biol Rev 68:413–471

Jacobs GH (1996) Primate photopigments and primate color vision. Proc Natl Acad Sci USA 93:577–581

Jacobs GH (1998) A perspective on color vision in platyrrhine monkeys. Vision Res 38:3307–3313

Jacobs GH (1999) Vision and behavior in primates. In: Archer SN, Djamgoz MBA, Loew ER et al (eds) Adaptive mechanisms in the ecology of vision. Kluwer, Dordrecht, pp 629–650

Jacobs GH (2007) New World monkeys and color. Int J Primatol 28:729–759

Jacobs GH, Deegan JF II (2001) Photopigments and colour vision in New World monkeys from the family Atelidae. Proc R Soc Lond B 268:695–702

Jacobs GH, Deegan JF II (2003) Cone pigment variations in four genera of new world monkeys. Vision Res 43:227–236

Jacobs GH, Deegan JF II (2005) Polymorphic New World monkeys with more than three M/L cone types. J Opt Soc Am A 22:2072–2080

Jacobs GH, Williams GA (2001) The prevalence of defective color vision in Old World monkeys and apes. Color Res Appl 26(suppl):S123–S127

Jacobs GH, Neitz M, Deegan JF et al (1996a) Trichromatic colour vision in New World monkeys. Nature (Lond) 382:156–158

Jacobs GH, Neitz M, Neitz J (1996b) Mutations in S-cone pigment genes and the absence of colour vision in two species of nocturnal primate. Proc R Soc Lond B 263:705–710

Jacobs GH, Deegan JF II, Tan Y et al (2002) Opsin gene and photopigment polymorphism in a prosimian primate. Vision Res 42:11–18

Jacobs GH, Williams GA, Cahill H et al (2007) Emergence of novel color vision in mice engineered to express a human cone photopigment. Science 315:1723–1725

Janzen DH (2002) Tropical dry forest: Area de Conservacion Guanacaste, northwestern Costa Rica. In: Perrow M, Davey A (eds) Handbook of ecological restoration: restoration in practice. Cambridge University Press, Cambridge, pp 559–583

Kawamura S, Kubotera N (2003) Absorption spectra of reconstituted visual pigments of a nocturnal prosimian, *Otolemur crassicaudatus*. Gene (Amst) 321:131–135

Kawamura S, Kubotera N (2004) Ancestral loss of short wave-sensitive cone visual pigment in lorisiform prosimians, contrasting with its strict conservation in other prosimians. J Mol Evol 58:314–321

Kawamura S, Hirai M, Takenaka O et al (2001) Genomic and spectral analyses of long to middle wavelength-sensitive visual pigments of common marmoset (*Callithrix jacchus*). Gene (Amst) 269:45–51

Kawamura S, Takenaka N, Hiramatsu C et al (2002) Y-chromosomal red-green opsin genes of nocturnal New World monkey. FEBS Lett 530:70–72

Kelber A, Vorobyev M, Osorio D (2003) Animal colour vision: behavioural tests and physiological concepts. Biol Rev 78:81–118

Levine JS, MacNichol EF Jr (1982) Color vision in fishes. Sci Am 246:140–149

Lucas PW, Darvell BW, Lee PKD et al (1998) Colour cues for leaf food selection by long-tailed macaques (*Macaca fascicularis*) with a new suggestion for the evolution of trichromatic colour vision. Folia Primatol 69:139–154

Lucas PW, Dominy NJ, Riba-Hernández P et al (2003) Evolution and function of routine trichromatic vision in primates. Evolution 57:2636–2643

Lythgoe JN (1979) The ecology of vision. Oxford University Press, Oxford

MacLeod DI, Boynton RM (1979) Chromaticity diagram showing cone excitation by stimuli of equal luminance. J Opt Soc Am 69:1183–1186

Martin PR (1998) Colour processing in the primate retina: recent progress. J Physiol 513(pt 3):631–638

Maximov VV (2000) Environmental factors which may have led to the appearance of colour vision. Philos Trans R Soc B 355:1239–1242

Melin AD, Fedigan LM, Hiramatsu C et al (2007) Effects of colour vision phenotype on insect capture by a free-ranging population of white-faced capuchins (*Cebus capucinus*). Anim Behav 73:205–214

Melin AD, Fedigan LM, Hiramatsu C et al (2008) Polymorphic color vision in white-faced capuchins (*Cebus capucinus*): is there foraging niche divergence among phenotypes? Behav Ecol Sociobiol 62:659–670

Melin AD, Fedigan LM, Hiramatsu C et al (2009) Fig foraging by dichromatic and trichromatic *Cebus capucinus* in a tropical dry forest. Int J Primatol 30:753–775

Melin AD, Fedigan LM, Young HC et al (2010) Can color vision variation explain sex differences in invertebrate foraging by capuchin monkeys? Curr Zool 56:300–312

Mollon JD (1989) "Tho' she kneel'd in that place where they grew…" The uses and origins of primate colour vision. J Exp Biol 146:21–38

Mollon JD, Bowmaker JK, Jacobs GH (1984) Variations of colour vision in a New World primate can be explained by polymorphism of retinal photopigments. Proc R Soc Lond B 222:373–399

Morgan MJ, Adam A, Mollon JD (1992) Dichromats detect colour-camouflaged objects that are not detected by trichromats. Proc R Soc Lond B 248:291–295

Morley RJ (2000) Origin and evolution of tropical rain forests. Wiley, Chichester

Mullen KT (1985) The contrast sensitivity of human colour vision to red-green and blue-yellow chromatic gratings. J Physiol 359:381–400

Nagao K, Takenaka N, Hirai M et al (2005) Coupling and decoupling of evolutionary mode between X- and Y-chromosomal red-green opsin genes in owl monkeys. Gene (Amst) 352:82–91

Nathans J (1987) Molecular biology of visual pigments. Annu Rev Neurosci 10:163–194

Nathans J, Thomas D, Hogness DS (1986) Molecular genetics of human color vision: the genes encoding blue, green, and red pigments. Science 232:193–202

Nei M, Kumar S (2000) Molecular evolution and phylogenetics. Oxford University Press, New York

Neitz M, Neitz J, Jacobs GH (1991) Spectral tuning of pigments underlying red-green color vision. Science 252:971–974

Onishi A, Koike S, Ida M et al (1999) Dichromatism in macaque monkeys. Nature (Lond) 402:139–140

Onishi A, Koike S, Ida-Hosonuma M et al (2002) Variations in long- and middle-wavelength-sensitive opsin gene loci in crab-eating monkeys. Vision Res 42:281–292

Osorio D, Vorobyev M (1996) Colour vision as an adaptation to frugivory in primates. Proc R Soc Lond B 263:593–599

Parraga CA, Troscianko T, Tolhurst DJ (2002) Spatiochromatic properties of natural images and human vision. Curr Biol 12:483–487

Perry GH, Martin RD, Verrelli BC (2007) Signatures of functional constraint at aye-aye opsin genes: the potential of adaptive color vision in a nocturnal primate. Mol Biol Evol 24:1963–1970

Pessoa DM, Araujo MF, Tomaz C et al (2003) Colour discrimination learning in black-handed tamarin (*Saguinus midas niger*). Primates 44:413–418

Pokorny J, Shevell SK, Smith VC (1991) Colour appearance and colour constancy. In: Cronly-Dillon JR (ed) Vision and visual dysfunction, vol 6. Macmillan, London, pp 43–61

Regan BC, Julliot C, Simmen B et al (1998) Frugivory and colour vision in *Alouatta seniculus*, a trichromatic platyrrhine monkey. Vision Res 38:3321–3327

Regan BC, Julliot C, Simmen B et al (2001) Fruits, foliage and the evolution of primate colour vision. Philos Trans R Soc B 356:229–283

Riba-Hernández P, Stoner KE, Osorio D (2004) Effect of polymorphic colour vision for fruit detection in the spider monkey *Ateles geoffroyi*, and its implications for the maintenance of polymorphic colour vision in platyrrhine monkeys. J Exp Biol 207:2465–2470

Robinson SR (1994) Early vertebrate color vision. Nature (Lond) 367:121

Saito A, Mikami A, Hasegawa T et al (2003) Behavioral evidence of color vision deficiency in a protanomalia chimpanzee (*Pan troglodytes*). Primates 44:171–176

Saito CA, da Silva Fiho M, Lee BB et al (2004) *Alouatta* trichromatic color vision: single-unit recording from retinal ganglion cells and microspectrophotometry. Invest Ophthalmol Vis Sci 45:E (abstract 4276)

Saito A, Kawamura S, Mikami A et al (2005a) Demonstration of a genotype-phenotype correlation in the polymorphic color vision of a non-callitrichine New World monkey, capuchin (*Cebus apella*). Am J Primatol 67:471–485

Saito A, Mikami A, Kawamura S et al (2005b) Advantage of dichromats over trichromats in discrimination of color-camouflaged stimuli in nonhuman primates. Am J Primatol 67:425–436

Schneider H (2000) The current status of the New World monkey phylogeny. An Acad Bras Ci 72:165–172

Sharpe LT, Stockman A, Jagle H et al (1999) Opsin genes, cone photopigments, color vision, and color blindness. In: Gegenfurtner KR, Sharpe LT (eds) Color vision: from genes to perception. Cambridge University Press, Cambridge, pp 3–51

Shyue SK, Hewett-Emmett D, Sperling HG et al (1995) Adaptive evolution of color vision genes in higher primates. Science 269:1265–1267

Shyue SK, Boissinot S, Schneider H et al (1998) Molecular genetics of spectral tuning in New World monkey color vision. J Mol Evol 46:697–702

Smallwood PM, Wang Y, Nathans J (2002) Role of a locus control region in the mutually exclusive expression of human red and green cone pigment genes. Proc Natl Acad Sci USA 99:1008–1011

Smith AC, Buchanan-Smith HM, Surridge AK et al (2003a) Leaders of progressions in wild mixed-species troops of saddleback (*Saguinus fuscicollis*) and mustached tamarins (*S. mystax*), with emphasis on color vision and sex. Am J Primatol 61:145–157

Smith AC, Buchanan-Smith HM, Surridge AK et al (2003b) The effect of colour vision status on the detection and selection of fruits by tamarins (*Saguinus* spp.). J Exp Biol 206:3159–3165

Smith AC, Buchanan-Smith HM, Surridge AK et al (2005) Factors affecting group spread within wild mixed-species troops of saddleback and mustached tamarins. Int J Primatol 26:337–355

Stoner KE, Riba-Hernández P, Lucas PW (2005) Comparative use of color vision for frugivory by sympatric species of platyrrhines. Am J Primatol 67:399–409

Sumner P, Mollon JD (2000) Catarrhine photopigments are optimized for detecting targets against a foliage background. J Exp Biol 203:1963–1986

Sumner P, Mollon JD (2003) Colors of primate pelage and skin: objective assessment of conspicuousness. Am J Primatol 59:67–91

Surridge AK, Mundy NI (2002) Trans-specific evolution of opsin alleles and the maintenance of trichromatic colour vision in callitrichine primates. Mol Ecol 11:2157–2169

Surridge AK, Smith AC, Buchanan-Smith HM et al (2002) Single-copy nuclear DNA sequences obtained from noninvasively collected primate feces. Am J Primatol 56:185–190

Surridge AK, Osorio D, Mundy NI (2003) Evolution and selection of trichromatic vision in primates. Trends Ecol Evol 18:198–205

Tajima F (1989) Statistical method for testing the neutral mutation hypothesis by DNA polymorphism. Genetics 123:585–595

Talebi MG, Pope TR, Vogel ER et al (2006) Polymorphism of visual pigment genes in the muriqui (Primates, Atelidae). Mol Ecol 15:551–558

Tan Y, Li WH (1999) Trichromatic vision in prosimians. Nature (Lond) 402:36

Tan Y, Yoder AD, Yamashita N et al (2005) Evidence from opsin genes rejects nocturnality in ancestral primates. Proc Natl Acad Sci USA 102:14712–14716

Terao K, Mikami A, Saito A et al (2005) Identification of a protanomalous chimpanzee by molecular genetic and electroretinogram analyses. Vision Res 45:1225–1235

Terborgh J (1986) Keystone plant resources in the tropical forest. In: Soule M (ed) Conservation biology: science of scarcity and diversity. Sinauer, Sunderland, pp 330–344

Veilleux CC, Bolnick DA (2009) Opsin gene polymorphism predicts trichromacy in a cathemeral lemur. Am J Primatol 71:86–90

Verrelli BC, Tishkoff SA (2004) Signatures of selection and gene conversion associated with human color vision variation. Am J Hum Genet 75:363–375

Verrelli BC, Lewis CM Jr, Stone AC et al (2008) Different selective pressures shape the molecular evolution of color vision in chimpanzee and human populations. Mol Biol Evol 25:2735–2743

Vogel ER, Neitz M, Dominy NJ (2007) Effect of color vision phenotype on the foraging of wild white-faced capuchins, *Cebus capucinus*. Behav Ecol 18:292–297

Vorobyev M (2004) Ecology and evolution of primate colour vision. Clin Exp Optom 87:230–238

Walls GL (1942) The vertebrate eye and its adaptive radiation. Cranbrook Institute of Science, Bloomfield Hills

Wang Y, Macke JP, Merbs SL et al (1992) A locus control region adjacent to the human red and green visual pigment genes. Neuron 9:429–440

Wang Y, Smallwood PM, Cowan M et al (1999) Mutually exclusive expression of human red and green visual pigment-reporter transgenes occurs at high frequency in murine cone photoreceptors. Proc Natl Acad Sci USA 96:5251–5256

Watterson GA (1975) On the number of segregating sites in genetical models without recombination. Theor Popul Biol 7:256–276

Yamada ES, Marshak DW, Silveira LC et al (1998) Morphology of P and M retinal ganglion cells of the bush baby. Vision Res 38:3345–3352

Yokoyama S (2000a) Molecular evolution of vertebrate visual pigments. Prog Retin Eye Res 19:385–419

Yokoyama S (2000b) Phylogenetic analysis and experimental approaches to study color vision in vertebrates. Methods Enzymol 315:312–325

Yokoyama S, Radlwimmer FB (1998) The "five-sites" rule and the evolution of red and green color vision in mammals. Mol Biol Evol 15:560–567

Yokoyama S, Radlwimmer FB (1999) The molecular genetics of red and green color vision in mammals. Genetics 153:919–932

Yokoyama S, Radlwimmer FB (2001) The molecular genetics and evolution of red and green color vision in vertebrates. Genetics 158:1697–1710

Yokoyama R, Yokoyama S (1990) Convergent evolution of the red- and green-like visual pigment genes in fish, *Astyanax fasciatus*, and human. Proc Natl Acad Sci USA 87:9315–9318

Yokoyama S, Yang H, Starmer WT (2008) Molecular basis of spectral tuning in the red- and green-sensitive (M/LWS) pigments in vertebrates. Genetics 179:2037–2043

Zhou YH, Li WH (1996) Gene conversion and natural selection in the evolution of X-linked color vision genes in higher primates. Mol Biol Evol 13:780–783

Part II
Genome Structure and Its Applications

Chapter 8
Human-Specific Changes in Sialic Acid Biology

Toshiyuki Hayakawa and Ajit Varki

Abbreviations

CMAH	CMP-*N*-acetylneuraminic acid hydroxylase
CMP-Kdn	Cytidine monophospho-2-keto-3 deoxynonulosonic acid
CMP-Neu5Ac	Cytidine monophospho-*N*-acetylneuraminic acid
CMP-Neu5Gc	Cytidine monophospho-*N*-glycolylneuraminic acid
Gal	Galactose
GlcNAc	*N*-Acetylglucosamine
ITIM	Immunoreceptor tyrosine-based inhibitory motif
Kdn	2-Keto-3 deoxynonulosonic acid
Neu5Ac	*N*-Acetylneuraminic acid
Neu5Gc	*N*-Glycolylneuraminic acid
Siglec	Sialic acid-binding immunoglobulin superfamily lectin

T. Hayakawa (✉)
Center for Human Evolution Modeling Research, Primate Research Institute,
Kyoto University, 41-2 Kanrin, Inuyama, Aichi 484-8506, Japan
e-mail: thayakawa@pri.kyoto-u.ac.jp

A. Varki
Center for Academic Research and Training in Anthropogeny, Glycobiology Research
and Training Center, Departments of Medicine and Cellular & Molecular Medicine,
University of California, San Diego, La Jolla, CA, USA
e-mail: a1varki@ucsd.edu

H. Hirai et al. (eds.), *Post-Genome Biology of Primates*, Primatology Monographs,
DOI 10.1007/978-4-431-54011-3_8, © Springer 2012

8.1 Introduction

Sialic acids are a family of nine-carbon sugars that are typically found at the terminal end of glycan chains on the cell surface and secreted molecules in the deuterostome lineage of animals (Angata and Varki 2002; Schauer 2000; Varki 2007) (Fig. 8.1a). The two most common forms of sialic acid are *N*-acetylneuraminic acid (Neu5Ac) and *N*-glycolylneuraminic acid (Neu5Gc), and Neu5Gc is enzymatically produced from Neu5Ac by adding a single oxygen atom (i.e., hydroxylation) (Fig. 8.1b). A third sialic acid, 2-keto-3 deoxynonulosonic acid (Kdn), is known but is less common in mammals. The amino group can also be rarely remained unmodified (Neu) and is regarded as a fourth sialic acid. These four forms of sialic acids can be further modified in various ways, and this diversity is increased by attachments to the underlying sugar chains by different linkages (α2-3, α2-6, and α2-8) (Angata and Varki 2002; Beyer et al. 1979; Traving and Schauer 1998). This diversity as presented on the cell surface and secreted molecules has been termed the "sialome" (Varki and Angata 2006).

Sialic acids are synthesized from certain precursor molecules or recycled from glycoconjugates located on the cell surface (Fig. 8.2). They can also be taken up from external sources and metabolically incorporated. The activation of free sialic acid to nucleotide donors (CMP-Neu5Ac, CMP-Neu5Gc, and CMP-Kdn) occurs in the nucleus, and nucleotide donors are then returned to the cytosol for transport into the lumen of the Golgi apparatus. In the Golgi, sialic acid residues are transferred from the CMP donors to newly synthesized glycoconjugates. Finally, sialic acid-containing glycoconjugates are transported to the cell surface, or secreted. Sialic acids on cell surfaces are often recognized by intrinsic and extrinsic molecules and play an important role in cell–cell communication and host–pathogen interaction. The intrinsic recognition is widely involved in many processes including the immune and nervous systems (Varki 2007). Elimination of sialic acid production results in embryonic lethality (Schwarzkopf et al. 2002). On the other hand, several pathogens use sialic acid recognition to initiate their infection (Angata and Varki 2002). More than 55 vertebrate genes are involved in these biochemical and biological processes (Altheide et al. 2006) (Fig. 8.2). A surprisingly large number of these have undergone genomic sequence, expression, or function changes uniquely in the human lineage.

8.2 Sialic Acid-Related Loci Changed in the Human Lineage

In the first year of the twenty-first century, the draft sequence of human genome was released, heralding a new era of genome biology (International Human Genome Sequencing Consortium 2001). The sequencing of the euchromatic region of the human genome was completed in 2004 (International Human Genome Sequencing Consortium 2004). Following these landmarks, a great deal of effort was made in the genome sequencing projects of non-human primates. This effort resulted in the release of the draft sequences of chimpanzee (*Pan troglodytes*) and rhesus monkey

8 Human-Specific Changes in Sialic Acid Biology 125

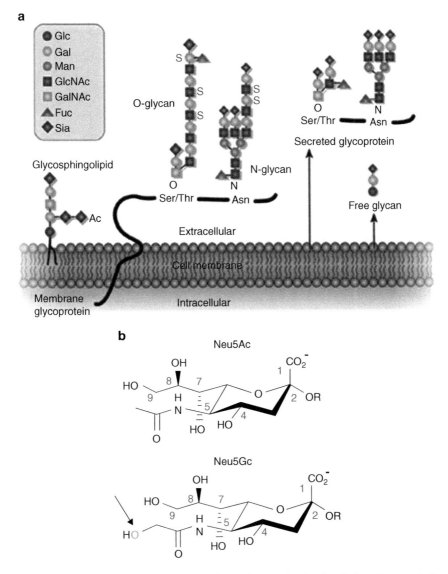

Fig. 8.1 Sialic acids. (**a**) Sialic acids on cell-surface and secreted molecules. Sialic acids are typically found at the terminal position of glycan chains on the cell surface and secreted molecules. *Ac*, *O*-acetyl ester; *Fuc*, fucose; *Gal*, galactose; *GalNAc*, N-acetylgalactosamine; *Glc*, glucose; *GlcNAc*, N-acetylglucosamine; *Man*, mannose; *S*, sulfate ester; *Sia*, sialic acid. (**b**) The two most common forms of sialic acids. The only difference between these two forms is the additional oxygen atom in the N-glycolylgroup of Neu5Gc (*arrow*). *Numbers* represent the serial numbers of the nine-carbon backbone. These sialic acids share various α-glycosidic linkages (α2-3, α2-6, and α2-8) to the underlying sugar chain (*R*) from the C2 position. (Reproduced with permission from Varki 2007)

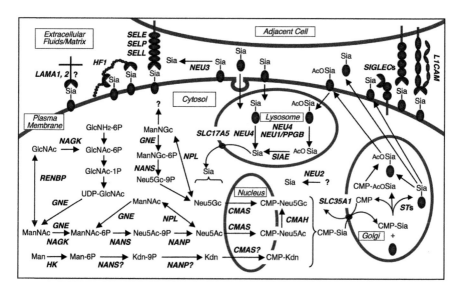

Fig. 8.2 Genes involved in sialic acid biology. All genes or groups of genes thought to be directly involved in sialic acid (Sia) biology are indicated in *bold italic font*. The *question marks* represent unknown or hypothetical pathways. *AcO*, *O*-acetyl ester; *Glc*, glucose; *GlcNAc*, *N*-acetylglucosamine; *Man*, mannose; *ManNAc*, *N*-acetylmannosamine; *STs*, sialyltransferases. (Modified from Altheide et al. 2006)

(*Macaca mulatta*) in 2005 and 2007, respectively (The Chimpanzee Sequencing and Analysis Consortium 2005; Rhesus Macaque Genome Sequencing and Analysis Consortium 2007). Following these genomic achievements, post-genome biology, in which transcriptome, proteome, and glycome are considered and integrated, has been accelerated in primatology. Such comparative analyses between human and non-human primates are expected to provide us certain hints about the genetic contribution to human uniqueness. Indeed, comparison of human and non-human primate genomes has uncovered the human-specific genomic changes in several genes (Varki and Altheide 2005). Moreover, the transcriptome analysis reveals differences in gene expression between humans and non-human primates (Caceres et al. 2003; Enard et al. 2002; Gilad et al. 2006; Khaitovich et al. 2004a,b, 2005, 2006; Preuss et al. 2004; Somel et al. 2009; Varki and Altheide 2005), as have proteomic and glycomic approaches.

The first example of human-specific change in sialic acid biology was discovered before the advent of genome biology. In 1998, it was found that Neu5Gc, one of the major sialic acids, was completely lacking in humans, but not in our closest evolutionary cousins, the great apes (that is, chimpanzees, bonobos, gorillas, and orangutans) (Muchmore et al. 1998). This complete lack of Neu5Gc was caused by a mutation in the *CMAH* gene (Chou et al. 1998; Irie et al. 1998) and was a starting point of the voyage to discover human-specific changes in the loci involved in sialic acid biology. At the time of this writing, human-specific changes have been found in 12 loci, equivalent to about 20% of the total known to be involved in sialic acid

8 Human-Specific Changes in Sialic Acid Biology

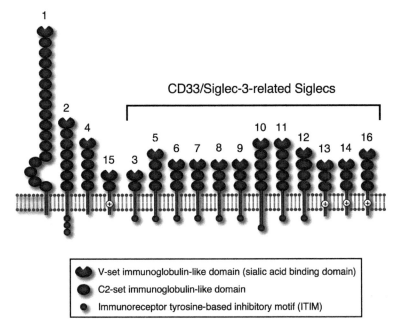

Fig. 8.3 Siglec proteins in hominids. Siglecs are type I transmembrane proteins, composed of one V-set immunoglobulin-like domain that binds sialic acids, variable numbers of C2-set immunoglobulin domains, transmembrane domain, and cytoplasmic tail. Many Siglecs have inhibitory motifs (ITIM) in their cytoplasmic tails. On the other hand, Siglecs-14, -15, and -16 have no ITIM in their cytoplasmic tails, but associate with activating adaptor molecules (e.g., DAP12) via charged residues in their transmembrane domains

biology (Varki 2007, 2009): *CMAH, SIGLEC1, SIGLEC5, SIGLEC6, SIGLEC7, SIGLEC9, SIGLEC11, SIGLEC12, SIGLEC13, SIGLEC14, SIGLEC16,* and *ST6GAL1*.

The *CMAH* locus encodes the CMP-*N*-acetylneuraminic acid hydroxylase (CMAH) enzyme that converts CMP-Neu5Ac to CMP-Neu5Gc in the cytosol (Fig. 8.2), and is essential for producing Neu5Gc (Kawano et al. 1994, 1995). *SIGLEC* loci encode sialic acid-binding immunoglobulin superfamily lectins (Siglecs), a family of sialic acid recognition proteins (Fig. 8.2). Siglecs are type I transmembrane proteins and mostly expressed in the cells involved in the innate and adaptive immune systems (Crocker 2005; Crocker et al. 2007; Crocker and Varki 2001; Varki and Angata 2006). Their extracellular regions have one V-set immunoglobulin-like domain that binds sialic acids and variable numbers of C2-set immunoglobulin-like domains (Fig. 8.3). An arginine residue in the V-set domain is essential for sialic acid binding (Crocker 2005; Crocker and Varki 2001; Varki and Angata 2006). Many Siglecs have inhibitory signaling motifs (immunoreceptor tyrosine-based inhibitory motif, ITIM) in their cytoplasmic tails (Fig. 8.3), and function as inhibitory receptors in signal transduction of the immune system (Carlin et al. 2009a,b; Crocker 2005; Crocker et al. 2007; Crocker and Varki 2001; Varki and Angata 2006). Other Siglecs lack signaling motifs in their cytoplasmic tails

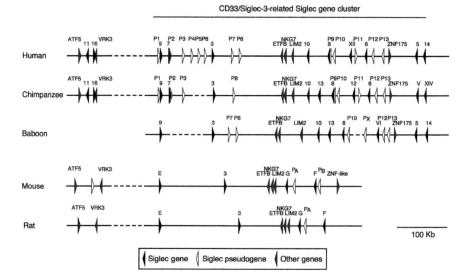

Fig. 8.4 Genomic regions containing CD33/Siglec-3-related Siglec genes in primates and rodents. Siglecs showing an "essential" arginine mutation are indicated by *roman numerals* (Modified from Angata et al. 2004)

(Fig. 8.3), and some of these can function as activating receptors by the association with adaptor molecules (Angata et al. 2006, 2007; Cao et al. 2008; Crocker et al. 2007). Each of the Siglecs displays a distinct preference for recognizing sialic acid-containing glycans (Crocker et al. 2007; Crocker and Varki 2001; Varki and Angata 2006), an important feature in their function (Carlin et al. 2009b).

Siglecs can be divided into two groups: an evolutionarily conserved subgroup (Siglec-1, -2, -4, and -15 in both primates and rodents) and a CD33/Siglec-3-related subgroup [Siglec-3, -5, -6, -7, -8, -9, -10, -11, -12, -13, -14, and -16 in primates; Fig. 8.3; Siglec-3, -E, -F, -G (10), and -H in rodents; Fig. 8.4] (Angata et al. 2004; Crocker et al. 2007; Varki and Angata 2006). The CD33/Siglec-3-related subgroup has experienced changes in gene number in each lineage of eutherian mammals (e.g., 12 genes in the chimpanzee and 5 genes in the mouse) (Angata et al. 2004; Cao et al. 2009; Crocker et al. 2007; Varki and Angata 2006). The *ST6GAL1* gene encodes one of the sialyltransferases that transfer sialic acid residues from CMP donor to glycoconjugates in the Golgi and is essential for the production of the α2-6 linkages of sialic acids to *N*-glycan chains (Martin et al. 2002; Weinstein et al. 1987). Interestingly, all loci that show certain human-specific changes are involved in diversification of sialic acid-containing glycan chains (i.e., sialome) or in sialic acid recognition (see Fig. 8.2). This finding suggests that sialic acid-mediated interactions (e.g., cell–cell communication) have been changed uniquely in the human lineage.

8.2.1 CMAH

CMAH, the key enzyme that generates variations of sialic acid-containing glycans by producing Neu5Gc (Fig. 8.2), has a Rieske iron–sulfur-binding region as an essential component for enzymatic activity (Schlenzka et al. 1996). Interestingly, the expression of *CMAH* is specifically suppressed in mammalian brain, which results in the near absence of Neu5Gc in the non-human brain even though Neu5Gc is abundant in other tissues (Gottschalk 1960; Mikami et al. 1998; Muchmore et al. 1998; Nakao et al. 1991; Rosenberg and Schengrund 1976; Schauer 1982; Tettamanti et al. 1965). This is a rather rare case wherein a gene widely expressed in many tissues is selectively downregulated only in brain. The reason why CMAH is downregulated in mammalian brain remains unknown, but some strong selection is likely expected to be behind this unusual situation.

CMAH is a single-copy gene that was inactivated by the deletion of a 92-bp exon (sixth exon) in the human lineage (Chou et al. 1998; Irie et al. 1998; Varki 2001). This deletion, which is found only in humans (Chou et al. 1998; Hayakawa et al. 2001; Irie et al. 1998; Varki 2001), resulted from the insertion of a human-specific Alu element [a SINE (short interspersed element); retroposon] (Hayakawa et al. 2001). As the exon deletion generates a new stop codon in the middle of the gene, humans have only the truncated CMAH protein (Chou et al. 1998). The truncated enzyme lacks the Reiske iron–sulfur cluster region and therefore cannot convert CMP-Neu5Ac to CMP-Neu5Gc (Chou et al. 1998).

Multiple approaches toward estimating the timing of this inactivation (i.e., the extraction of sialic acids from fossils, calculation of pseudogenization time, and estimation of timing of Alu insertion that caused the exon deletion) showed that CMAH was inactivated approximately 3 million years ago (Chou et al. 2002). This calculated timing is compatible with the fact that the CMAH inactivation is universal to modern humans that appeared about 0.2 million years ago (White et al. 2003). Interestingly, the transition from genus *Australopithecus* to genus *Homo* occurred about 2–3 million years ago (Carroll 2003; McHenry 1994; Wood 2002; Wood and Collard 1999). Thus, it is possible that transition between these two genera might have involved CMAH inactivation.

One of the major phenotypic differences between the two genera is brain size (McHenry 1994; Wood and Collard 1999). Considering that CMAH is selectively downregulated in the mammalian brain, one possible hypothesis is that the low level of brain Neu5Gc has a suppressive role for brain expansion in other mammals and that CMAH inactivation released our ancestor from this constraint (Chou et al. 2002). However, a Cmah-null mouse did not show any gross increase in brain size (Hedlund et al. 2007). This hypothesis is currently under further evaluation.

Many microbial pathogens initiate their infection by recognizing sialic acids on host cells. Some of them exert a distinct preference for Neu5Gc. For example, enterotoxigenic *Escherichia coli* K99 adheres to ganglioside GM3 (Neu5Gc), but not to GM (Neu5Ac), in intestinal epithelial cells (Smit et al. 1984). Other pathogens that prefer Neu5Gc for binding include the S40 virus (Campanero-Rhodes et al. 2007), the transmissible gastoenteritis coronavirus (Angata and Varki 2002), and the great ape malaria parasite *Plasmodium reichenowi* (see below).

As compared with genus *Australopithecus*, the range of genus *Homo* expanded widely throughout the Old World, and *Homo sapiens* to every corner of the globe. After the emergence from Africa, our ancestors undoubtedly met new pathogens in their Out-of-Africa journey. Thus, another hypothesis is that CMAH inactivation protected against the infection by new non-human animal pathogens that prefer Neu5Gc, thereby providing an advantage for the adaptation to a new environment (Hayakawa et al. 2001).

The transgenic mouse that mimics the human inactivation of CMAH, that is, the *Cmah*-deficient mouse, shows a variety of phenotypes: age-associated decrease in hearing, histological abnormalities in the inner ears, and defects in wound healing (Hedlund et al. 2007). Of these phenotypes, the age-associated decrease in hearing reminds us of the age-associated hearing loss that is common in humans. In addition, abnormality in the inner ears could potentially cause a mild deterioration in balance sense (Hedlund et al. 2007). In this regard, it is interesting that the shift from a mixed arboreal (climbing) and terrestrial (walking) behavior to a primarily terrestrial lifestyle seems to occur in the transition to genus *Homo* (Bramble and Lieberman 2004; Klein 1999).

The foregoing hypotheses are currently waiting for some supporting evidence from further studies and are based on an assumption that inactivation of CMAH was advantageous for our ancestors to survive. The distant inactivation time (~3 million years ago) of CMAH makes it difficult to detect positive selection on the inactivated *CMAH* locus. However, the time back to the most recent common ancestor (TMRCA; 2.9 million years) of all human *CMAH* haplotypes is very close to the CMAH inactivation time (3.2 million years) with a short duration time, which indicates that inactivated *CMAH* allele may have fixed quickly in our ancestral population after its emergence (Hayakawa et al. 2006). This point is far from adequate to support a positive selection on the inactivated *CMAH* locus but it is suggestive.

Under the concept of an "arms race" between hosts and pathogens ("Red Queen" effects) (Hamilton et al. 1990; Van Valen 1974), the human-specific inactivation of CMAH (human-specific loss of Neu5Gc) is expected to result in certain adaptive changes in sialic acid recognition molecules of pathogens. A human malaria parasite, *Plasmodium falciparum*, uses the sialic acid recognition molecule EBA-175 to bind host sialic acids in their invasion of red blood cells (Baum et al. 2003; Camus and Hadley 1985; Tomita et al. 1978). Interestingly, the sialic acid preference of EBA-175 is different between *Plasmodium falciparum* and its most closely related chimpanzee parasite, *Plasmodium reichenowi*: that is, EBA-175 of *Plasmodium falciparum* prefers Neu5Ac to Neu5Gc, but that of *Plasmodium reichenowi* prefers Neu5Gc (Martin et al. 2005). This Neu5Ac preference of *Plasmodium falciparum* may be regarded as an adaptation to the human-specific loss of Neu5Gc. Indeed, recent studies have shown that all extant strains of *Plasmodium falciparum* are likely derived from a single strain of *Plasmodium reichenowi* (Rich et al. 2009). In this scenario, human ancestors may have escaped from the common ape malaria by eliminating Neu5Gc, but then later became susceptible to a variant that evolved to now recognize the Neu5Ac-rich red blood cells of humans (Varki and Gagneux 2009).

Shiga toxigenic *Escherichia coli*, which causes serious gastrointestinal disease in humans, can also secrete a subtilase cytotoxin (SubAB) (Paton et al. 2004). The pentameric B subunit of SubAB, which directs target cell uptake after the binding of surface glycans, has a strong preference for Neu5Gc rather than Neu5Ac, and binds Neu5Gc that is incorporated from the diet into human gut epithelia and kidney vasculature (Byres et al. 2008). This mechanism likely confers susceptibility to the gastrointestinal and systemic toxicities of SubAB in humans (Byres et al. 2008). The CMAH inactivation causes the lack of protective Neu5Gc-containing glycoproteins in serum and other body fluids and may thus make humans kidneys hypersusceptible to the toxin. The susceptibility to SubAB expressing *Escherichia coli* may be a case where CMAH inactivation increases the risk of infectious disease in humans.

8.2.2 *SIGLEC1*

SIGLEC1 belongs to the Siglec gene family and encodes Siglec-1 (sialoadhesin). It is composed of one V-set domain, sixteen C2-set domains, a transmembrane domain, and a short cytoplasmic tail without any signaling motifs (Crocker 2005; Crocker et al. 2007; Crocker and Varki 2001; Hartnell et al. 2001) (Fig. 8.3). As compared with other Siglecs that have at most six C2-set domains, Siglec-1 displays a very long extracellular component (Fig. 8.3). In addition to Siglec-1, there are three known Siglecs having no signaling motif in their cytoplasmic tails: Siglec-14, Siglec-15, and Siglec-16 (Angata et al. 2006, 2007; Cao et al. 2008) (Fig. 8.3). Even though these three Siglecs lack signaling motifs in their cytoplasmic tails, they can function as receptors in signal transduction by the association with an adaptor molecule (Angata et al. 2006, 2007; Cao et al. 2008) (see the "*SIGLEC5/SIGLEC14*" and "*SIGLEC11* and *SIGLEC16*" sections for details; see Fig. 8.3). In contrast to these Siglecs, Siglec-1 has no such intracellular partner molecules and seems to predominantly act as a simple adhesion molecule without signaling properties. These structural features make Siglec-1 unique among Siglecs.

Siglec-1 is evolutionarily conserved in all mammals examined (Crocker et al. 2007; Crocker and Varki 2001; Varki and Angata 2006). Similar to mouse Siglec-1, human Siglec-1 strongly prefers binding Neu5Ac to Neu5Gc (Brinkman-Van der Linden et al. 2000). However, considering that the loss of Neu5Gc results in the enrichment of Neu5Ac in humans, human Siglec-1 has been exposed to an increase of endogenous ligand density. Siglec-1 is expressed in macrophages (Hartnell et al. 2001), but the expression pattern of human Siglec-1 appears unique. In rats and chimpanzees, Siglec-1 on macrophages is found in both the perifollicular zone and marginal zone of the spleen (Brinkman-Van der Linden et al. 2000). In contrast, human Siglec-1 is found only in perifollicular zone (Brinkman-Van der Linden et al. 2000). Furthermore, almost all CD68+ macrophages in the human spleen are Siglec-1 positive, but only a subpopulation of macrophages expresses Siglec-1 in the chimpanzee spleen (Brinkman-Van der Linden et al. 2000). These human-specific changes may have certain implications in Siglec-1 biology.

Macrophages phagocytose cellular debris and pathogens and play an important role in innate immunity. Siglec-1 on macrophages binds to sialic acids on pathogens such as *Trypanosoma cruzi* and *Neisseria meningitidis* and increases the pathogen uptake by macrophages (Jones et al. 2003; Monteiro et al. 2005). If the striking expression of Siglec-1 on almost all macrophages in the spleen means the advanced ability of phagocytosis of human macrophages, human Siglec-1 might be a direct case wherein Siglec evolution improved the ability of protection against certain pathogens. Interestingly, most of the Neu5Ac-expressing pathogens are human specific, and Neu5Gc-synthesizing pathogens have never been recognized (Vimr et al. 2004).

8.2.3   *SIGLEC5/SIGLEC14*

SIGLEC5 and *SIGLEC14* are members of the CD33/Siglec-3-related subset of Siglec genes and make a primate gene cluster with other members (*SIGLEC3, SIGLEC6, SIGLEC7, SIGLEC8, SIGLEC9, SIGLEC10, SIGLEC12*, and *SIGLEC13*) (Angata et al. 2004, 2006; Cao et al. 2009; Varki and Angata 2006) (Figs. 8.3 and 8.4). *SIGLEC5* and *SIGLEC14* are positioned side-by-side in a tail-to-head orientation on the most telomeric end in the CD33/Siglec-3-related Siglec gene cluster (Fig. 8.4) and are conserved in all primates examined (Angata et al. 2004, 2006; Cao et al. 2009; Varki and Angata 2006).

The *SIGLEC5* gene consists of nine exons (Yousef et al. 2002), and the *SIGLEC14* is composed of seven exons (Angata et al. 2006). The approximately 1.5-kb region including the first four exons of *SIGLEC5* shows homology to that of *SIGLEC14*, which suggests that gene duplication was involved in the emergence of this Siglec gene pair (Angata et al. 2006). Interestingly, the 5'-end (~1.3-kb part; designated as A/A') of the similar region shows extremely high similarity (>99%) between *SIGLEC5* and *SIGLEC14* in all primate lineages studied (Angata et al. 2006). In contrast, the rest of the similar region (~0.2 kb; designated as B/B') shows a lesser identity (78%) (Angata et al. 2006). The high similarity in region A/A' cannot be explained by the paralogous relationship between two primate genes. The only possible explanation is that recurrent gene conversion events have been occurring in the primate lineages (Angata et al. 2006).

Siglec-5 is composed of one V-set domain, three C2-set domains, a transmembrane domain, and a cytoplasmic tail containing an inhibitory signaling motif (ITIM) (Fig. 8.3), and thus functions as an inhibitory receptor (Crocker 2005; Crocker et al. 2007; Crocker and Varki 2001; Varki and Angata 2006). Siglec-14 is a smaller molecule that consists of one V-set domain, two C2-set domains, a transmembrane domain, and a short cytoplasmic tail with no signaling motif

(Angata et al. 2006) (Fig. 8.3). However, Siglec-14 has a positively charged residue (arginine) in its transmembrane domain and associates with activating adaptor molecule DAP12 (*TYROBP*) via its charged residue (Angata et al. 2006) (Fig. 8.3). Unlike Siglec-5, Siglec-14 therefore functions as an activating receptor. In addition to extreme sequence similarity, sialic acid recognition properties and tissue distribution are very similar between human Siglec-5 and Siglec-14 (Angata et al. 2006; Yamanaka et al. 2009). Taken together, Siglec-5 and Siglec-14 are regarded as "paired receptors" (Angata et al. 2006). The role of paired receptors is elusive, but it is proposed that they are involved in fine-tuning of immune responses via a balance in the ligand binding by activating and inhibitory pair. Because the A/A′ region contains the upstream region and an exon that codes a sialic acid-binding domain (i.e., V-set domain), the gene conversion is the most likely a genomic mechanism that ensures that Siglec-5 and Siglec-14 are the paired receptors.

Interestingly, independent gene conversion between *SIGLEC5* and *SIGLEC14* (S5–S14 gene conversion) is also found in each of the non-human primates studied, such as chimpanzee, gorilla, orangutan, and baboon, which indicates a high frequency of the S5–S14 gene conversion (Angata et al. 2006). A short (TG)n tract is located between the A/A′ and B/B′ regions (Angata et al. 2006) and is an interesting sequence feature in the terms of the high frequency of the S5–S14 gene conversion. Regardless, the high frequency of S5–S14 gene conversion indicates the necessity of gene conversion in the Siglec-5/Siglec-14 pair as paired receptors. More importantly, the species-specific gene conversion suggests that sialic acid recognition properties and tissue distributions of Siglec-5 and Siglec-14 are similar but become unique in each primate species. In addition, the "essential" arginine residue that confers optimal sialic acid recognition to the sialic acid-binding domain is mutated in the great ape Siglec-5 and Siglec-14 (see Fig. 8.4), which results in the reduction of sialic acid binding in the great ape Siglecs (Angata et al. 2006). It is therefore likely that the biological function of Siglec-5/Siglec-14 pair in the human differs from those in the great apes.

A striking contrast of Siglec-5/Siglec-14 expression has been found between human and chimpanzee T cells. Although the majority of chimpanzee CD4+ T cells express Siglec-5 and/or Siglec-14, CD4+ T cells in humans are mostly negative (Nguyen et al. 2006; Yamanaka et al. 2009). The CD4+ T cells are involved in the pathology of common human diseases such as acquired immunodeficiency syndrome (AIDS), bronchial asthma, rheumatoid arthritis, and type I diabetes. In this regard, human immunodeficiency virus (HIV) progression to AIDS is common in humans but rare in chimpanzees (Novembre et al. 1997; Olson and Varki 2003). Moreover, rheumatoid arthritis, bronchial asthma, and type I diabetes have not been reported in great apes (Olson and Varki 2003; Varki and Altheide 2005). The lack of Siglec-5 and Siglec-14 expression on CD4+ T cells may therefore contribute to T-cell overreactivity in these common human diseases.

A functional deletion of *SIGLEC14*, which is caused by a fusion between *SIGLEC5* and *SIGLEC14* loci, has recently been found in human populations from around the world (Yamanaka et al. 2009). As mentioned earlier, the approximately 1.3-kb A/A′ region of human *SIGLEC14* represents extreme identity with the corresponding region of human *SIGLEC5*. The gene fusion caused the lack of region unique to *SIGLEC14* (i.e., the fusion boundary is in A/A' region), resulting in the lack of Siglec-14 expression (Yamanaka et al. 2009) but continued expression of Siglec-5-like protein. The Siglec-14 null individuals are apparently healthy, which shows that the loss of Siglec-14 is not deleterious in healthy human populations (Yamanaka et al. 2009). However, the frequency of the SIGLEC14 null allele is significantly higher in Asians than in Africans and Europeans (Yamanaka et al. 2009), suggesting unknown selective pressures. In this regard, it is interesting that group B *Streptococcus*, a common cause of sepsis and meningitis in human newborns, binds to Siglec-5 via cell wall-anchored β protein and thereby impairs leukocyte phagocytosis, oxidative burst, and extracellular trap production (Carlin et al. 2009a). This interaction increases bacterial resistance to phagocytosis and killing by human leukocytes and is considered as one of the strategies of microbial innate immune evasion (Carlin et al. 2009a). Siglec-14 may be involved in this interaction as a partner molecule of Siglec-5, and its functional deletion may have certain implication on pathogenic infection.

8.2.4 SIGLEC6

Human *SIGLEC6* is located in the CD33/Siglec-3-related Siglec gene cluster on the long arm of chromosome 19 with orthologues in chimpanzee, baboon, and rhesus monkey (Angata et al. 2004; Cao et al. 2009; Varki and Angata 2006; Yousef et al. 2002) (Fig. 8.4). Siglec-6 is composed of one V-set domain, two C2-set domains, a transmembrane domain, and cytoplasmic tail having an ITIM, and is a structurally typical member of CD33/Siglec-3-related Siglecs (Crocker 2005; Crocker et al. 2007; Crocker and Varki 2001; Patel et al. 1999; Varki and Angata 2006) (Fig. 8.3). However, its expression pattern and binding ability are unusual in humans. Human Siglec-6 is found not only on B cells but also on the trophoblast of the placenta, and binds not only to sialic acids but also to leptin, a non-sialic acid-containing protein (hormone) secreted by adipose tissue and placenta (Patel et al. 1999). The placental expression and leptin binding imply that human Siglec-6 may play an important role in placenta biology.

Interestingly, even though Siglec-6 is expressed on B cells of great apes, its placental expression is unique to humans (Brinkman-Van der Linden et al. 2007). Potential ligands (sialic acid-containing glycoproteins) of Siglec-6 are also expressed in placenta, which indicates that Siglec-6 was recruited to placental expression in the human lineage (Brinkman-Van der Linden et al. 2007). Human pregnancy and parturition are unique as compared with that of other mammals, including non-human primates. Human parturition is prolonged as a result of the negotiation between the larger fetal head and small birth canal in the compensation for the gain of the large brain and erect bipedalism (Lovejoy 2005). Human labor tends to be long in duration, whereas chimpanzees give birth more rapidly (Keeling and Roberts 1972). Siglec-6 expression increases with the onset and progression of labor

(Brinkman-Van der Linden et al. 2007). By its role as an inhibitory receptor, Siglec-6, whose expression is recruited in the human placenta, might potentially contribute to the prolongation of the birth process in humans. Siglec-6 also binds to leptin, and they colocalize in the placenta (Brinkman-Van der Linden et al. 2007). The plasma leptin concentration increases during labor (Nuamah et al. 2004), and leptin-deficiency mice represent increase of parturition time (Mounzih et al. 1998). Even though leptin is not a dominant ligand, the leptin–Siglec-6 binding might be important in the prolongation of parturition. Interestingly, recent studies have shown that Siglec-6 expression is further upregulated in the placenta of patients with preeclampsia, a uniquely human disease (Winn et al. 2009).

8.2.5 *SIGLEC7 and SIGLEC9*

Siglec-7 and Siglec-9 are members of the CD33/Siglec-3-related Siglec subgroup (Crocker 2005; Crocker et al. 2007; Crocker and Varki 2001; Varki and Angata 2006) (Fig. 8.3). The orthologue of Siglec-7 is not found in rodents, but that of Siglec-9 is probably rodent Siglec-E because of their similar gene location and expression pattern (Angata et al. 2004; Varki and Angata 2006) (see Fig. 8.4). Siglec-9 is therefore considered as an ancient Siglec. Because of the close phylogenetic relationship and identical genomic/domain structure (Angata et al. 2004; Yousef et al. 2002), Siglec-7 might have emerged from Siglec-9 via a gene duplication event in the Siglec expansion in primates. Siglec-7 is dominantly expressed on natural killer (NK) cells and weakly on monocytes (Crocker et al. 2007; Nicoll et al. 1999). On the other hand, the expression level of Siglec-9 is high on monocytes, neutrophils, and conventional dendritic cells, but low on NK cells (Crocker et al. 2007; Zhang et al. 2000). Indeed, although expression intensity is different between Siglec-7 and Siglec-9, both Siglecs are found on the same set of immune cells, which suggests that Siglec-7 and Siglec-9 genes retained the similar set of regulatory elements via gene duplication. Siglec-7 and Siglec-9 can both function as inhibitory receptors in these sets of immune cells because of the presence of ITIM in their cytoplasmic tails (Carlin et al. 2009b; Crocker 2005; Crocker et al. 2007; Crocker and Varki 2001; Nicoll et al. 1999; Varki and Angata 2006; Zhang et al. 2000).

Although human Siglec-9 has nonpreferred binding to Neu5Ac and Neu5Gc, chimpanzee and gorilla Siglec-9 represent a strong preference for Neu5Gc-containing ligands (Sonnenburg et al. 2004). This preference indicates that Siglec-9 has changed its sialic acid preference to bind Neu5Ac in the human lineage after the loss of Neu5Gc. Interestingly, the greatest differences of sequence are found in the V-set domains (i.e., sialic acid-binding domains) of human and great ape Siglec-9 (Sonnenburg et al. 2004). Moreover, the nonsynonymous substitution rate is higher than synonymous substitution rate in the exon encoding the V-set domain in the human lineage (Sonnenburg et al. 2004). These findings indicate that accelerated evolution has occurred in the V-set domain of human Siglec-9. As mentioned in the foregoing "*CMAH*" section, the loss of Neu5Gc occurred in the human lineage

(Muchmore et al. 1998; Varki 2001). The acquisition of Neu5Ac binding in human Siglec-9 is therefore thought to be a consequence of adaptive evolution to the human-specific loss of Neu5Gc. Additionally, Siglec-7 shows similar differences in sequence and sialic acid binding between humans and chimpanzees, as does Siglec-9: multiple amino acid changes in the sialic acid-binding domain, chimpanzee Siglec-7 binding to Neu5Gc, and a human Siglec-7 accommodation of Neu5Ac (Sonnenburg et al. 2004). Similar to Siglec-9, human Siglec-7 may have evolved under an adaptation to the loss of Neu5Gc.

Human Siglec-7 and Siglec-9 seem to present interesting cases of likely molecular coevolution between ligands and endogenous receptors in sialic acid biology. Based on this strong coevolutionary relationship, endogenous sialic acids are considered as functional ligands of Siglec molecules and direct the evolution of sialic acid-binding specificity (Sonnenburg et al. 2004). Human-specific loss of Neu5Gc might have also caused changes in sialic acid-binding specificity of other Siglecs (Sonnenburg et al. 2004).

The alteration of sialic acid preference of human Siglec-7 and Siglec-9 might also be implicated in pathogen infection. *Campylobacter jejuni*, a human pathogen commonly responsible for gastroenteritis, can express a variety of different sialyloligosaccharides in its lipo-oligosaccharides (Avril et al. 2006; Crocker et al. 2007). Siglec-7 binds to Neu5Acα2-8Neu5Acα2-3Gal in lipo-oligosaccharides on this pathogen and increases pathogen binding to NK cells and monocytes (Avril et al. 2006; Crocker et al. 2007). *Campylobacter jejuni* might thus be exploiting the Neu5Ac accommodation of human Siglec-7 in its infection. On the other hand, Siglec-9 is dominantly expressed on neutrophils, specialized granulocytes that recognize and kill microorganisms, and binds Neu5Acα2-3Galβ1-4GlcNAc units (Carlin et al. 2009b). Interestingly, the human-specific pathogen group B *Streptococcus* has the same units on its sialylated capsule and can suppress neutrophil function using its sialic acid-rich capsule to engage human Siglec-9 (Carlin et al. 2009b).

8.2.6 SIGLEC11 and SIGLEC16

Human *SIGLEC11* and *SIGLEC16* are members of CD33/Siglec-3-related Siglec genes, but located about 1 Mb centromeric of the CD33/Siglec-3-related Siglec gene cluster (Angata et al. 2002; Crocker 2005; Crocker et al. 2007; Varki and Angata 2006) (Fig. 8.4). They are found in a head-to-head orientation about 9 kb apart (Angata et al. 2002; Cao et al. 2008) (Fig. 8.4). The *SIGLEC11* gene is also found in non-human primates such as the great apes and rhesus monkey, and *SIGLEC16* is clearly identified at least from the chimpanzee (Angata et al. 2004; Cao et al. 2008; Hayakawa et al. 2005; Varki and Angata 2006) (see Fig. 8.4). On the other hand, mouse, dog, and cow have a single Siglec gene or a Siglec-like pseudogene in the orthologous region of their genomes (Cao et al. 2008). The approximately 3-kb genomic part containing the first eight exons of *SIGLEC11* represents sequence similarity to *SIGLEC16*, which indicates an evolutionary kinship between *SIGLEC11* and *SIGLEC16* (Angata et al. 2002; Hayakawa et al. 2005). Thus, it has been proposed that *SIGLEC11* and *SIGLEC16* genes emerged via a gene duplication event in the primate lineage (Angata et al. 2002; Cao et al. 2008).

Interestingly, the approximately 2-kb region (designated A/A′) including the first five exons represents extreme identity (99.3%) between human *SIGLEC11* and *SIGLEC16*, whereas the rest of the similar part (designated B/B′) shows 94.6% identity (Hayakawa et al. 2005). In contrast to human genes, chimpanzee orthologues of *SIGLEC11* and *SIGLEC16* present no such extreme identity in the similar region (Hayakawa et al. 2005). The phylogenetic trees show that the clustering of human *SIGLEC11* and *SIGLEC16* is found only in the A/A′ region, not in the B/B′ region (Hayakawa et al. 2005). These findings indicate that gene conversion occurred in the A/A′ region exclusively in the human lineage (Hayakawa et al. 2005). In addition, the genetic distance analysis and the phylogenetic tree constructed by adding bonobo, gorilla, and orangutan *SIGLEC11* sequences clearly show that *SIGLEC16* converted *SIGLEC11* (S16→S11 gene conversion) only in the human lineage (Hayakawa et al. 2005).

The converted part of human *SIGLEC11* contains an exon encoding the sialic acid-binding domain (Hayakawa et al. 2005). Indeed, human Siglec-11 shows a different binding ability from chimpanzee Siglec-11 (Hayakawa et al. 2005). Interestingly, Neu5Gc binding is dramatically reduced in human Siglec-11 (Hayakawa et al. 2005). If chimpanzee Siglec-11 shows the ancestral situation of sialic acid binding, this may again be a consequence of adaptive evolution to the human-specific loss of Neu5Gc, as in the case of Siglec-7/Siglec-9.

In addition to the coding region, the upstream region was involved in the S16→S11 gene conversion (Hayakawa et al. 2005). In contrast to other CD33/Siglec-3-related Siglec genes, the *SIGLEC11* gene shows expression in the human brain (Angata et al. 2002); that is, human brain microglia show a positive staining with anti Siglec-11 antibody, but chimpanzee and orangutan brain microglia do not (Hayakawa et al. 2005). Considering that Siglec-11 binds to sialic acids enriched in the brain, *SIGLEC11* seems to have been recruited to brain expression in the human lineage. In this regard, it is also interesting to note that human Siglec-11 binds to oligo sialic acids [$(Neu5Ac\alpha2\text{-}8)_{2\text{-}3}$], which are enriched in the brain (Hayakawa et al. 2005).

The S16→S11 gene conversion makes an impact on human Siglec-11 evolution. This gene conversion results in the extreme sequence similarity between the extracellular parts of human Siglec-11 and Siglec-16, which suggests that human Siglec-11 and Siglec-16 have the same sialic acid-binding specificity. Human *SIGLEC16* is also expressed in the brain (Cao et al. 2008). It appears that human Siglec-16 functions as an activating receptor by the association with an activating adaptor molecule, DAP12, via a positively charged lysine residue in the transmembrane domain (Cao et al. 2008) (Fig. 8.3). As Siglec-11 is an inhibitory receptor because of the presence of ITIM in the cytoplasmic tail (Angata et al. 2002; Crocker 2005; Crocker et al. 2007; Varki and Angata 2006), the S16→S11 gene conversion may have assured that Siglec-11 became an inhibitory partner of Siglec-16. In other words, human Siglec-11 and Siglec-16 became human-specific paired receptors via the $S16 \rightarrow S11$ gene conversion. The emergence of paired receptors may possibly have contributed to the human brain evolution via alteration of microglial function such as the interaction with neural cells.

In this regard, the cytoplasmic tail of Siglec-11 is known to recruit the protein tyrosine phosphatase SHP-1 (Src homology domain 2-containing phosphatase 1) (Angata et al. 2002). The SHP-1-deficient mice show a marked decrease of the number of microglial cells and a slightly smaller brain size than littermate controls (Wishcamper et al. 2001). It is therefore possible to hypothesize that the Siglec-11 recruitment to brain expression is involved in human brain expansion via association with SHP-1 in microglial cells.

As the human *SIGLEC16* sequence released by the human genome project has a 4-bp deletion (4-bpΔ) in exon 2, human *SIGLEC16* was originally proposed as a pseudogene (*SIGLECP16*) (Angata et al. 2002; Hayakawa et al. 2005); this is attributed to the high frequency of the 4-bpΔ allele (null allele) of *SIGLEC16* in human populations (e.g., ~40% in the UK population) (Cao et al. 2008). Because the chimpanzee *SIGLEC16* released by the chimpanzee genome project has no 4-bp deletion, it is supposed that the 4-bpΔ allele appeared uniquely in the human lineage. Surprisingly, the same situation, that is, the presence of a null allele of the activating Siglec gene, is also found in *SIGLEC14* (see the "*SIGLEC5/SIGLEC14*" section) (Yamanaka et al. 2009), and both Siglec-16 and Siglec-14 are the activating partners of paired receptors. These findings may provide some hint about the role of Siglec paired receptors.

Siglec-11 has been recently proposed as a molecule related to Alzheimer's disease, a progressive neurodegenerative disorder (Salminen and Kaarniranta 2009). Alzheimer's disease is characterized by the continuous increase in the numbers and size of β-amyloid plaques (Salminen and Kaarniranta 2009). In this hypothesis, Siglec-11 induces an antiinflammatory response in microglial cells by recognizing sialylated glycolipids or glycoproteins that bind to β-amyloid plaques, which allow plaques to evade the immune surveillance of microglia (Salminen and Kaarniranta 2009). This is interesting, because microglial expression of Siglec-11 is found only in humans (Angata et al. 2002; Hayakawa et al. 2005), and Alzheimer's disease is common only in humans (Olson and Varki 2003; Varki and Altheide 2005).

8.2.7 *SIGLEC12*

In contrast to other Siglecs, Siglec-12 has two V-set domains (Angata et al. 2001; Yousef et al. 2002) (Fig. 8.3). However, the "essential" arginine residue, which confers the optimal sialic acid recognition, is conserved only in the first V-set domain, indicating that only the first V-set domain functions as a sialic acid-binding domain (Angata et al. 2001). This unique second V-set domain does not define Siglec-12 as an evolutionary lone Siglec because the genomic structure of SIGLEC12 is very similar to that of SIGLEC7 (Angata et al. 2001; Yousef et al. 2002). As the Siglec-7 gene has an exon fossil that corresponds to an exon encoding the second V-set domain of Siglec-12 (Angata et al. 2001), it is suggested that Siglec-12 and Siglec-7 are sibling molecules generated via a gene duplication event; this is also supported by the phylogenetic tree analysis (Angata et al. 2001, 2004). On the other hand, despite the close evolutionary kinship with Siglec-7 expressed on immune cells (NK cells and monocytes), Siglec-12 displays a very different expression pattern, that is, expression on the luminal edge of epithelial cells in organs such as the stomach and tonsils (Angata et al. 2001). It seems that the set of regulatory elements was not conserved in the gene duplication generating Siglec-12 and Siglec-7.

The "essential" arginine residue of Siglec-12 is conserved among the great apes but is changed to a cysteine residue in humans (Angata et al. 2001) (see Fig. 8.4). This "essential" arginine mutation causes the loss of sialic acid binding in human Siglec-12 (Angata et al. 2001). Meanwhile, chimpanzee Siglec-12 binds to both Neu5Ac and Neu5Gc but shows strong preference for Neu5Gc (Angata et al. 2001). Sonnnenburg et al. proposed that the lack of the sialic acid-binding ability in human Siglec-12 reflects a "retirement" caused by the human-specific loss of Neu5Gc (Sonnenburg et al. 2004).

8.2.8 SIGLEC13

The sequence comparison of the CD33/Siglec-3-related Siglec gene cluster between human, chimpanzee, and baboon shows a few species-specific gene losses (Angata et al. 2004) (Fig. 8.4). In humans, the *SIGLEC13* locus is completely deleted (Angata et al. 2004; Varki 2007) (Fig. 8.4). To understand the impact of *SIGLEC13* deletion on human evolution, further analysis on sialic acid preference and expression pattern in non-human primates is under way.

8.2.9 ST6GAL1

Sialic acids have three common linkages to acceptor sugars: $\alpha2$-3, $\alpha2$-6, and $\alpha2$-8 linkages (Angata and Varki 2002; Beyer et al. 1979; Traving and Schauer 1998). A striking contrast in the distribution of $\alpha2$-6-linked sialic acids is found between humans and great apes (Gagneux et al. 2003). The $\alpha2$-6-linked sialic acids are abundant in the epithelium lining of the human trachea and lung airways, but not in the epithelial goblet cells that secrete heavily sialylated soluble mucins to the lumen of airways (Gagneux et al. 2003). On the other hand, great apes have no $\alpha2$-6-linked sialic acids in the former cell types but show an abundance in the latter (Gagneux et al. 2003). These findings clearly indicate that human airway epithelium underwent a concerted bidirectional switch in the expression of $\alpha2$-6-linked sialic acids, that is, a likely upregulation of *ST6GAL1* in the epithelial lining and downregulation in the goblet cells and secreted mucins.

Human influenza virus A and B are known as pathogens that show strong preference for α2-6-linked sialic acids during their infection and target the human respiratory epithelia (Rogers and Paulson 1983). In contrast, avian and other mammalian influenza viruses prefer α2-3-linked sialic acids in their infection (Webster et al. 1992). It is also reported that chimpanzees show attenuation of human influenza infection (Murphy et al. 1992; Snyder et al. 1986; Subbarao et al. 1995). In humans, α2-6-linked sialic acids are abundant in the target cell surfaces but not in mucins that act as potential soluble decoys for viruses. Thus, the uniquely human distribution of α2-6-linked sialic acids in airway epithelium is considered to correlate to the susceptibility of human influenza viruses.

The upregulation of α2-6-linked sialic acids in the epithelial lining in humans is probably explained by the increase in the common α2-6 linkage of sialic acid to galactose on the *N*-glycan chain (Gagneux et al. 2003; Martin et al. 2002; Weinstein et al. 1987). This sialic acid structure is primarily produced by a sialyltransferase, ST6Gal-I. It is therefore considered that the unique human distribution of α2-6-linked sialic acids results from the altered spatial regulation of ST6Gal-I expression. However, further work is needed to confirm this hypothesis.

8.3 Overall Evolution of Sialic Acid Biology in the Primate Lineage

Based on their roles, genes involved in sialic acid biology can be divided into five categories: "biosynthesis"; "activation, transport, and transfer"; "modification"; "recognition"; and "recycling and degradation" (Altheide et al. 2006) (see Fig. 8.2). Biosynthesis genes encode several enzymes involved in the production of sialic acid residues from precursor molecules. Activation, transport, and transfer genes encode many enzymes that activate free sialic acid into the nucleotide donor CMP-sialic acid and transport it into the Golgi in which sialic acid residues are transferred to newly synthesized glycoconjugates from the CMP donors. ST6Gal-I falls into this category. The modification category consists of one known member to date, *CMAH*. Recognition refers to the loci involved in sialic acid recognition, and the major group of this category is the Siglecs. Recycling and degradation genes encode several molecules that function in the release of sialic acid residues from the glycan chain, the breakdown of sialic acid into acylmannosamines and pyruvate, the removal of 9-*O*-acetyl esters from sialic acid residues, the lysosomal transport of free sialic acid residues to the cytosol, and the stabilization of the sialidase that releases sialic acid residues.

Evolutionary analysis of these represents an overall acceleration across all categories in primates compared with rodents (Altheide et al. 2006), which suggests that the biological functions based on sialic acid biology have evolved more rapidly in the primate lineage.

8 Human-Specific Changes in Sialic Acid Biology

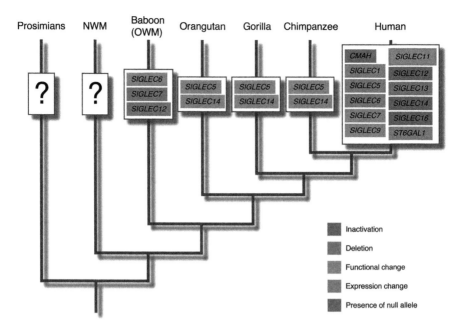

Fig. 8.5 Evolution of loci involved in sialic acid biology in the primate lineage. *NWM*, New World monkeys; *OWM*, Old World monkeys

8.3.1 Rapid Evolution of Recognition Molecules

Further comparison among categories shows that the recognition category has evolved more rapidly than the others (Altheide et al. 2006). Notably, Siglecs display a higher rate of evolution than other members in the recognition category (Altheide et al. 2006). Indeed, many human-specific changes occurred in Siglecs, especially, in CD33/Siglec-3-related Siglecs. This finding is consistent with human–chimpanzee genomic comparison, which reveals that CD33/Siglec-3-related Siglecs are one of the fastest evolving groups of genes in the entire genome (The Chimpanzee Sequencing and Analysis Consortium 2005). In addition to the human, non-human primates also display some changes in CD33/Siglec-3-related Siglecs (Fig. 8.5). The *SIGLEC7* and *SIGLEC12* loci are completely deleted in the baboon (Angata et al. 2004). Moreover, the "essential" arginine residue of Siglec-6 changed to leucine in the baboon, suggesting the loss of optimal sialic acid binding in baboon Siglec-6 (Angata et al. 2004). In addition to baboon Siglec-6, the great apes (chimpanzee, gorilla, and orangutan) show the "essential" arginine changed to histidine or tyrosine in both Siglec-5 and Siglec-14 (Angata et al. 2006). Interestingly, these Siglecs also show human-specific changes and might be most changeable in the primate Siglecs. The CD33/Siglec-3-related Siglec subgroup has been uniquely changed in each primate lineage (see Fig. 8.5).

Even within the components of Siglec molecules, the sialic acid-binding domain is the most rapidly evolving (Altheide et al. 2006). It has been proposed that this high rate of evolution of the sialic acid-binding domains is a secondary "Red Queen" effect (Varki and Angata 2006). Namely, the rapid evolution of sialome is caused by the arms race with sialic acid-binding pathogens (primary "Red Queen" effect), and this rapid sialome evolution then forces the sialic acid-binding domains to evolve quickly to keep up with recognizing the changing sialome as "self" (secondary "Red Queen" effect). Even though sialyltransferase sequences are highly conserved in primates, specific gains and losses of sialic acid expression, for example, human unique changes of the distribution of α2-3- and α2-6-linked sialic acids, are found in each primate species (Altheide et al. 2006; Gagneux et al. 2003). Thus, sialic acid-binding domains may have evolved uniquely in each primate species.

8.3.2 *More Rapid Evolution in the Human Lineage?*

The human sialome was changed dramatically by the loss of Neu5Gc (Muchmore et al. 1998; Varki 2001), suggesting that the recognition molecules have experienced more sialome changes in the human than in non-human primates. Considering secondary "Red Queen" effects (Varki and Angata 2006), it is supposed that sialic acid-binding domains of human Siglecs have evolved more rapidly than those of non-human primates. This conjecture is consistent with the highest rate of evolution of the sialic acid-binding domain in the human compared with non-human primates (Altheide et al. 2006) and is supported by the human unique alteration in sialic acid-binding ability shown by many Siglecs (i.e., Siglec-5, Siglec-7, Siglec-9, Siglec-11, Siglec-12, and Siglec-14) (see Fig. 8.5).

The loss of Neu5Gc, that is, the CMAH inactivation, occurred about 3 million years ago (Chou et al. 2002). If the changes of sialic acid-binding ability are a consequence of this event, such changes should have occurred after the loss of Neu5Gc. The timing estimation of genomic event (e.g., gene conversion) involved in the change of sialic acid-binding ability would be helpful to test this hypothesis. At this time, it is supposed that the changes of sialic acid-binding ability in human Siglec-5 and Siglec-14 occurred about 3 million years ago (Angata et al. 2006), which is consistent with the hypothesis. Timing estimations for other Siglecs are currently under way. In addition, the emergence of many human-specific pathogens that express Neu5Ac might have also occurred because of the change in inhibitory human Siglecs toward binding Neu5Ac rather than Neu5Gc (Carlin et al. 2009b).

8.4 Future Directions and Perspectives

An unusually large number of human-specific changes have been found in genomic loci involved in sialic acid biology (Fig. 8.5), suggesting that certain phenotypes expressed by sialic acid biology have changed uniquely in the human lineage. Based on the biological and biochemical roles of the changed molecules (i.e., CMAH, Siglecs, and ST6Gal-I), it is possible that these human-specific changes have contributed to human uniqueness through the alteration in the interaction between sialome and recognition molecules. Because this interaction is known to be important in the immunity and host–pathogen interaction, a major impact of the detected human-specific changes may have been in these biological systems. Additionally, further findings on these human-specific changes might give some explanation on the background of human unique/common diseases such as Alzheimer's disease and AIDS (Olson and Varki 2003; Varki and Altheide 2005).

Even though the detected changes are undoubtedly unique in the human, we are still speculating on their significance in human biology. Several hypotheses have been proposed to explain the role of each human-specific change. So far, we have difficulty in testing these hypotheses because human and great apes can never be subjects of genetic manipulation. On the other hand, mice may be too distantly related to primates to use them as models. However, genetic manipulating techniques were recently established for the common marmoset (*Callithrix jacchus*), a species of New World monkeys (Sasaki et al. 2009).

What makes us human? This is a popular question for mankind, and has been posed by many a researcher. Exploration of human uniqueness in sialic acid biology is one of the scientific approaches to answering this question. Scientists could not directly deal with this question in the twentieth century. However, genome biology, which heralded the first year of the twenty-first century (International Human Genome Sequencing Consortium 2001), has given us many useful tools and hints to consider this question, and post-genome biology will allow for further progress.

References

Altheide TK, Hayakawa T, Mikkelsen TS, Diaz S, Varki N, Varki A (2006) System-wide genomic and biochemical comparisons of sialic acid biology among primates and rodents: evidence for two modes of rapid evolution. J Biol Chem 281:25689–25702

Angata T, Varki A (2002) Chemical diversity in the sialic acids and related alpha-keto acids: an evolutionary perspective. Chem Rev 102:439–469

Angata T, Varki NM, Varki A (2001) A second uniquely human mutation affecting sialic acid biology. J Biol Chem 276:40282–40287

Angata T, Kerr SC, Greaves DR, Varki NM, Crocker PR, Varki A (2002) Cloning and characterization of human Siglec-11. A recently evolved signaling that can interact with SHP-1 and SHP-2 and is expressed by tissue macrophages, including brain microglia. J Biol Chem 277:24466–24474

Angata T, Margulies EH, Green ED, Varki A (2004) Large-scale sequencing of the CD33-related Siglec gene cluster in five mammalian species reveals rapid evolution by multiple mechanisms. Proc Natl Acad Sci USA 101:13251–13256

Angata T, Hayakawa T, Yamanaka M, Varki A, Nakamura M (2006) Discovery of Siglec-14, a novel sialic acid receptor undergoing concerted evolution with Siglec-5 in primates. FASEB J 20:1964–1973

Angata T, Tabuchi Y, Nakamura K, Nakamura M (2007) Siglec-15: an immune system Siglec conserved throughout vertebrate evolution. Glycobiology 17:838–846

Avril T, Wagner ER, Willison HJ, Crocker PR (2006) Sialic acid-binding immunoglobulin-like lectin 7 mediates selective recognition of sialylated glycans expressed on *Campylobacter jejuni* lipooligosaccharides. Infect Immun 74:4133–4141

Baum J, Pinder M, Conway DJ (2003) Erythrocyte invasion phenotypes of *Plasmodium falciparum* in The Gambia. Infect Immun 71:1856–1863

Beyer TA, Rearick JI, Paulson JC, Prieels JP, Sadler JE, Hill RL (1979) Biosynthesis of mammalian glycoproteins. Glycosylation pathways in the synthesis of the nonreducing terminal sequences. J Biol Chem 254:12531–12534

Bramble DM, Lieberman DE (2004) Endurance running and the evolution of *Homo*. Nature (Lond) 432:345–352

Brinkman-Van der Linden EC, Sjoberg ER, Juneja LR, Crocker PR, Varki N, Varki A (2000) Loss of N-glycolylneuraminic acid in human evolution. Implications for sialic acid recognition by siglecs. J Biol Chem 275:8633–8640

Brinkman-Van der Linden EC, Hurtado-Ziola N, Hayakawa T, Wiggleton L, Benirschke K, Varki A, Varki N (2007) Human-specific expression of Siglec-6 in the placenta. Glycobiology 17:922–931

Byres E, Paton AW, Paton JC, Lofling JC, Smith DF, Wilce MC, Talbot UM, Chong DC, Yu H, Huang S, Chen X, Varki NM, Varki A, Rossjohn J, Beddoe T (2008) Incorporation of a non-human glycan mediates human susceptibility to a bacterial toxin. Nature (Lond) 456:648–652

Caceres M, Lachuer J, Zapala MA, Redmond JC, Kudo L, Geschwind DH, Lockhart DJ, Preuss TM, Barlow C (2003) Elevated gene expression levels distinguish human from non-human primate brains. Proc Natl Acad Sci USA 100:13030–13035

Campanero-Rhodes MA, Smith A, Chai W, Sonnino S, Mauri L, Childs RA, Zhang Y, Ewers H, Helenius A, Imberty A, Feizi T (2007) N-Glycolyl GM1 ganglioside as a receptor for simian virus 40. J Virol 81:12846–12858

Camus D, Hadley TJ (1985) A *Plasmodium falciparum* antigen that binds to host erythrocytes and merozoites. Science 230:553–556

Cao H, Lakner U, de Bono B, Traherne JA, Trowsdale J, Barrow AD (2008) SIGLEC16 encodes a DAP12-associated receptor expressed in macrophages that evolved from its inhibitory counterpart SIGLEC11 and has functional and non-functional alleles in humans. Eur J Immunol 38:2303–2315

Cao H, de Bono B, Belov K, Wong ES, Trowsdale J, Barrow AD (2009) Comparative genomics indicates the mammalian CD33r Siglec locus evolved by an ancient large-scale inverse duplication and suggests all Siglecs share a common ancestral region. Immunogenetics 61:401–417

Carlin AF, Chang YC, Areschoug T, Lindahl G, Hurtado-Ziola N, King CC, Varki A, Nizet V (2009a) Group B *Streptococcus* suppression of phagocyte functions by protein-mediated engagement of human Siglec-5. J Exp Med 206:1691–1699

Carlin AF, Uchiyama S, Chang YC, Lewis AL, Nizet V, Varki A (2009b) Molecular mimicry of host sialylated glycans allows a bacterial pathogen to engage neutrophil Siglec-9 and dampen the innate immune response. Blood 113:3333–3336

Carroll SB (2003) Genetics and the making of *Homo sapiens*. Nature (Lond) 422:849–857

Chou HH, Takematsu H, Diaz S, Iber J, Nickerson E, Wright KL, Muchmore EA, Nelson DL, Warren ST, Varki A (1998) A mutation in human CMP-sialic acid hydroxylase occurred after the *Homo–Pan* divergence. Proc Natl Acad Sci USA 95:11751–11756

Chou HH, Hayakawa T, Diaz S, Krings M, Indriati E, Leakey M, Paabo S, Satta Y, Takahata N, Varki A (2002) Inactivation of CMP-N-acetylneuraminic acid hydroxylase occurred prior to brain expansion during human evolution. Proc Natl Acad Sci USA 99:11736–11741

Crocker PR (2005) Siglecs in innate immunity. Curr Opin Pharmacol 5:431–437

Crocker PR, Varki A (2001) Siglecs, sialic acids and innate immunity. Trends Immunol 22:337–342

Crocker PR, Paulson JC, Varki A (2007) Siglecs and their roles in the immune system. Nat Rev Immunol 7:255–266

Enard W, Khaitovich P, Klose J, Zollner S, Heissig F, Giavalisco P, Nieselt-Struwe K, Muchmore E, Varki A, Ravid R, Doxiadis GM, Bontrop RE, Paabo S (2002) Intra- and interspecific variation in primate gene expression patterns. Science 296:340–343

Gagneux P, Cheriyan M, Hurtado-Ziola N, van der Linden EC, Anderson D, McClure H, Varki A, Varki NM (2003) Human-specific regulation of alpha 2-6-linked sialic acids. J Biol Chem 278:48245–48250

Gilad Y, Oshlack A, Smyth GK, Speed TP, White KP (2006) Expression profiling in primates reveals a rapid evolution of human transcription factors. Nature (Lond) 440:242–245

Gottschalk A (1960) The chemistry and biology of sialic acids and related substance. Cambridge University Press, Cambridge

Hamilton WD, Axelrod R, Tanese R (1990) Sexual reproduction as an adaptation to resist parasites (a review). Proc Natl Acad Sci USA 87:3566–3573

Hartnell A, Steel J, Turley H, Jones M, Jackson DG, Crocker PR (2001) Characterization of human sialoadhesin, a sialic acid binding receptor expressed by resident and inflammatory macrophage populations. Blood 97:288–296

Hayakawa T, Satta Y, Gagneux P, Varki A, Takahata N (2001) Alu-mediated inactivation of the human CMP-N-acetylneuraminic acid hydroxylase gene. Proc Natl Acad Sci USA 98:11399–11404

Hayakawa T, Angata T, Lewis AL, Mikkelsen TS, Varki NM, Varki A (2005) A human-specific gene in microglia. Science 309:1693

Hayakawa T, Aki I, Varki A, Satta Y, Takahata N (2006) Fixation of the human-specific CMP-N-acetylneuraminic acid hydroxylase pseudogene and implications of haplotype diversity for human evolution. Genetics 172:1139–1146

Hedlund M, Tangvoranuntakul P, Takematsu H, Long JM, Housley GD, Kozutsumi Y, Suzuki A, Wynshaw-Boris A, Ryan AF, Gallo RL, Varki N, Varki A (2007) N-Glycolylneuraminic acid deficiency in mice: implications for human biology and evolution. Mol Cell Biol 27:4340–4346

International Human Genome Sequencing Consortium (2001) Initial sequencing and analysis of the human genome. Nature (Lond) 409:860–921

International Human Genome Sequencing Consortium (2004) Finishing the euchromatic sequence of the human genome. Nature (Lond) 431:931–945

Irie A, Koyama S, Kozutsumi Y, Kawasaki T, Suzuki A (1998) The molecular basis for the absence of N-glycolylneuraminic acid in humans. J Biol Chem 273:15866–15871

Jones C, Virji M, Crocker PR (2003) Recognition of sialylated meningococcal lipopolysaccharide by siglecs expressed on myeloid cells leads to enhanced bacterial uptake. Mol Microbiol 49:1213–1225

Kawano T, Kozutsumi Y, Kawasaki T, Suzuki A (1994) Biosynthesis of N-glycolylneuraminic acid-containing glycoconjugates. Purification and characterization of the key enzyme of the cytidine monophospho-N-acetylneuraminic acid hydroxylation system. J Biol Chem 269:9024–9029

Kawano T, Koyama S, Takematsu H, Kozutsumi Y, Kawasaki H, Kawashima S, Kawasaki T, Suzuki A (1995) Molecular cloning of cytidine monophospho-N-acetylneuraminic acid hydroxylase. Regulation of species- and tissue-specific expression of N-glycolylneuraminic acid. J Biol Chem 270:16458–16463

Keeling MR, Roberts JR (1972) The chimpanzee. In: Bourne GH (ed) Histology, reproduction and restraint, vol 5. Karger, New York, pp 143–150

Khaitovich P, Muetzel B, She X, Lachmann M, Hellmann I, Dietzsch J, Steigele S, Do HH, Weiss G, Enard W, Heissig F, Arendt T, Nieselt-Struwe K, Eichler EE, Paabo S (2004a) Regional patterns of gene expression in human and chimpanzee brains. Genome Res 14:1462–1473

Khaitovich P, Weiss G, Lachmann M, Hellmann I, Enard W, Muetzel B, Wirkner U, Ansorge W, Paabo S (2004b) A neutral model of transcriptome evolution. PLoS Biol 2:E132

Khaitovich P, Hellmann I, Enard W, Nowick K, Leinweber M, Franz H, Weiss G, Lachmann M, Paabo S (2005) Parallel patterns of evolution in the genomes and transcriptomes of humans and chimpanzees. Science 309:1850–1854

Khaitovich P, Enard W, Lachmann M, Paabo S (2006) Evolution of primate gene expression. Nat Rev Genet 7:693–702

Klein RG (1999) The human career: human biology and cultural origins. The University of Chicago, Chicago

Lovejoy CO (2005) The natural history of human gait and posture. Part 1. Spine and pelvis. Gait Posture 21:95–112

Martin LT, Marth JD, Varki A, Varki NM (2002) Genetically altered mice with different sialyltransferase deficiencies show tissue-specific alterations in sialylation and sialic acid 9-*O*-acetylation. J Biol Chem 277:32930–32938

Martin MJ, Rayner JC, Gagneux P, Barnwell JW, Varki A (2005) Evolution of human–chimpanzee differences in malaria susceptibility: relationship to human genetic loss of *N*-glycolylneuraminic acid. Proc Natl Acad Sci USA 102:12819–12824

McHenry HM (1994) Tempo and mode in human evolution. Proc Natl Acad Sci USA 91:6780–6786

Mikami T, Kashiwagi M, Tsuchihashi K, Daino T, Akino T, Gasa S (1998) Further characterization of equine brain gangliosides: the presence of GM3 having *N*-glycolyl neuraminic acid in the central nervous system. J Biochem 123:487–491

Monteiro VG, Lobato CS, Silva AR, Medina DV, de Oliveira MA, Seabra SH, de Souza W, DaMatta RA (2005) Increased association of *Trypanosoma cruzi* with sialoadhesin-positive mice macrophages. Parasitol Res 97:380–385

Mounzih K, Qiu J, Ewart-Toland A, Chehab FF (1998) Leptin is not necessary for gestation and parturition but regulates maternal nutrition via a leptin resistance state. Endocrinology 139:5259–5262

Muchmore EA, Diaz S, Varki A (1998) A structural difference between the cell surfaces of humans and the great apes. Am J Phys Anthropol 107:187–198

Murphy BR, Hall SL, Crowe J, Collins P, Subbarao K, Connors M, London WT, Chanock R (1992) In: Corwin J, Landon JC (eds) Chimpanzee conservation and public health. Diagnon/Bioqual, Rockville, MD, pp 21–27

Nakao T, Kon K, Ando S, Hirabayashi Y (1991) A NeuGc-containing trisialoganglioside of bovine brain. Biochim Biophys Acta 1086:305–309

Nguyen DH, Hurtado-Ziola N, Gagneux P, Varki A (2006) Loss of Siglec expression on T lymphocytes during human evolution. Proc Natl Acad Sci USA 103:7765–7770

Nicoll G, Ni J, Liu D, Klenerman P, Munday J, Dubock S, Mattei MG, Crocker PR (1999) Identification and characterization of a novel siglec, siglec-7, expressed by human natural killer cells and monocytes. J Biol Chem 274:34089–34095

Novembre FJ, Saucier M, Anderson DC, Klumpp SA, O'Neil SP, Brown CR 2nd, Hart CE, Guenthner PC, Swenson RB, McClure HM (1997) Development of AIDS in a chimpanzee infected with human immunodeficiency virus type 1. J Virol 71:4086–4091

Nuamah MA, Yura S, Sagawa N, Itoh H, Mise H, Korita D, Kakui K, Takemura M, Ogawa Y, Nakao K, Fujii S (2004) Significant increase in maternal plasma leptin concentration in induced delivery: a possible contribution of pro-inflammatory cytokines to placental leptin secretion. Endocr J 51:177–187

Olson MV, Varki A (2003) Sequencing the chimpanzee genome: insights into human evolution and disease. Nat Rev Genet 4:20–28

Patel N, Brinkman-Van der Linden EC, Altmann SW, Gish K, Balasubramanian S, Timans JC, Peterson D, Bell MP, Bazan JF, Varki A, Kastelein RA (1999) OB-BP1/Siglec-6: a leptin- and sialic acid-binding protein of the immunoglobulin superfamily. J Biol Chem 274:22729–22738

Paton AW, Srimanote P, Talbot UM, Wang H, Paton JC (2004) A new family of potent AB(5) cytotoxins produced by Shiga toxigenic *Escherichia coli*. J Exp Med 200:35–46

Preuss TM, Caceres M, Oldham MC, Geschwind DH (2004) Human brain evolution: insights from microarrays. Nat Rev Genet 5:850–860

Rhesus Macaque Genome Sequencing and Analysis Consortium (2007) Evolutionary and biomedical insights from the rhesus macaque genome. Science 316:222–234

Rich SM, Leendertz FH, Xu G, Lebreton M, Djoko CF, Aminake MN, Takang EE, Diffo JL, Pike BL, Rosenthal BM, Formenty P, Boesch C, Ayala FJ, Wolfe ND (2009) The origin of malignant malaria. Proc Natl Acad Sci USA 106(35):14902–14907

Rogers GN, Paulson JC (1983) Receptor determinants of human and animal influenza virus isolates: differences in receptor specificity of the H3 hemagglutinin based on species of origin. Virology 127:361–373

Rosenberg A, Schengrund C (1976) Biological roles of sialic acids. Plenum, New York and London

Salminen A, Kaarniranta K (2009) Siglec receptors and hiding plaques in Alzheimer's disease. J Mol Med 87:697–701

Sasaki E, Suemizu H, Shimada A, Hanazawa K, Oiwa R, Kamioka M, Tomioka I, Sotomaru Y, Hirakawa R, Eto T, Shiozawa S, Maeda T, Ito M, Ito R, Kito C, Yagihashi C, Kawai K, Miyoshi H, Tanioka Y, Tamaoki N, Habu S, Okano H, Nomura T (2009) Generation of transgenic non-human primates with germline transmission. Nature (Lond) 459:523–527

Schauer R (1982) Sialic acids: chemistry, metabolism and function. Springer, New York

Schauer R (2000) Achievements and challenges of sialic acid research. Glycoconj J 17:485–499

Schlenzka W, Shaw L, Kelm S, Schmidt CL, Bill E, Trautwein AX, Lottspeich F, Schauer R (1996) CMP-*N*-acetylneuraminic acid hydroxylase: the first cytosolic Rieske iron–sulphur protein to be described in *Eukarya*. FEBS Lett 385:197–200

Schwarzkopf M, Knobeloch KP, Rohde E, Hinderlich S, Wiechens N, Lucka L, Horak I, Reutter W, Horstkorte R (2002) Sialylation is essential for early development in mice. Proc Natl Acad Sci USA 99:5267–5270

Smit H, Gaastra W, Kamerling JP, Vliegenthart JF, de Graaf FK (1984) Isolation and structural characterization of the equine erythrocyte receptor for enterotoxigenic *Escherichia coli* K99 fimbrial adhesin. Infect Immun 46:578–584

Snyder MH, London WT, Tierney EL, Maassab HF, Murphy BR (1986) Restricted replication of a cold-adapted reassortant influenza A virus in the lower respiratory tract of chimpanzees. J Infect Dis 154:370–371

Somel M, Franz H, Yan Z, Lorenc A, Guo S, Giger T, Kelso J, Nickel B, Dannemann M, Bahn S, Webster MJ, Weickert CS, Lachmann M, Paabo S, Khaitovich P (2009) Transcriptional neoteny in the human brain. Proc Natl Acad Sci USA 106:5743–5748

Sonnenburg JL, Altheide TK, Varki A (2004) A uniquely human consequence of domain-specific functional adaptation in a sialic acid-binding receptor. Glycobiology 14:339–346

Subbarao K, Webster RG, Kawaoka Y, Murphy BR (1995) Are there alternative avian influenza viruses for generation of stable attenuated avian-human influenza A reassortant viruses? Virus Res 39:105–118

Tettamanti G, Bertona L, Berra B, Zambotti V (1965) Glycolyl-neuraminic acid in ox brain gangliosides. Nature (Lond) 206:192

The Chimpanzee Sequencing and Analysis Consortium (2005) Initial sequence of the chimpanzee genome and comparison with the human genome. Nature (Lond) 437:69–87

Tomita M, Furthmayr H, Marchesi VT (1978) Primary structure of human erythrocyte glycophorin A. Isolation and characterization of peptides and complete amino acid sequence. Biochemistry 17:4756–4770

Traving C, Schauer R (1998) Structure, function and metabolism of sialic acids. Cell Mol Life Sci 54:1330–1349

Van Valen L (1974) Two modes of evolution. Nature (Lond) 252:298–300

Varki A (2001) Loss of *N*-glycolylneuraminic acid in humans: mechanisms, consequences, and implications for hominid evolution. Am J Phys Anthropol Suppl 33:54–69

Varki A (2007) Glycan-based interactions involving vertebrate sialic-acid-recognizing proteins. Nature (Lond) 446:1023–1029

Varki A (2009) Multiple changes in sialic acid biology during human evolution. Glycoconj J 26:231–245

Varki A, Altheide TK (2005) Comparing the human and chimpanzee genomes: searching for needles in a haystack. Genome Res 15:1746–1758

Varki A, Angata T (2006) Siglecs: the major subfamily of I-type lectins. Glycobiology 16:1R–27R

Varki A, Gagneux P (2009) Human-specific evolution of sialic acid targets: explaining the malignant malaria mystery? Proc Natl Acad Sci USA 106:14739–14740

Vimr ER, Kalivoda KA, Deszo EL, Steenbergen SM (2004) Diversity of microbial sialic acid metabolism. Microbiol Mol Biol Rev 68:132–153

Webster RG, Bean WJ, Gorman OT, Chambers TM, Kawaoka Y (1992) Evolution and ecology of influenza A viruses. Microbiol Rev 56:152–179

Weinstein J, Lee EU, McEntee K, Lai PH, Paulson JC (1987) Primary structure of beta-galactoside alpha 2,6-sialyltransferase. Conversion of membrane-bound enzyme to soluble forms by cleavage of the NH$_2$-terminal signal anchor. J Biol Chem 262:17735–17743

White TD, Asfaw B, DeGusta D, Gilbert H, Richards GD, Suwa G, Howell FC (2003) Pleistocene *Homo sapiens* from Middle Awash, Ethiopia. Nature (Lond) 423:742–747

Winn VD, Gormley M, Paquet AC, Kjaer-Sorensen K, Kramer A, Rumer KK, Haimov-Kochman R, Yeh RF, Overgaard MT, Varki A, Oxvig C, Fisher SJ (2009) Severe preeclampsia-related changes in gene expression at the maternal–fetal interface include sialic acid-binding immunoglobulin-like lectin-6 and pappalysin-2. Endocrinology 150:452–462

Wishcamper CA, Coffin JD, Lurie DI (2001) Lack of the protein tyrosine phosphatase SHP-1 results in decreased numbers of glia within the motheaten (me/me) mouse brain. J Comp Neurol 441:118–133

Wood B (2002) Hominid revelations from Chad. Nature (Lond) 418:133–135

Wood B, Collard M (1999) The human genus. Science 284:65–71

Yamanaka M, Kato Y, Angata T, Narimatsu H (2009) Deletion polymorphism of SIGLEC14 and its functional implications. Glycobiology 19:841–846

Yousef GM, Ordon MH, Foussias G, Diamandis EP (2002) Genomic organization of the siglec gene locus on chromosome 19q13.4 and cloning of two new siglec pseudogenes. Gene (Amst) 286:259–270

Zhang JQ, Nicoll G, Jones C, Crocker PR (2000) Siglec-9, a novel sialic acid binding member of the immunoglobulin superfamily expressed broadly on human blood leukocytes. J Biol Chem 275:22121–22126

Chapter 9
Duplicated Gene Evolution of the Primate Alcohol Dehydrogenase Family

Hiroki Oota and Kenneth K. Kidd

Abbreviations

ADH   Alcohol dehydrogenase
ALDH   Aldehyde dehydrogenase
BP   Before present
cDNA   Complementary DNA
kb   Kilobase pairs
K_m Inverse of enzyme affinity
mRNA   Messenger RNA
My   Million years
NWM   New World monkey
OWM   Old World monkey
PCR   Polymerase chain reaction
WGS   Whole genome sequencing

9.1 Alcohol Metabolism in Humans

Alcohol metabolism and the genes coding enzymes for ethanol digestion have been well studied in many species from bacteria through vertebrates, in part because of the medical consequences of excessive alcohol consumption by humans (Barth and Kunkel 1979; Canestro et al. 2000; Chen et al. 2009; Fischer and Maniatis 1985;

H. Oota (✉)
Graduate School of Frontier Sciences, The University of Tokyo, Kashiwa, Chiba, Japan

Present address: Kitasato University School of Medicine, Sagamihara, Kanagawa, Japan
e-mail: hiroki_oota@med.kitasato-u.ac.jp

K.K. Kidd
Department of Genetics, Yale University School of Medicine, New Haven, CT, USA

H. Hirai et al. (eds.), *Post-Genome Biology of Primates*, Primatology Monographs,
DOI 10.1007/978-4-431-54011-3_9, © Springer 2012

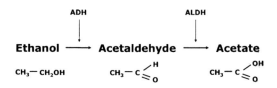

Fig. 9.1 The steps in ethanol metabolism

Guagliardi et al. 1996; Han et al. 2005; Martinez et al. 1996; Oota et al. 2004, 2007; Osier et al. 2002b; Reimers et al. 2004). Following alcohol ingestion, ethanol is first digested to acetaldehyde by alcohol dehydrogenase (ADH), and secondarily acetaldehyde is digested to acetate by aldehyde dehydrogenase (ALDH) 2, primarily in the liver (Fig. 9.1). The intermediate, acetaldehyde, is pharmacologically poisonous, and the accumulation of acetaldehyde causes aversive reactions such as facial flush, palpitations, and nausea. These unpleasant symptoms are thought to be protective against developing alcoholism (Lu et al. 2005; Peng et al. 1999, 2002; Slutske et al. 1995). It is well known that there exist polymorphisms changing the functions of ADH and ALDH2 enzymes, and the alleles of higher ADH activity and deficient ALDH2 have been found at high frequencies in East Asians (Harada et al. 1999; Li et al. 2008, 2009; Lu et al. 2005; Oota et al. 2004; Osier et al. 1999, 2002a; Sherman et al. 1994; Yoshida 1990, 1994). A few previous studies using global human population samples have shown that positive selection might have operated at these loci or at close genomic regions, although the selective pressure is unclear (Han et al. 2007; Oota et al. 2004; Li et al. 2008).

9.2 ADH Gene Cluster in Vertebrates

The vertebrate *ADH* genes are classified into five classes that tandemly cluster on the genome across more than 350 kb (Kent et al. 2002; Lander et al. 2001; Venter et al. 2001). The *ADH* clusters of human and mouse are located on chromosomes 4 and 3, respectively (Ceci et al. 1987; Szalai et al. 2002; Waterston et al. 2002). In humans, the *ADH* gene family contains seven genes: three of these genes (*ADH1A, -1B, -1C*, coding alpha, beta, and gamma subunits, respectively), called Class I *ADH*, produce the most active enzymes in ethanol digestion (Table 9.1) (Duester et al. 1986; Hur and Edenberg 1992; Matsuo and Yokoyama 1989; Satre et al. 1994; von Bahr-Lindstrom et al. 1991; Yasunami et al. 1991; Yokoyama et al. 1992). Each subunit forms homo- and heterodimers that have different K_m values and can best digest various concentrations of ethanol (Eklund et al. 1976a, b; Hoog et al. 2001; von Bahr-Lindstrom et al. 1986). The human Class I genes are expressed mainly in liver and differentially in kidney, stomach, small intestine, and skin. The mouse has only one Class I gene, which shows less tissue-specific expression (Table 9.1) (Engeland and Maret 1993). The human Class II and IV genes are expressed in liver and stomach, respectively, suggesting those enzymes may support digesting ethanol at a lower level than Class I enzymes. The expression patterns of mouse are similar to, but generally less tissue specific than, those of the human. The Class III gene is

9 Duplicated Gene Evolution of the Primate Alcohol Dehydrogenase Family 151

Table 9.1 Main substrates and expressed tissues of AHD enzymes in human and mouse

	Human			Mouse	
Class	Gene	Enzyme subunit (main substrate)	Expression	Gene	Expression
I	*ADH1A* *ADH1B* *ADH1C*	α (Ethanol) β (Ethanol) γ (Ethanol)	Liver, kidney, stomach, small intestine, skin	*Adh1*	Liver, adrenal, small intestine at high level, kidney at low level, and detectable in ovary, uterus, seminal vesicle
II	*ADH4*	π (Ethanol lower than Class I)	Liver	*Adh2*	Liver and less in kidney
III	*ADH5*	χ (Formaldehyde)	All tissues including red blood cells	*Adh3*	All tissues
IV	*ADH7*	σ (Retinol)	Stomach	*Adh4*	Stomach, esophagus, skin, and ovary, uterus, seminal vesicle at low level
V	*ADH6*	Not isolated	mRNAs are detected in stomach, liver	*Adh5a* *Adh5b* *Adh5ps*	Unknown

expressed in all tissues in both human and mouse. For the mouse, the entire *ADH* cluster was sequenced and the Class V genes were found (Fig. 9.2a) (Szalai et al. 2002). The mouse has multiple Class V genes whereas the human has multiple Class I genes. Class V mRNA is detected in stomach and liver in the human, although the expression pattern is unknown in the mouse. The *ADH* genes from the same class have very high similarity (>80%) in amino acid sequence but extremely high differences (<60%) in the nucleotide sequences of the introns between human and mouse. The similarity of amino acid sequences in each class is conservative throughout vertebrates. The similarity of the genes from the different classes is also high (60–70%) in the same species. In addition, the exon–intron structure of each gene is quite similar, although the lengths of the introns differ slightly from one another (Fig. 9.2b). Overall, the *ADH* genes are evolutionarily conserved, suggesting the importance of these ADH enzymes in vertebrates.

The conservativeness of the *ADH* genes also suggests that the gene family has evolved by multiple gene duplications (Oota et al. 2007). Figure 9.3 is the phylogenetic tree of vertebrate ADH based on amino acid sequences, derived from the international database, Genebank/EMBL/DDBJ. The sequence from *Escherichia coli* plasmid ADH (Hayashi et al. 2001) is added into the trees as an outgroup. Only one *ADH* gene is found in amphioxus and fish (Canestro et al. 2000; Funkenstein and Jakowlew 1996), which come to the very root of the tree together with human, horse, and mouse Class III, suggesting the Class III *ADH* gene is the ancestral type of the gene family and that the others have evolved by gene duplication events from the Class III *ADH* gene. Because no other *ADH* gene has been found in amphioxus

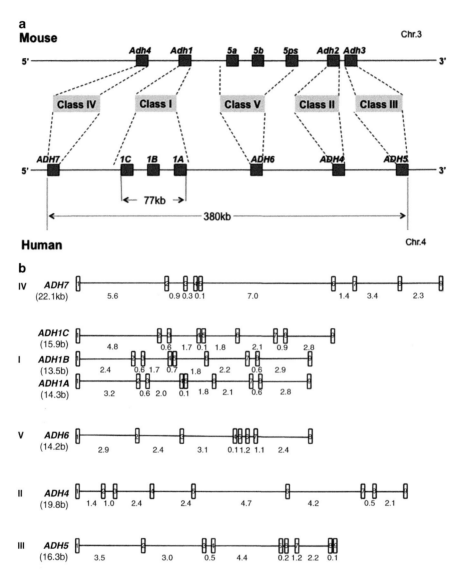

Fig. 9.2 (**a**) The *ADH* gene family cluster on chromosomes (*Chr.*) 3 and 4 for mouse and human, respectively. (**b**) The exon–intron structure of human *ADH* genes. For historical reasons the naming of the human genes does not correspond to the enzyme class nomenclature. Many species are now following the human nomenclature for the genes; therefore, one cannot extrapolate from enzyme class to gene name and vice versa

and fish so far, it is possible that the duplication events occurred as early as the tetrapod ancestor. The phylogenetic analyses based on amino acid and cDNA sequences give an insight into how the *ADH* gene family evolved by gene duplication events. Following the initial duplication event, the Class III gene evolved separately and did not duplicate again. The next two duplications gave rise to Class II and then

9 Duplicated Gene Evolution of the Primate Alcohol Dehydrogenase Family   153

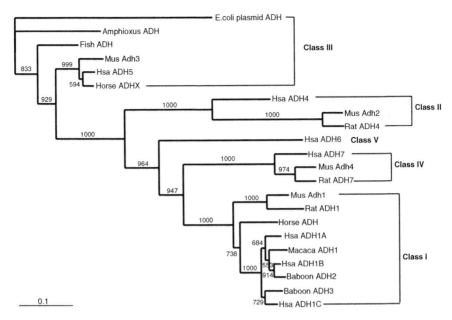

Fig. 9.3 Neighbor-joining (NJ) tree (Saitou and Nei 1987) of the *ADH* genes based on amino acid sequences from vertebrates. Abbreviations (accession numbers or CCDS ID) are the following: *E. coli* plasmid ADH (NP-308438), *Escherichia coli*; Amphioxus (AF154331), *Branchiostoma floridae* (lancelets); Fish (U84791), *Sparus aurata* (bream); Horse (M64865 for Class I ADH and P19854 for ADHX), *Equus caballus*; Mus (CCDS17867 for Adh1, 38653 for Adh2, 17868 for Adh3, 38651 for Adh4), *Mus musculus* (mouse); Rat (BC062403 for ADH1, BC127504 for ADH4, P41682 for ADH7), *Rattus norvegicus*; Hsa (P07327 for ADH1A, P00325 for ADH1B, P00326 for ADH1C, M15943 for ADH4, M29872 for ADH5, AK092768 for ADH6, X76342 for ADH7), *Homo sapiens* (human); Macaca (M81807 for ADH1), *Macaca mulatta* (rhesus monkey); Baboon (M25035 for ADH2, L30113 for ADH3), *Papio hamadryas*. Numbers on the branches represent the bootstrap values (1,000 resampling)

Class V genes. The final duplication separated Class IV and Class I (Fig. 9.3). The synteny of the *ADH* gene cluster in human and mouse (Fig. 9.2a) clearly supports these duplications occurring before the mammalian radiation. Data on the nonmammalian tetrapods will help determine exactly when in evolution these duplications occurred.

9.3 Puzzle About the Number of the ADH Genes

The whole genome sequencing (WGS) projects should give the complete picture of the duplication history of the *ADH* gene family. Using the human *ADH* gene sequences as queries, we looked for the orthologous and paralogous *ADH* genes in ten non-human primates and nine non-primate mammals (Table 9.2). We downloaded assemblies of shotgun sequences across the genome regions that should contain at

Table 9.2 The Number of the ADH Genes Estimated from the Database

Species		Chromosome	Class IV	I	V	II	III	Reference
Primates								
Apes	Human	4	1	3	1	1	1 §	WGS
	Chimpanzee	4	1	3	1	1	1	WGS
	Bonobo		?	3 ¶	?	?	?	Oota et al. (2007)
	Gorilla	3	1	3 ¶	1	?	?	Oota et al. (2007); WGS
	Orangutan	4	1	3 ¶	1	1	1	Oota et al. (2007); WGS
OWMs	Baboon		?	3 ¶	?	?	?	Oota et al. (2007)
	Macaque	5	1	5	1	1	1	WGS
NWMs	Tarsier		1	?	1	1	?	WGS
	Marmoset		1	?	1	1	1	WGS
Prosimians	Mouse lemur		1	?	1	1	1	WGS
	Bushbaby		1	?	?	1	1	WGS
Non-primate mammals								
	Tree shrew		1	?	1	1	1	WGS
	Mouse	3	1	1	3 ¿	1 §	1	Szalai et al. (2002); WGS
	Rat	2	1	1	3	1	1	WGS; Holmes (2009)
	Rabbit		0	2	1	2	1	Yasunami et al. (1990)
	Horse		?	2	?	?	1	Yasunami et al. (1990)
	Sheep		?	2	?	?	?	Yasunami et al. (1990)
	Megabat		1	?	?	1	1	WGS
	Microbat		1	?	1	1	1	WGS
	Opposum	5	3	1	1	2	1	WGS; Holmes (2009)

WGS, the whole genome sequencing project in September 2009

?, unclear numbers in both WGS and the literature

§, one pseudogene exists on the other chromosome

¶, at least three genes exist, based on PCR-direct sequencing analysis by Oota et al. (2007);

¿, it is known that one of the genes is a pseudogene

least one gene and tried to count the number of open reading frames. However, we found it difficult for many species because there are still many gaps in the assemblies, especially in the Class I region. We assume that the Class I genes are quite similar to one another and the shotgun sequences cannot therefore be assembled accurately. Some Southern hybridization data from a previous study (Yasunami et al. 1990) are available, but it is almost impossible to count the exact number of the Class I genes by such nonsequencing methods. Additional sequencing data, ideally using the clone-by-clone method, are required in the near future to obtain a complete assembly of sequences covering the entire *ADH* cluster. Whenever WGS data gave unclear

9 Duplicated Gene Evolution of the Primate Alcohol Dehydrogenase Family 155

assemblies or contained large gaps, we searched for literature about the species and used the published data on the sequence of this particular genomic region.

The number of Class I *ADH* genes varies among species, but the numbers of the other classes are relatively constant across species in both primates and non-primate mammals (Table 9.2). All apes have at least three Class I *ADH* genes. Oota et al. (2007) found three Class I genes in the baboon by polymerase chain reaction (PCR) direct sequencing, but based on our in silico research in this review using the WGS data, the rhesus macaque has five Class I *ADH* genes. As another Old World monkey (OWM), the baboon might also have five or more Class I genes. For OWMs, New World monkeys (NWMs), and prosimians, overall the assemblies of WGS are, unfortunately, not yet of sufficiently high quality. Marmoset WGS is of relatively high quality but has no distinguishable annotation of the Class I genes. The question mark in Table 9.2 represents uncertain numbers, even searching both WGS and the literature (Holmes 2009; Oota et al. 2007; Szalai et al. 2002; Yasunami et al. 1990). Under the assumption that multiple similar genes make shotgun sequences difficult to assemble in the genomic region, the question marks imply that more than one *ADH* gene exists. If this is true, NWMs, prosimians, and several non-primates also have multiple (and perhaps very similar to each other) Class I *ADH* genes.

Now we have two questions about the primate *ADH* gene cluster: When did gene duplication occur in Class I *ADH*? Why do such similar genes exist? There are three possibilities for the timing of the duplication events. One is that the Class I gene duplications occurred independently after the divergence between primates and the other mammalian lineages. Another is that the duplications occurred in the common ancestor of mammals and some non-primate lineages lost copies. A third is that successive duplications occurred during mammalian radiation leaving some lineages with one, others with two, and the primates with three. A previous study argued that gene conversion has homogenized three Class I *ADH* genes, based on the exon sequence data determined in OWMs (rhesus macaque and baboon) (Cheung et al. 1999). Oota et al. (2007) determined the intron sequences of Class I *ADH* genes in great apes (chimpanzee, bonobo, gorilla, and orangutan) and OWM (baboon), and showed no evidence of gene conversion(s) by the maximum likelihood tree constructed using the intron nucleotide sequences (Fig. 9.4). The estimated divergence times assuming the linear evolutionary rate and baboon divergence at 25 My before present (BP) are 40–70 My between *ADH1A* and *ADH1B*, and 54–84 My between *ADH1C* and *ADH1A/ADH1B*. Fossil data show that the first primates emerged 66–98 My BP, the first prosimians emerged before 66 My BP, and the divergence time between NWMs and OWMs is 36–55 My BP (Martin 1993). Thus, we hypothesize that the first duplication event probably occurred during the mammalian radiation, and the second duplication event probably occurred around or just before prosimians emerged. Some studies argue the OWM divergence time is at least 10 My older (Martin 1993; Takahata and Satta 1997) (for details of origins and phylogeny of early primates, see Chaps. 16 and 17). When we assume the divergence time of 35 My, the estimates obtained are pushed back to much earlier.

For the second question, one of the possible answers is positive selection that operated on the Class I *ADH* genes. In general, duplicated genes are homogenized by gene conversions (Ohta 1988, 1993, 2000). That the duplicated genes diverge,

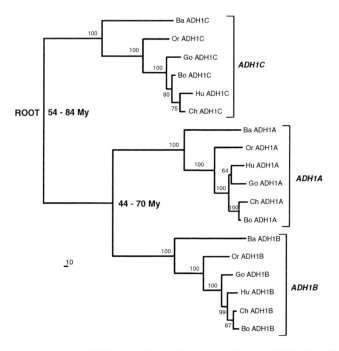

Fig. 9.4 Maximum likelihood (ML) tree (Cavalli-Sforza and Edwards 1967) of the Class I *ADH* genes based on nucleotide sequences of introns 2, 3, and 8 from primates. Hu *Homo sapiens* (human), Bo *Pan paniscus* (bonobo), Ch *Pan troglodytes* (common chimpanzee), Go *Gorilla gorilla* (gorilla), Or *Pongo pygmaeus* (orangutan), Ba *Papio anubis* (olive baboon). *Numbers on the branches* represent the ML bootstrap values (100 resamplings). The detailed rooting method is described in Oota et al. (2007)

escaping from homogenization, positive selection, or neutral fixation must cause accumulation different mutations at a rate that exceeds homogenization by gene conversion (Innan 2003; Teshima and Innan 2004). Oota et al. (2007) found there were inconsistencies in the Class I *ADH* genes: the analysis based on the intron sequences of *ADH* in apes shows no evidence for gene conversions but clear evidence for negative selection. Namely, the genes from different species cluster with the orthologues in the phylogenetic tree, indicating no gene conversion (Fig. 9.4). The topology of the tree also seemed to be nonneutral because the generally accepted phylogenetic clustering (i.e., *ADH1B*, chimpanzee, and bonobo cluster, and that branch clusters with human; then gorilla and orangutan come outside of the human and chimpanzee/bonobo cluster) was not observed in *ADH1A* and *ADH1C*: human *ADH1A* clusters with gorilla, and bonobo *ADH1C* clusters with the human–chimp branch (Fig. 9.4), suggesting the three Class I genes have evolved independently without concerning each other. However, all the pairwise d_N/d_S (Nei and Gojobori 1986) values were less than 1, suggesting that negative selection has operated on the Class I *ADH* genes, when comparing the exon sequences from human, chimpanzee, rhesus, and mouse (Oota et al. 2007). This is another puzzle in the evolution of this duplicated gene family.

9.4 Frugivory Hypothesis for Class I Evolution

A hint of the answer might be taken in the d_N/d_S tests on the catalytic and the coenzyme-binding domains of *ADH* genes. Overall, lower rates of nonsynonymous substitution were observed in the coenzyme binding domain than within the catalytic domain (Oota et al. 2007). The highly conserved sequences in the coenzyme-binding domains suggest that the selection for heterodimer stability and the acquisition of regulatory elements determining the organ-specific expression levels could be keys to evolution of the Class I *ADH* genes. All the human Class I ADH enzymes are expressed in liver and various organs (Table 9.1) (Engeland and Maret 1993). The Class I ADH subunits combine randomly and form homo- and heterodimers; the different combinations have different ethanol catalytic efficiencies (von Bahr-Lindstrom et al. 1986). To metabolize various concentrations of ethanol in different organs, various combinations of the homo- and heterodimers of the Class I subunits would be necessary, and the stabilities of the heterodimers would be very important in any subunit combinations. The highly conserved binding domain is probably maintained by such functional constraints.

The duplication of the Class I *ADH* genes may be associated with dietary adaptation of primates. Some current leaf-eaters (i.e., *Colobus*) ferment leaves in their foregut (Kay and Davies 1994). In the process of fermentation, the ADH enzymes must play an important role in digesting any alcohols generated by the fermentation. Furthermore, low-level dietary exposure to ethanol via ingestion of fermenting fruit has probably characterized the predominantly frugivorous anthropoid lineage for about 40 My (Dudelly 2002). Ripe and overripe fruits of the palm contain ethanol within the pulp at concentrations averaging 0.9% and 4.5%, respectively, in the Neotropical region (Dudelly 2004). To detect the location of the ripe fruit, the aroma of ethanol is possibly effective for frugivorous anthropoid lineage (Dudelly 2004) (for details about importance of olfactory and taste receptors in primates, see Chaps. 4 and 6). Figure 9.5 shows the phylogeny of current primates and the ratios of fruit in diet based on the data from Fleagle (1999), imitating Figs. 1 and 3 of Dudelly (2002, 2004), respectively. The smaller primates including prosimians or fossil primates are mostly insectivorous, although some prosimians (i.e., *Microcebus murinus*, *Cheirogaleus medius*, *Galagoides alleni*) are highly frugivorous. On the other hand, most NWMs and OWMs are highly frugivorous, although they eat leaves and insects as well. For apes, the ratios of fruit are quite high (Fleagle 1999). Recent studies of the paleoecological surroundings where *Ardipithecus ramidus* lived reported the habitat may be woodland and *A. ramidus* could have been frugivorous (Louchart et al. 2009; Suwa et al. 2009; White et al. 2009; WoldeGabriel et al. 2009). Thus, current diet ratios and paleoecological data suggest dietary adaptation concerning frugivorous behavior should have been important for primate (especially for anthropoid) evolution.

If our hypothesis is true, the number of Class I *ADH* genes varies depending on the dietary behavior of the species: frugivorous animals and leaf-eaters have more Class I genes, and even if the insectivores have multiple Class I genes as well as the

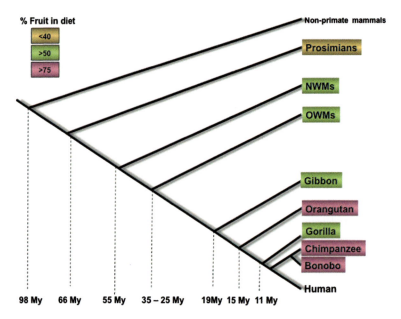

Fig. 9.5 Phylogeny, approximate divergence times, and frugivory for primates and the other mammals. The divergence times and the proportions of fruit in the diets are approximations described in Fleagle (1999). *My* million years, *NWM* New World monkeys, *OWM* Old World monkeys

frugivores and leaf-eaters, the insectivore Class I genes could be homogenized by gene conversion and not divergent as independent genes. Currently, prosimians are not so frugivorous, but there are not enough data to characterize the diet of the ancestor of prosimians that could be close to the ancestor of primates. To prove how and why the *ADH* gene cluster evolved in primates, we need more sequences of the entire *ADH* cluster from many species, especially including prosimians.

Acknowledgments We must thank Drs. H. Hirai and H. Imai for kindly giving us the opportunity to write this book chapter.

References

Barth G, Kunkel W (1979) Alcohol dehydrogenase (ADH) in yeasts. II. NAD+−and NADP+−dependent alcohol dehydrogenases in *Saccharomycopsis lipolytica*. Z Allg Mikrobiol 19:381–390

Canestro C, Hjelmqvist L, Albalat R et al (2000) *Amphioxus* alcohol dehydrogenase is a class 3 form of single type and of structural conservation but with unique developmental expression. Eur J Biochem 267:6511–6518

Cavalli-Sforza LL, Edwards AW (1967) Phylogenetic analysis. Models and estimation procedures. Am J Hum Genet 19:233–257

Ceci JD, Zheng YW, Felder MR (1987) Molecular analysis of mouse alcohol dehydrogenase: nucleotide sequence of the Adh-1 gene and genetic mapping of a related nucleotide sequence to chromosome 3. Gene 59:171–182

Chen YC, Peng GS, Tsao TP et al (2009) Pharmacokinetic and pharmacodynamic basis for over-coming acetaldehyde-induced adverse reaction in Asian alcoholics, heterozygous for the variant ALDH2*2 gene allele. Pharmacogenet Genomics 19:588–599

Cheung B, Holmes RS, Easteal S et al (1999) Evolution of class I alcohol dehydrogenase genes in catarrhine primates: gene conversion, substitution rates, and gene regulation. Mol Biol Evol 16:23–36

Dudelly R (2002) Fermenting fruit and the historical ecology of ethanol ingestion: is alcoholism in modern humans an evolutionary hangover? Addiction 97:381–388

Dudelly R (2004) Ethanol, fruit ripening, and the historical origins of human alcoholism in primate frugivory. Integr Comp Biol 44:315–323

Duester G, Smith M, Bilanchone V et al (1986) Molecular analysis of the human class I alcohol dehydrogenase gene family and nucleotide sequence of the gene encoding the beta subunit. J Biol Chem 261:2027–2033

Eklund H, Branden CI, Jornvall H (1976a) Structural comparisons of mammalian, yeast and bacillar alcohol dehydrogenases. J Mol Biol 102:61–73

Eklund H, Nordstrom B, Zeppezauer E et al (1976b) Three-dimensional structure of horse liver alcohol dehydrogenase at 2–4 resolution. J Mol Biol 102:27–59

Engeland K, Maret W (1993) Extrahepatic, differential expression of four classes of human alcohol dehydrogenase. Biochem Biophys Res Commun 193:47–53

Fischer JA, Maniatis T (1985) Structure and transcription of the *Drosophila mulleri* alcohol dehydrogenase genes. Nucleic Acids Res 13:6899–6917

Fleagle JG (1999) Primate adaptation and evolution. Academic, London

Funkenstein B, Jakowlew SB (1996) Molecular cloning of fish alcohol dehydrogenase cDNA (Amst). Gene 174:159–164

Guagliardi A, Martino M, Iaccarino I et al (1996) Purification and characterization of the alcohol dehydrogenase from a novel strain of *Bacillus stearothermophilus* growing at 70 degrees C. Int J Biochem Cell Biol 28:239–246

Han Y, Oota H, Osier MV et al (2005) Considerable haplotype diversity within the 23 kb encompassing the ADH7 gene. Alcohol Clin Exp Res 29:2091–2100

Han Y, Gu S, Oota H et al (2007) Evidence of positive selection on a class I ADH locus. Am J Hum Genet 80:441–456

Harada S, Okubo T, Nakamura T et al (1999) A novel polymorphism (−357G/A) of the ALDH2 gene: linkage disequilibrium and an association with alcoholism. Alcohol Clin Exp Res 23:958–962

Hayashi T, Makino K, Ohnishi M et al (2001) Complete genome sequence of enterohemorrhagic *Escherichia coli* O157:H7 and genomic comparison with a laboratory strain K-12. DNA Res 8:11–22

Holmes RS (2009) Opossum alcohol dehydrogenases: sequences, structures, phylogeny and evolution. Evidence for the tandem location of ADH genes on opossum chromosome 5. Chem Biol Interact 178:8–15

Hoog JO, Hedberg JJ, Stromberg P et al (2001) Mammalian alcohol dehydrogenase: functional and structural implications. J Biomed Sci 8:71–76

Hur MW, Edenberg HJ (1992) Cloning and characterization of the ADH5 gene encoding human alcohol dehydrogenase 5, formaldehyde dehydrogenase. Gene 121:305–311

Innan H (2003) A two-locus gene conversion model with selection and its application to the human RHCE and RHD genes. Proc Natl Acad Sci USA 100:8793–8798

Kay RNB, Davies AG (1994) Colobinem monkeys: their ecology, behaviour and evolution. In: Oate JF (ed) Davis AG. Cambridge University Press, Cambridge, pp 229–250

Kent WJ, Sugnet CW, Furey TS et al (2002) The human genome browser at UCSC. Genome Res 12:996–1006

Lander ES, Linton LM, Birren B et al (2001) Initial sequencing and analysis of the human genome. Nature (Lond) 409:860–921

Li H, Gu S, Cai X et al (2008) Ethnic related selection for an ADH Class I variant within East Asia. PLoS One 3:e1881

Li H, Borinskaya S, Yoshimura K et al (2009) Refined geographic distribution of the oriental ALDH2*504Lys (nee 487Lys) variant. Ann Hum Genet 73:335–345

Louchart A, Wesselman H, Blumenschine RJ et al (2009) Taphonomic, avian, and small-vertebrate indicators of *Ardipithecus ramidus* habitat. Science 326:66e61–66e64

Lu RB, Ko HC, Lee JF et al (2005) No alcoholism-protection effects of ADH1B*2 allele in antisocial alcoholics among Han Chinese in Taiwan. Alcohol Clin Exp Res 29:2101–2107

Martin RD (1993) Primate origins: plugging the gaps. Nature (Lond) 363:223–234

Martinez MC, Achkor H, Persson B et al (1996) *Arabidopsis* formaldehyde dehydrogenase. Molecular properties of plant class III alcohol dehydrogenase provide further insights into the origins, structure and function of plant class P and liver class I alcohol dehydrogenases. Eur J Biochem 241:849–857

Matsuo Y, Yokoyama S (1989) Molecular structure of the human alcohol dehydrogenase 1 gene. FEBS Lett 243:57–60

Nei M, Gojobori T (1986) Simple methods for estimating the numbers of synonymous and nonsynonymous nucleotide substitutions. Mol Biol Evol 3:418–426

Ohta T (1988) Evolution by gene duplication and compensatory advantageous mutations. Genetics 120:841–847

Ohta T (1993) Pattern of nucleotide substitutions in growth hormone-prolactin gene family: a paradigm for evolution by gene duplication. Genetics 134:1271–1276

Ohta T (2000) Evolution of gene families. Gene (Amst) 259:45–52

Oota H, Pakstis AJ, Bonne-Tamir B et al (2004) The evolution and population genetics of the ALDH2 locus: random genetic drift, selection, and low levels of recombination. Ann Hum Genet 68:93–109

Oota H, Dunn CW, Speed WC et al (2007) Conservative evolution in duplicated genes of the primate Class I ADH cluster. Gene (Amst) 392:64–76

Osier M, Pakstis AJ, Kidd JR et al (1999) Linkage disequilibrium at the ADH2 and ADH3 loci and risk of alcoholism. Am J Hum Genet 64:1147–1157

Osier MV, Pakstis AJ, Goldman D et al (2002a) A proline-threonine substitution in codon 351 of ADH1C is common in Native Americans. Alcohol Clin Exp Res 26:1759–1763

Osier MV, Pakstis AJ, Soodyall H et al (2002b) A global perspective on genetic variation at the ADH genes reveals unusual patterns of linkage disequilibrium and diversity. Am J Hum Genet 71:84–89

Peng GS, Wang MF, Chen CY et al (1999) Involvement of acetaldehyde for full protection against alcoholism by homozygosity of the variant allele of mitochondrial aldehyde dehydrogenase gene in Asians. Pharmacogenetics 9:463–476

Peng GS, Yin JH, Wang MF et al (2002) Alcohol sensitivity in Taiwanese men with different alcohol and aldehyde dehydrogenase genotypes. J Formos Med Assoc 101:769–774

Reimers MJ, Hahn ME, Tanguay RL (2004) Two zebrafish alcohol dehydrogenases share common ancestry with mammalian class I, II, IV, and V alcohol dehydrogenase genes but have distinct functional characteristics. J Biol Chem 279:38303–38312

Saitou N, Nei M (1987) The neighbor-joining method: a new method for reconstructing phylogenetic trees. Mol Biol Evol 4:406–425

Satre MA, Zgombic-Knight M, Duester G (1994) The complete structure of human class IV alcohol dehydrogenase (retinol dehydrogenase) determined from the ADH7 gene. J Biol Chem 269:15606–15612

Sherman DI, Ward RJ, Yoshida A et al (1994) Alcohol and acetaldehyde dehydrogenase gene polymorphism and alcoholism. EXS 71:291–300

Slutske WS, Heath AC, Madden PA et al (1995) Is alcohol-related flushing a protective factor for alcoholism in Caucasians? Alcohol Clin Exp Res 19:582–592

Suwa G, Kono RT, Simpson SW et al (2009) Paleobiological implications of the *Ardipithecus ramidus* dentition. Science 326:94–99

Szalai G, Duester G, Friedman R et al (2002) Organization of six functional mouse alcohol dehydrogenase genes on two overlapping bacterial artificial chromosomes. Eur J Biochem 269:224–232

Takahata N, Satta Y (1997) Evolution of the primate lineage leading to modern humans: phylogenetic and demographic inferences from DNA sequences. Proc Natl Acad Sci USA 94:4811–4815

Teshima KM, Innan H (2004) The effect of gene conversion on the divergence between duplicated genes. Genetics 166:1553–1560

Venter JC, Adams MD, Myers EW et al (2001) The sequence of the human genome. Science 291:1304–1351

von Bahr-Lindstrom H, Hoog JO, Heden LO et al (1986) cDNA and protein structure for the alpha subunit of human liver alcohol dehydrogenase. Biochemistry 25:2465–2470

von Bahr-Lindstrom H, Jornvall H, Hoog JO (1991) Cloning and characterization of the human ADH4 gene. Gene 103:269–274

Waterston RH, Lindblad-Toh K, Birney E et al (2002) Initial sequencing and comparative analysis of the mouse genome. Nature (Lond) 420:520–562

White TD, Ambrose SH, Suwa G et al (2009) Macrovertebrate paleontology and the Pliocene habitat of *Ardipithecus ramidus*. Science 326:87–93

WoldeGabriel G, Ambrose SH, Barboni D et al (2009) The geological, isotopic, botanical, invertebrate, and lower vertebrate surroundings of *Ardipithecus ramidus*. Science 326:65e61–65e65

Yasunami M, Chen CS, Yoshida A (1990) Multiplication of the class I alcohol dehydrogenase locus in mammalian evolution. Biochem Genet 28:591–599

Yasunami M, Chen CS, Yoshida A (1991) A human alcohol dehydrogenase gene (ADH6) encoding an additional class of isozyme. Proc Natl Acad Sci USA 88:7610–7614

Yokoyama S, Matsuo Y, Rajasekharan S et al (1992) Molecular structure of the human alcohol dehydrogenase 3 gene. Jpn J Genet 67:167–171

Yoshida A (1990) Isozymes of human alcohol dehydrogenase and aldehyde dehydrogenase. Prog Clin Biol Res 344:327–340

Yoshida A (1994) Genetic polymorphisms of alcohol metabolizing enzymes related to alcohol sensitivity and alcoholic diseases. Alcohol Alcohol 29:693–696

Chapter 10
Genome Structure and Primate Evolution

Yoko Satta

Abbreviations

AMEL	Amelogenin
AMY1	Amylase 1
CD	Chromodomain
CMAH	CMP-N-acetylneuraminic acid hydroxylase
CNV	Copy number variation
CSP1	Chimpanzee-specific palindrome 1
DAZ	Deleted in azoospermia
ELN	Tropoelastin
HSAY	Human (*Homo sapiens*) Y chromosome
HSF	Heat-shock transcription factor
KAL	Kallman syndrome
LINEs	Long interspersed elements
MHC	Major histocompatibility complex
Myr	Million years
NWMs	New World monkeys
OWMs	Old World monkeys
PAR1	Pseudo-autosomal region 1
PTRY	Chimpanzee (*Pan troglodytes*) Y chromosome
RBM	RNA-binding motif
SINEs	Short interspersed elements
VC	Variable charge
XKR	X Kell blood-related

Y. Satta (✉)
Department of Evolutionary Studies of Biosystems, The Graduate University
for Advanced Studies, Sokendai Hayama, Hayama, Kanagawa 240-0193, Japan
e-mail: satta@soken.ac.jp

H. Hirai et al. (eds.), *Post-Genome Biology of Primates*, Primatology Monographs,
DOI 10.1007/978-4-431-54011-3_10, © Springer 2012

10.1 Structure in the Human Genome

With the preliminary report of the human genome sequence in 2001 (International Human Genome Sequencing Consortium 2001), Li et al. (2001) presented several evolutionary features of the human genome. They showed that approximately 43% of the genome was composed of interspersed repetitive elements of the four major types shown in Table 10.1. Among these, *Alu* (one of the short interspersed elements, or *SINEs*) and *L1* (one of the long interspersed elements, or *LINEs*) are most frequent. Because a repetitive element insertion into a gene usually destroys the gene function or causes a structural change in the gene, insertions are mostly deleterious. Deletion also results in deterioration of genes; for example, *Alu* is involved in an exon deletion of CMP-*N*-acetylneuraminic acid hydroxylase (*CMAH*; Hayakawa et al. 2001) and tropoelastin gene (*ELN*; Szabó et al. 1999). In *CMAH*, the *Alu*-mediated exon deletion causes a frameshift that renders the gene functionless in humans. In *ELN*, *Alu*-mediated recombination caused the independent loss of exons in an ancestor of Catarrhini [Old World monkeys (OWMs) and hominoids] and in humans. Although insertions into genes are not observed frequently, insertion of repetitive elements into protein-coding genes and the translation of these elements as parts of genes were identified. Approximately 5% (more than 2,000 cases) of predicted or hypothetical genes are found to contain some repetitive elements, whereas the average proportion of inserted elements in the genome is approximately 40%. This lower frequency (5%) of insertions in genes suggests a deleterious effect on protein-coding genes. Nevertheless, inserted elements in proteins may play a role in generating new gene functions. Although the function or biological meaning of repetitive elements in host genes is unclear, there are several recent reports of the effect of inserted repetitive elements (or transposons) on the activity of transcriptional regulation of nearby genes (Sinzelle et al. 2009 and references therein).

The Human Genome Sequencing project revealed frequent segmental duplications in the genome. Segmental duplications show nearly identical (>90% of identity) copies of genomic DNA with sizes ranging from 1 to 300 kb. They are present in at least two locations in the human genome, and the duplicates occur within the same chromosome (intrachromosomal duplication) or on different chromosomes (interchromosomal duplication). The average proportion of duplicates is about 5% of the genome, and these duplications occurred during the past 35 million years (Myr), after the divergence of OWMs and hominoids (Samonte and Eichler 2002). In particular, on human chromosome 22, a region with a length of more than 400 kb has shown to be the result of a human-specific duplication of chromosome 14 (Bailey et al. 2002). This human-specific duplication occurred after the divergence

Table 10.1 The content of repetitive elements in the human genome

Types	Percentage in the genome
SINEs (short interspersed elements)	12.5
LINEs (long interspersed elements)	18.9
Elements with long terminal repeats	7.9
DNA transposons	2.7

10 Genome Structure and Primate Evolution 165

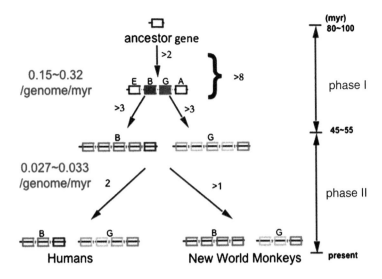

Fig. 10.1 Gene duplication of major histocompatibility complex (*MHC*) Class I genes during primate evolution. The evolutionary process is divided into phases I and II. Phase I corresponds to the time before the divergence of humans and New World Monkeys (NWMs), whereas phase II corresponds to the time after divergence. The duplication rate has been estimated from phylogenetic studies of the NWM and human *MHC* Class I genes. The rate in phase I is approximately ten times faster than that in phase II. *B* and *G*, two groups of *MHC* class I genes

of humans and chimpanzees, with the nucleotide divergence between two paralogues being 0.2–0.6%. This divergence, however, suggests that the possibility of the duplication being polymorphic in the extant human population is very low.

The pattern of shared duplication in primate genomes reveals that the rate of segmental duplication appears to slow down in the human lineage (Marques-Bonet et al. 2009). Based on the number of shared duplicates between different primate species and the divergence time estimation of segmental duplicates, Marques-Bonet et al. (2009) estimated the rates of duplications among ancestors of different branches in hominoid evolution: the ancestor of humans/chimpanzees/gorillas had a relatively high rate (~60 duplicates/Myr/genome), whereas the rate slowed down in the human lineage (~20 duplicates/Myr/genome).

Although the analysis of primate major histocompatibility complex (*MHC*) class I coding regions may not reflect the whole genome (Sawai et al. 2004), it reveals that the gene duplication burst occurred before the divergence of New World monkeys (NWMs) and Catarrhini (OWMs and hominoids) (Fig. 10.1). Similar differences in the gene duplication rate have been observed in the evolution of killer T-cell receptors and scavenger receptor genes (R. Dawkins, personal communication). However, because it is unlikely that the *de novo* duplication rate varies depending on time, the apparent differences in rate might be the result of environmental conditions that allow the presence of duplications.

When segmental duplication occurs in an adjacent region, it creates tandem duplications or inverted repeats. These structures often affect copy number variation

(CNV), not only within a species but also among primate species. A typical example of CNV in humans is variation in the number of green and red color vision genes on the X chromosome. Every X chromosome has a single red gene, but there is polymorphism for the number of green genes. Each X chromosome has one, two, three, or more (up to five) green genes (Drummond-Borg et al. 1989). Some CNV is positively correlated with physiological variation not only within humans (*AMY1*, *CYP2D6*, etc.) but also between humans and chimpanzees (*aquaporin 7*, *NBRF15*, etc.) (Marques-Bonet et al. 2009).

10.2 Evolution of X and Y Chromosome and Genome Structure

The paired sex chromosomes in mammals originated from a pair of autosomes. The pair of autosomes differentiated into X and Y chromosomes after several (at least two) sex-determining genes differentiated on one of the autosomes, the proto-Y chromosome. The stable transmission and segregation of the sexes in the offspring is essential for the sex development in individuals and for the maintenance of species. Because of this, recombination between proto-X and proto-Y chromosomes has been arrested. This halt in recombination allows the accumulation of X- and Y-specific mutations and, consequently, the differentiation of the pair of sex chromosomes has proceeded. As the Y chromosome, in particular, always exists as a hemizygous or haploid, there is no way to remove deleterious mutations from the chromosome, and consequently the chromosome has deteriorated.

Despite of the deterioration of the Y chromosome, a comparison of the nucleotide sequences of the X and Y chromosomes shows 19 pairs of homologous genes, which is consistent with the autosome pair origin. The extent of synonymous divergences (d_s) of these pairs varies greatly, from 5% at *ARSE/ARSEP* to 125% at *SRY/SOX3*. The divergence d_s decreases by position from the tip of the long arm to the tip of the short arm on the X chromosome. Based on this position and the d_s value, 19 pairs were classified into four groups, each containing genes in a block or a segment where recombination stopped simultaneously. Based on this observation, Lahn and Page (1999) proposed the four-strata scenario of mammalian sex chromosome evolution. This scenario places the generation of the youngest stratum, stratum 4 with d_s =about 10%, after the divergence of prosimian and simian primates and before the divergence of NWMs and Catarrhini. The stratum covers more than half of the short arm of the X chromosome.

To examine whether the nucleotide divergence (d) is about 10% within stratum 4, in general, the nucleotide sequence in the non-coding or intergenic regions in stratum 4 was compared with the corresponding region on the Y chromosome (Fig. 10.2; Iwase et al. 2003). The compared sequence in the stratum 4 about 2.7 Mb encompasses from the tip to the middle of intron 2 of amelogenin X (*AMELX*) and the region shows almost 10% of d. However, in the middle of stratum 4, there is a subregion in which divergence is significantly lower (0–5%) than in neighboring regions. The region covers about 100 kb and ranges from Kallman syndrome (*KAL*)

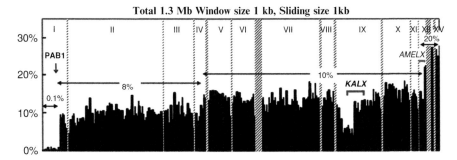

Fig. 10.2 Window analysis of nucleotide differences (*y*-axis) between human X and Y chromosomes along the X chromosome (*x*-axis). Each *bar* corresponds to the nucleotide difference (*p*-distance) in a non-overlapping 1,000-bp window. The *Roman numerals* at the top of the graph indicate contiguous segments of DNA that can be compared between the X and Y chromosomes. The *two-headed arrows and numbers* indicate a region showing somewhat constant *p*-distance over the region. The *grey bracket* indicates a region with lower *p*-distance region than the neighboring region. *KALX* is located in the low *p*-distance region. A *grey bar* in region XII indicates the position of *AMELX*, in which the boundary of strata 3 and 4 is localized

to variable charge X (*VCX*) genes. This low divergence might be caused by gene conversion. We are examining the molecular causes of these low *d* regions and the evolutionary history of the region in primates (Iwase et al. 2010).

In contrast to the X chromosome, gametologues of genes in stratum 4 are interspersed on the entire Y chromosome as a result of the frequent inversions on the chromosome after the divergence of NWMs and Catarrhini. Comparing the gene order on the human Y (HSAY) to the chimpanzee Y (PTRY) shows at least a single inversion in each of the lineages during the 5–7 Myr after the divergence of humans from chimpanzees (Kuroki et al. 2006), resulting in different relative locations of the pseudo-autosomal region (*PAR1*) between the two species. *PAR1* is located at the tip of the short arm in HSAY but at the tip of the long arm in PTRY.

The male-specific region, which does not recombine with the X chromosome, on HSAY contains three kinds of sequences related to the origin and history of sequences: X-transposed, X-degenerated, and ampliconic sequences (Skaletsky et al. 2003; Ross et al. 2005). The X-transposed sequences originated from a transposition from X to Y after the divergence of humans and chimpanzees, approximately 3–5 Myr ago. These sequences are located on the short arm of the Y chromosome and are unique to humans, although the region contains only two genes (*TGIF2LY* and *PCDH11Y*). The X-degenerated sequences are derived from sequences in proto-sex chromosomes, as already mentioned, and have a homologue (gametologue) on the X chromosome. The ampliconic sequences are composed of palindromes, inverted repeats, and arrays of "no long open reading frame" and testis-specific-Y repeats.

HSAY possesses eight palindromic structures (P1–P8) and PTRY possesses four (P6–P8, and CSP1), although the nucleotide sequences of the PTRY short arm have not been fully examined (Skaletsky et al. 2003; Kuroki et al. 2006) In 2010, the chimpanzee Y chromosome has been completely sequenced (Hughes et al. 2010). The sequence revealed that chimpanzee Y chromosome has 19 palindromes. Among

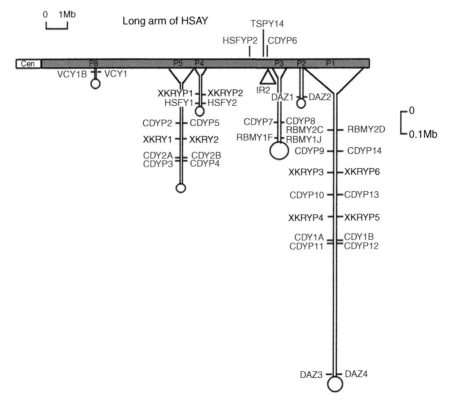

Fig. 10.3 Positions of ampliconic genes on palindromes. Palindromes are shown as hairpin-like structures. *Circles*, loop region. *XKRY* and *XKRP*, the history of which is mentioned in the text, are **boldface**. *HSAY*, human (*Homo sapiens*) Y chromosome

the 19, only three are orthologous to humans (P6–P8). Others are either chimpanzee specific (CSP1) or partial duplicates of human P1–P3. P7 on HSAY and PTRY are relatively small and their structures are nearly identical (Kuroki et al. 2006). Among eight palindromes in humans, six (P1, P2, P3, P4, P5, and P8) contain ampliconic genes: the locations are shown in Fig. 10.3.

On the other hand, there are structural differences in P6. PTRY P6 is 760 kb long and is larger than HSAY P6, whose size is about one-third (~280 kb) that of its counterpart on PTRY. The difference in size of PTRY P6 is probably the result of an extension of P6 in chimpanzees. P8 in both PTRY and HSAY contain variable charge Y (*VCY*) and *VCY1B*, which form a relatively small gene family with *VCX*. In humans, four members of *VCX* and two of *VCY* compose the family. The 1,881-bp *VCX* gene consists of four exons, whereas the 742-bp *VCY* has two exons. All members of *VCX* and *VCY* are expressed exclusively in male germ cells (Lahn and Page 2000). However, *VCX*s and *VCY*s show high degrees of sequence similarity within each gene family as well as between gametologues in humans and chimpanzees (Bhowmick et al. 2007), suggesting frequent gene conversion within and between the X and Y chromosomes. It is clear that conversion in *VCY* is enhanced by a palindromic structure in both humans and chimpanzees; however, on the X chromosome, relationships between the structures and gene conversion have not been fully investigated.

Our interest in the structures on the Y chromosome is focused on identifying when palindrome structures were generated. Palindrome structures contain several gene families. They are "deleted in azoospermia" (*DAZ*), chromodomain Y (*CDY*), X Kell blood-related Y (*XKRY*), RNA-binding motif Y (*RBMY*), heat-shock transcription factor Y (*HSFY*), and *VCY*. Some members of these gene families are located on palindromic arms (Fig. 10.3), and others are included in a different palindrome. As expected, all pairs of copies at symmetrical arm positions within a palindrome are identical or nearly identical. However, different copies at different (nonsymmetrical) positions within a palindrome or in different palindromes differ greatly. Our way to estimate the origin of a palindrome structure is to estimate the time when members of these gene families are distributed at nonsymmetrical positions within a palindrome. If the time of dispersion of members in different families is similar to each other, we can infer the time of the dispersion as the time of the origin of the palindrome arm.

Concerning *XKRY*, there are two functional copies (*XKRY1–XKRY2*) and six copies of pseudogenes (*XKRYP1–XKRYP6*). *XKRY1* and *XKRY2* are located on P5; *XKRYP1–XKRYP6* are on P1 and P4. A search of homologous sequences over the genome of non-human primates and mammals showed that homologues are located on the X chromosome (*XKRX*) and on an autosome (*XKRYL* on chromosome 22). A tree constructed with homologues of *XKRY* (Fig. 10.4) on the Y chromosome, X chromosome, and chromosome 22 formed reciprocal monophyletic clusters to each other. When a homologue (*XK*) from chickens is used as the outgroup sequence, the copy on the X chromosome (*XKRX*) and that on the Y chromosome (*XKRY*) show divergence at the node XY (in Fig. 10.4). The timing corresponds to a point before the radiation of eutherian mammals. A copy of *XKRY* was transposed and generated an autosomal copy (*XKRYL*) on chromosome 22 at the node of YL. Although a homologue of *XKRYL* has been detected in the orangutan genome with unknown chromosomal location (http://genome.ucsc.edu/), the possibility that the detected sequence might be an orthologue of *XKRY* cannot be ruled out. On the other hand, homologues of *XKRY* and *XKRYL* are not detected in the rhesus monkey genome (http://genome.ucsc.edu/), which is inconsistent with the nucleotide divergence between *XKRYL* and *XKRY* in humans of 12%, which is greater than the divergence between humans and rhesus monkeys (5–7%). Thus, we could not rule out the possibility of both genes being present in both species genomes but simply not detected in the rhesus monkey genome due to some technical problems. Alternatively, the generation of *XKRYL* could have occurred before the divergence of OWMs and hominoids, and both *XKRYL* and *XKRY* have been lost from the rhesus monkey genome. Subsequently, at the node of YO, *XKRY* on different palindromes and that at different positions in a single palindrome were generated simultaneously, and therefore the tree suggests that YO is the time at which several palindrome arms were generated.

A similar diversification pattern of other gene family members on palindromes is observed. If we assume the synonymous nucleotide substitution rate per site per

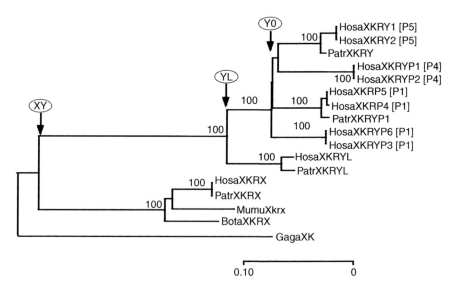

Fig. 10.4 Phylogenetic relationship of ampliconic genes (*XKRY* and *XKRYP*) and copies on autosomal chromosome 22 (*XKRYL*) and the X chromosome (*XKRX*). *Numbers at nodes* indicate bootstrap probability. Palindromes where a copy is located are given in *brackets*. Encircled Y0, YL, and XY show the divergence for XKRY genes, the gene translocation to autosome, and the divergence between XKRX and XKRY, respectively (see text). *Scale bar on bottom right* indicates the number of nucleotides divergence per site. Hosa, *Homo sapiens*; Patr, *Pan troglodytes*; Mumu, *Mus musculus*; Bota, *Bos taurus*; Gaga, *Galus galus*

year for the Y-chromosomal gene of $r_Y = 1.6 \times 10^{-9}$ (Ebersberger et al. 2002), we can estimate the time of diversification for each gene family (Bhowmick et al. 2007; Table 10.2). Roughly speaking, the estimates are close to each other and suggest that palindromes P1, P3, P4, and P5 were generated before the divergence of NWMs and Catarrhini; this implies that palindromes could also be detected in the rhesus monkey Y chromosome.

However, the divergence time of genes on P2 and P8 are different from that on other palindromes. Aside from the *DAZ* genes, flanking *DAZ* sequences are also similar between P1 and P2 (Kuroda-Kawaguchi et al. 2001), and the sequence shows 0.1% differences between the arms of P1 and P2, whereas the loop region shows approximately 1% differences between P1 and P2 (Bhowmick et al. 2007). Although we cannot deny the possibility of gene conversion between P1 and P2, 1% sequence divergence between loop regions suggests that the P2 has probably been generated after the divergence of humans and chimpanzees.

Table 10.2 Time for generation of palindrome

Gene family	XKRRY	CDY	RBMY	HSFY	DAZ
Palindrome	P1, P4, P5	P1, P3, P5	P1, P3	P4	P1, P2
Y0 (myr)	37 ± 6	87 ± 9	43 ± 5	46 ± 6	3~5

10.3 Gene Expression and Genomic Structure on Autosomes

Compared with the X and Y chromosomes, the meaning of genomic structures on autosomes has not been fully examined in the evolutionary process because genomic information of non-human primates has not been fully available. However, recent improvements in *in situ* technologies (for example, fiber fluorescence *in situ* hybridization, FISH) enable us to directly compare the genomic structure between humans and non-human primates. Recent focus on genomic structures on autosomes has elucidated the relationship of phenotypic differences among species, especially differences in gene expression (Khaitovich et al. 2004; Cheng et al. 2005; Blekhman et al. 2009).

Blekhman et al. (2009), comparing the gene expression patterns of humans and chimpanzees, found a slight enrichment of genes that are differentially expressed between the two species on chromosomes 2, 16, and 18. These three chromosomes are known to have large-scale chromosomal rearrangements between the two species, and the enrichment of genes is found to be within a 10-Mb region on either side of the rearrangement breakpoints. However, there are five other chromosomes that have large-scale rearrangements specific to humans, and no enrichment of these genes is observed there. The relationship between differences in gene expression among species and large-scale genomic structures still needs to be investigated.

In contrast to large-scale rearrangement, gene expression differences and CNV in humans are relatively well understood. The relationship is as can be intuitively expected. CNV is the variation in the number of copies of a gene and, consequently, the amount of RNA or proteins expressed would vary in proportion to the number of gene copies (Fu et al. 2007). For example, amylase 1 (*AMY1*) on chromosome 1 illustrates well the relationship between CNV and amount of protein expression. The amylase gene family contains two protein-coding genes (*AMY1* and *AMY2*) and one pseudogene (*AMYP*). The two protein-coding genes encode amylase for the salivary gland and pancreas, respectively, with salivary amylases catalyzing dietary starch at the first step of digestion in the mouth. *AMY1* shows wide variation in the copy number, ranging from 2 to 15 in diploids (Groot et al. 1991; Iafrate et al. 2004; Perry et al. 2007), and AMY1 protein quantities show a significant positive correlation with gene copy number. This study shows that individuals or populations with high-starch diets have more gene copies than those with low-starch diets (Perry et al. 2007), suggesting that natural selection maintains a large number of *AMY1* copies.

The case of color vision genes is different from amylase genes. Here, gene copy number does not affect gene expression. The number of green genes varies in the human population, and these are arranged in tandem next to a single copy of a red gene. However, only the most proximal (5′) green gene in the arrangement is expressed in the retina, even if several copies of green genes are present in the genome. In this case, the increased number of copies does not contribute to an increase in the expression of the gene or proteins.

10.4 Conclusions

After the completion of the human genome sequencing project, several approaches to reveal human specificity, especially when compared with other non-human primates, have been undertaken. Advances in technology for sequencing and structural comparison allow the comparison of human and non-human primate genomes *in situ* and *in silico*. These studies have identified relationships among genome structure, gene expression and gene evolution. In future studies, the interrelationships among structure, gene expression, and gene evolution will be further elucidated.

Acknowledgments The author thanks Drs. Mineyo Iwase and Hielim Kim for their help in the analysis and discussion. The findings presented here are from research supported in part by a grant (16107001) from the Japan Science Promotion Society (JSPS) and in part by a grant (17018032) from the Ministry of Education, Culture, Sports, Science and Technology (MEXT).

References

Bailey JA, Yavor AM, Viggiano L et al (2002) Human-specific duplication and mosaic transcripts: the recent paralogous structure of chromosome 22. Am J Hum Genet 70:83–100

Bhowmick BK, Satta Y, Takahata N (2007) The origin and evolution of human ampliconic gene families and ampliconic structure. Genome Res 17:441–450

Blekhman R, Oshlack A, Gilad Y (2009) Segmental duplications contribute to gene expression differences between humans and chimpanzees. Genetics 182:627–630

Cheng Z, Ventura M, She X et al (2005) A genome-wide comparison of recent chimpanzee and human segmental duplications. Nature (Lond) 437:88–93

Drummond-Borg M, Deeb SS, Motulsky AG (1989) Molecular patterns of X chromosome-linked color vision genes among 134 men of European ancestry. Proc Natl Acad Sci USA 86:983–987

Ebersberger I, Metzler D, Schwarz C et al (2002) Genomewide comparison of DNA sequences between humans and chimpanzees. Am J Hum Genet 70:1490–1497

Fu N, Drinnenberg I, Kelso J et al (2007) Comparison of protein and mRNA expression evolution in humans and chimpanzees. PLoS One 2(2):e216

Groot PC, Mager WH, Frants RR (1991) Interpretation of polymorphic DNA patterns in the human alpha-amylase multigene family. Genomics 10:779–785

Hayakawa T, Satta Y, Gagneux P et al (2001) Alu-mediated inactivation of the human CMP-*N*-acetylneuraminic acid hydroxylase gene. Proc Natl Acad Sci USA 98:11399–11140

Hughes JF, Skaletsky H, Pyntikova T et al (2010) Chimpanzee and human Y chromosomes are remarkably divergent in structure and gene content. Nature 463:536–539

Iafrate AJ, Feuk L, Rivera MN et al (2004) Detection of large-scale variation in the human genome. Nat Genet 36:949–951

International Human Genome Sequencing Consortium (2001) Initial sequencing and analysis of the human genome. Nature (Lond) 409:860–921

Iwase M, Satta Y, Hirai Y et al (2003) The amelogenin loci span an ancient pseudoautosomal boundary in diverse mammalian species. Proc Natl Acad Sci USA 100:5258–5263

Iwase M, Satta Y, Hirai H et al (2010) Frequent gene conversion events between the X and Y homologous chromosomal regions in primates. BMC Evol Biol 10:225

Khaitovich P, Muetzel B, She X et al (2004) Regional patterns of gene expression in human and chimpanzee brains. Genome Res 14:1462–1473

Kuroda-Kawaguchi T, Skaletsky H, Brown LG et al (2001) The AZFc region of the Y chromosome features massive palindromes and uniform recurrent deletions in infertile men. Nat Genet 29(3):279–286

Kuroki Y, Toyoda A, Noguchi H et al (2006) Comparative analysis of chimpanzee and human Y chromosomes unveils complex evolutionary pathway. Nat Genet 38:158–167

Lahn BT, Page DC (1999) Four evolutionary strata on the human X chromosome. Science 286:964–967

Lahn BT, Page DC (2000) A human sex-chromosomal gene family expressed in male germ cells and encoding variably charged proteins. Hum Mol Genet 9:311–319

Li WH, Gu Z, Wang H et al (2001) Evolutionary analyses of the human genome. Nature (Lond) 409:847–849

Marques-Bonet T, Kidd JM, Ventura M et al (2009) A burst of segmental duplications in the genome of the African great ape ancestor. Nature (Lond) 457:877–881

Perry GH, Dominy NJ, Claw KG et al (2007) Diet and the evolution of human amylase gene copy number variation. Nat Genet 39:1256–1260

Ross MT, Grafham DV, Coffey AJ et al (2005) The DNA sequence of the human X chromosome. Nature (Lond) 434:325–337

Samonte RV, Eichler EE (2002) Segmental duplications and the evolution of the primate genome. Nat Rev Genet 3:65–72

Sawai H, Kawamoto Y, Takahata N et al (2004) Evolutionary relationships of major histocompatibility complex class I genes in simian primates. Genetics 166:1897–1907

Sinzelle L, Izsvák Z, Ivics Z (2009) Molecular domestication of transposable elements: from detrimental parasites to useful host genes. Cell Mol Life Sci 66:1073–1093

Skaletsky H, Kuroda-Kawaguchi T, Minx PJ et al (2003) The male-specific region of the human Y chromosome is a mosaic of discrete sequence classes. Nature (Lond) 423:825–837

Szabó Z, Levi-Minzi SA, Christiano AM et al (1999) Sequential loss of two neighboring exons of the tropoelastin gene during primate evolution. J Mol Evol 49:664–671

Chapter 11
Contribution of DNA-Based Transposable Elements to Genome Evolution: Inferences Drawn from Behavior of an Element Found in Fish

Akihiko Koga

Abbreviations

DTE DNA-based transposable element
Mya Million years ago
RTE RNA-mediated transposable element
TE Transposable element

11.1 Lifespans of DTEs Are Relatively Short

DNA-based transposable elements (DTEs) are genetic elements that are, or were at some time in the past, capable of changing their chromosomal locations, usually present as dispersed repetitive sequences. They are transposed mostly in a cut-and-paste fashion. The enzyme that mediates the transposition reaction is called a transposase, which is encoded by a gene lying in the element itself. The enzyme recognizes terminal regions of the element, excises the element from the double-stranded DNA, and then inserts the element into DNA at another location.

When DTEs are transposed, they cause changes in genetic information at their excision and insertion sites. Even without transposition, DTEs cause, because of their repetitive nature, chromosomal rearrangements such as inversions, deletions, duplications, and translocations. For example, a homologous recombination between two elements on different chromosomes may result in a reciprocal translocation of the chromosomes.

A. Koga (✉)
Primate Research Institute, Kyoto University, 41-2 Kanrin, Inuyama, Aichi 484-8506, Japan
e-mail: koga@pri.kyoto-u.ac.jp

H. Hirai et al. (eds.), *Post-Genome Biology of Primates*, Primatology Monographs,
DOI 10.1007/978-4-431-54011-3_11, © Springer 2012

Although a small fraction of DTE-caused new mutations might be beneficial to their host organisms, the vast majority of them are thought to be deleterious to the hosts or selectively neutral. The average effect of DTE-caused mutations on the fitness of the hosts is thought to be negative. It is thus likely that natural selection acts against the transposition activity of DTEs. In accordance with this inference, the number of DTE copies that carry an intact transposase gene is, in many DTE families and in many host organisms, smaller than that of DTE copies that have suffered mutational changes giving rise to inactivation of the transposase (Fedoroff et al. 1983; Streck et al. 1986; Warren et al. 1994). Another factor affects survival of DTEs in their host organisms: stochastic change in the copy number (Lohe et al. 1995). Once the copy number of the intact DTE becomes, by chance, zero, defective copies, even if present in the genome, are no longer transposable. It is a widely accepted idea that, because of these factors that work to eliminate active DTEs from the genomes, the lifespans of DTEs in their host organisms are shorter than those of RTEs that are transposed through copy-and-paste mechanisms (IHGSC 2001).

11.2 Most DTEs Are Dead in Mammals

The DTEs first found as natural mutagens in eukaryotes are the *P* element of *Drosophila* (Rubin and Spradling 1982) and the *Activator* element of maize (Fedoroff et al. 1983). Since then, in various organisms, many DTEs have been identified as insertion sequences in mutant genes. Such mutations, however, had not been observed in mammals, which led to the speculation that mammalian genomes do not contain DTEs. This supposition was found to be false as soon as the genome sequencing era began. Approximately 3% of the human genome consists of DTEs (MGSC 2002). Analyses of sequences of these DTEs, however, have shown that they are all or mostly defective copies: their transposase genes had already become pseudogenes. With these analysis results, biologists revised their speculation: TEs may have worked as mutagens in the remote past, but their effects seem to be negligible in recent and modern mammals. In the current post-genome era, this view is being further revised, as described next.

11.3 DTEs Were Highly Active in Mammals

Recent bioinformatic studies of genomes of humans and other mammals have revealed intensive activity of DTEs during the mammalian radiation and early primate evolution (Pace and Feschotte 2007). Utilizing their own approach, which employs a much higher degree of precision than ever before, the authors have shown that approximately 74,000 of the DTE copies now fixed in the human genome (~33 Mb) were integrated in the period between the divergence of a primate ancestor from other mammalian clades they analyzed, such as the mouse (~75–85 Mya), and the emergence of prosimian primates (~63 Mya). They also estimated that about 23,000 DTE copies (~5 Mb) were inserted and fixed in the human genome in the period between the split of prosimians (~63 Mya) and the emergence of New World monkeys (~40 Mya). The authors suggest that DTEs had a strong impact on the evolution of mammalian genomes during these periods.

Another significant finding about DTE activity in mammals is the repeated horizontal transfer of an element called *SPACE INVADER* (*SPIN*). This element exhibits a patchy distribution among taxa of mammals and other tetrapods, with extremely high nucleotide sequence identities among their copies that would not be expected from the phylogeny of their host species (Pace et al. 2008). Pace et al. estimated that the element invaded multiple species independently around the same evolutionary time (46–15 Mya). The element was found in the rat and mouse (murine rodents), bushbaby (prosimian primates), little brown bat (laurasiatherian), tenrec (afrotherian), opossum (marsupial), and some other tetrapods. The *SPIN* element is thought to have been highly active because their copy numbers are as high as about 100,000 in tenrecs and about 30,000 in bushbabies.

11.4 Live DTEs Were Identified in Fish

The lack of any apparent sign of the presence of currently active DTEs had been longstanding, not only in mammals but also in other classes of the subphylum Vertebrata. The first clear signs were shown in fish about 10 years ago. The medaka fish is a freshwater fish species native to eastern Asia, including China, Korea, and Japan. Mutant fish exhibiting various degrees of albino phenotypes were found in stocks collected from natural populations. Tyrosinase is the key enzyme for melanin biosynthesis, and mutations of its gene were known to be one of the main causes of albinism in humans and mice (Oetting and King 1999). We first cloned the wild-type tyrosinase gene of medaka fish and then analyzed the structures of the gene of the mutant fish, identifying two TEs inserted there (Koga et al. 1995, 1996) (Fig. 11.1). These TEs were named *Tol1* and *Tol2* (*Tol* = transposable element of *Oryzias latipes*). A few hundred *Tol1* copies and 10–20 *Tol2* copies are present in the haploid genome of the medaka fish. Although these DTEs are not from mammals but from fish, results of their analyses, especially of the *Tol2* element, coupled with results of extensive analyses of mammalian genomes by bioinformatics researchers, gave new insights into how mammalian DTEs contributed to genome evolution.

11.5 Fish DTE Is a Recent Invader of Its Host Genome

The *Tol2* element is highly homogeneous in nucleotide sequences, while its host species exhibits, throughout the genome, high degrees of sequence polymorphisms even in essential nuclear genes. Our quantitative analyses gave the conclusion that

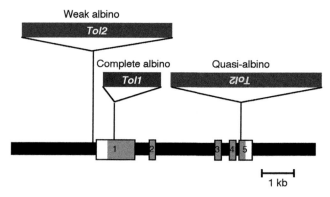

Fig. 11.1 Copies of *Tol1* and *Tol2* identified as insertion sequences in the tyrosinase gene. The *scale* is in kilobase pairs, 0 being assigned to the major transcription initiation site of the tyrosinase gene. Exons are shown as *boxes with numerals*, in which the coding regions are *lightly shaded*. The *heavily shaded boxes* toward the top of the figure are transposable elements. The names of the elements are shown *inside*, their orientation being that of the transcription of their transposase gene. The phenotypes of mutants caused by the respective insertions are described *above* the elements. For the actual colors of these mutant fish, see Koga and Hori (1997, 2001) and Iida et al. (2005)

this TE invaded the fish genome recently and was amplified in a short time span (Koga et al. 2000). It is not known from where and how the DTE was introduced into the fish genome.

DTEs are subject to natural selection against the transposition activity and, in addition, stochastic change in copy number works as a factor to eliminate DTEs from host genomes. Nevertheless, DTEs are present in most known organisms. Thus, there should be some opposing factors against the negative selection pressure and stochastic loss. Horizontal transfer is a primary candidate for being one such factor (Lohe et al. 1995). Transmission to a new genome may give a DTE an opportunity to proliferate there because of the absence of mechanisms for suppression of transposition of this particular TE.

11.6 DTE Provides Genetic Variation but Tends to Be Overlooked as a Cause

The *Tol2* element was first identified as an insertion sequence of the tyrosinase gene of an albino mutant fish. It was subsequently demonstrated to be the sole cause of the albino phenotype: excision of this particular copy from the gene resulted in a wild-type body color (Iida et al. 2005).

Analysis of this reversion mutation revealed that a precise excision of the DTE had occurred in this particular fish. Our subsequent large-scale screening, however, showed that precise excision is rather rare. We observed highly frequent secondary mutations leading to various pigmentation patterns, and their tyrosinase genes exhibited different nucleotide sequences consequent to imprecise excision of the

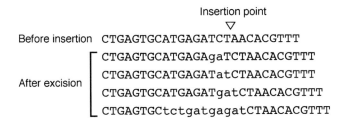

Fig. 11.2 Nucleotide sequences of the region around the *Tol2* insertion point (*arrowhead*) in new alleles caused by excision. The *lowercase letters* indicate nucleotides newly added to, or altered from, the original sequence

DTE copy (Koga et al. 2006). In all the cases, most of the element had been excised, with 1–8 nucleotides left over, or the whole element had been deleted together with 1–4 nucleotides originally belonging to the tyrosinase gene (Fig. 11.2). Excision mutations of the former type are considered, when compared with the wild-type gene, to be insertions of some nucleotides, and those of the latter type are considered deletions of some nucleotides. Because the DTE is no longer present there, one cannot recognize participation of the DTE in their generation. Thus, a young DTE does contribute to genetic variation of the host genome, but the contribution appears to be overlooked because of its transposition mechanisms that are involved in the "cutting" process.

11.7 Impact of DTEs on Mammalian Genomes May Be Larger than Commonly Supposed

The *Tol2* element contributes to genetic variation, as already described, leaving no recognizable traces in nucleotide sequences. The gene we analyzed in this study was the tyrosinase gene that regulates the melanin biosynthesis, and the new mutations we found by screening exhibited various pigmentation patterns. Thus, these DTE-caused mutations are not "loss of function" or simple "gain of function" but rather "diversification of function."

The *Tol2* element is a young DTE in its host, the medaka fish. Resembling this element in the modern fish, huge numbers of DTEs that infiltrated mammals probably contributed to the genome evolution while the respective elements were young, and their impact may have been greater than what we can now estimate by analysis of sequence data.

References

Fedoroff NV, Wessler S, Shure M (1983) Isolation of the transposable maize controlling elements *Ac* and *Ds*. Cell 35:235–242

Iida A, Takamatsu N, Hori H et al (2005) Reversion mutation of i^b oculocutaneous albinism to wild-type pigmentation in medaka fish. Pigment Cell Res 18:382–384

International Human Genome Sequencing Consortium (2001) Initial sequencing and analysis of the human genome. Nature (Lond) 409:860–921

Koga A, Inagaki H, Bessho Y et al (1995) Insertion of a novel transposable element in the tyrosinase gene is responsible for an albino mutation in the medaka fish, *Oryzias latipes*. Mol Gen Genet 249:400–405

Koga A, Hori H (1997) Albinism due to transposable element insertion in fish. Pigment Cell Res 10:377–381

Koga A, Hori H (2001) The *Tol2* transposable element of the medaka fish: an active DNA-based element naturally occurring in a vertebrate genome. Genes Genet Syst 76:1–8

Koga A, Suzuki M, Inagaki H et al (1996) Transposable element in fish. Nature (Lond) 383:30

Koga A, Shimada A, Shima A et al (2000) Evidence for recent invasion of the medaka fish genome by the *Tol2* transposable element. Genetics 155:273–281

Koga A, Iida A, Hori H et al (2006) Vertebrate DNA transposon as a natural mutator: the medaka fish *Tol2* element contributes to genetic variation without recognizable traces. Mol Biol Evol 23:1414–1419

Lohe AR, Moriyama EN, Lidholm DA et al (1995) Horizontal transmission, vertical inactivation, and stochastic loss of *mariner*-like transposable elements. Mol Biol Evol 12:62–72

Mouse Genome Sequencing Consortium (MGSC) (2002) Initial sequencing and comparative analysis of the mouse genome. Nature (Lond) 420:520–562

Oetting WS, King RA (1999) Molecular basis of albinism: mutations and polymorphisms of pigmentation genes associated with albinism. Hum Mutat 13:99–115

Pace JK 2nd, Feschotte C (2007) The evolutionary history of human DNA transposons: evidence for intense activity in the primate lineage. Genome Res 17:422–432

Pace JK 2nd, Gilbert C, Clark MS et al (2008) Repeated horizontal transfer of a DNA transposon in mammals and other tetrapods. Proc Natl Acad Sci USA 105:17023–17028

Rubin GM, Spradling AC (1982) Genetic transformation of *Drosophila* with transposable element vectors. Science 218:348–353

Streck RD, MacGaffey JE, Beckendorf SK (1986) The structure of hobo transposable elements and their insertion sites. EMBO J 5:3615–3623

Warren WD, Atkinson PW, O'Brochta DA (1994) The *Hermes* transposable element from the house fly, *Musca domestica*, is a short inverted repeat-type element of the *hobo*, *Ac*, and *Tam3* (*hAT*) element family. Genet Res 64:87–97

Chapter 12
Application of Phylogenetic Network

Takashi Kitano

Abbreviations

cDNA	Complementary DNA
mtDNA	Mitochondrial DNA
OXTR	Oxytocin receptor
Ts	Transition
Tv	Transversion
UPGMA	Unweighted pair group method with arithmetic means

12.1 Phylogenetic Tree and Phylogenetic Network

A phylogenetic tree is a diagram that illustrates the evolutionary pathway of organisms or genes. It can be described as a tree-like structure (dendrogram) that has separate diverging branches without reticulations (or loops). However, some evolutionary pathways of organisms, genes, or genomes cannot be illustrated by a phylogenetic tree, which has simple separating branches. Such complex evolutionary pathways can, however, be illustrated by a phylogenetic network.

A phylogenetic network can show incompatible phylogenetic information (or character states) on a single diagram using reticulations, a feature unavailable in phylogenetic trees. If incompatible phylogenetic information is not included in the data, a phylogenetic network automatically becomes an unrooted phylogenetic tree, because

---

T. Kitano (✉)
Department of Biomolecular Functional Engineering, College of Engineering,
Ibaraki University, 4-12-1 Nakanarusawa-cho, Hitachi 316-8511, Japan
e-mail: tkitano@mx.ibaraki.ac.jp

H. Hirai et al. (eds.), *Post-Genome Biology of Primates*, Primatology Monographs,
DOI 10.1007/978-4-431-54011-3_12, © Springer 2012

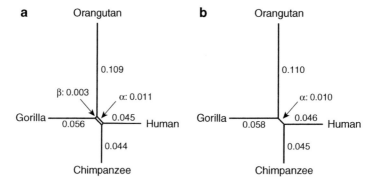

Fig. 12.1 A phylogenetic network constructed by the neighbor-net method (**a**) and a phylogenetic tree constructed by the neighbor-joining method (**b**). The evolutionary distances for both were calculated using Kimura's (1980) two-parameter method. *Numbers on branches* indicate evolutionary distances. α indicates the branch separating human and chimpanzee from gorilla and orangutan; β indicates the branch separating chimpanzee and gorilla from human and orangutan

Table 12.1 Numbers of parsimony informative sites of human, chimpanzee, gorilla, and orangutan mitochondrial genomes excluding D-loop regions

Topology	Substitution	Ts	Tv	Ts/Tv
α [(hum, chi), (gor, ora)]	303	280	23	12.17
β [(chi, gor), (hum, ora)]	209	204	5	40.80
γ [(hum, gor), (chi, ora)]	170	164	4	27.33

Ts transition, *Tv* transversion, *hum* human, *chi* chimpanzee, *gor* gorilla, *ora* orangutan

the presence of reticulations in a phylogenetic network indicates incompatible phylogenetic information. Therefore, a phylogenetic network is a generalization of the concept of a phylogenetic tree.

The phylogenetic network of human (*Homo sapiens*), chimpanzee (*Pan troglodytes*), gorilla (*Gorilla gorilla*), and orangutan (*Pongo pygmaeus*) mitochondrial genomes excluding D-loop regions was constructed by the neighbor-net method (Bryant and Moulton 2004) (Fig. 12.1a) using the SplitsTree4 program (Huson and Bryant 2006). A reticulation, indicating incompatible phylogenetic information, was observed. The branch labeled α separated human and chimpanzee from gorilla and orangutan, thereby reflecting species relationship. In contrast, the branch labeled β separated chimpanzee and gorilla from human and orangutan; this pattern was incompatible with the first one. Because recombination does not occur in mitochondrial genomes, these incompatibilities could have been produced by parallel nucleotide substitutions. The numbers of parsimony informative sites are summarized in Table 12.1. The third topology, γ [(human, gorilla), (chimpanzee, orangutan)], has not been illustrated in the phylogenetic network (Fig. 12.1a). Because parallel transversions are very rare, transition/transversion (Ts/Tv) values of the topologies β and γ were two to three times higher than that of α. When the phylogenetic tree was constructed using the same data by the neighbor-joining method (Saitou and Nei 1987), the β branch was eliminated as noise (Fig. 12.1b). MEGA version 4 (Tamura et al. 2007) was used to construct the neighbor-joining tree. Although reticulations in the phylogenetic network of mitochondrial DNA data were treated as parallel, the phylogenetic network method can be considered a powerful tool for analyzing recombination events occurring in nuclear genomes.

12.2 Variation in Phylogenetic Network Constructing Methods

The several methods for constructing phylogenetic networks can be roughly divided into two groups depending on the usage of data (Table 12.2). One group comprises character state-based methods, which use data directly to construct a phylogenetic network as character states, in a fashion similar to the maximum parsimony and maximum likelihood tree-making methods. The other group comprises distance matrix-based methods, which calculate a distance matrix from data to construct a phylogenetic network, in a fashion similar to the neighbor-joining and UPGMA tree-making methods. In principle, phylogenetic network construction by a distance matrix-based method is depicted by a two-dimensional diagram, whereas the dimensions of a character state-based phylogenetic network depend on the data. A series of studies (Kitano and Saitou 1999; Kitano et al. 2000, 2004, 2009; Liu et al. 2008; Noda et al. 2000; Sumiyama et al. 2000) conducted by our group often used the character state-based method following Saitou and Yamamoto (1997), in which all the possible nodes are connected. This method is an enhancement of the concept of the median network method (Bandelt et al. 1995) that contains all equally parsimonious trees.

Table 12.2 Variation in phylogenetic network constructing methods

Character state-based methods

  Parsimony splits (Bandelt and Dress 1992)

  Spectral splits (Hendy and Penny 1993)

  Median network (Bandelt et al. 1995)

  QNet[a] (Grünewald et al. 2007)

Distance matrix-based methods

  Split decomposition (Bandelt and Dress 1992)

  Neighbor-net (Bryant and Moulton 2004)

  T-REX (Makarenkov 2001)

  TOM-networks (Willson 2006)

[a]It can be viewed as compromise between character state-based and distance matrix-based methods

12.3 Phylogenetic Network of *OXTR* Genes for Human and Great Apes

Oxytocin receptor (OXTR) belongs to the G protein-coupled receptor family (Gimpl and Fahrenholz 2001) and acts as a receptor for the peptide hormone oxytocin. cDNA and genome sequences of *OXTR* were reported by Kimura et al. (1992) and Inoue et al. (1994). The oxytocin-OXTR system plays an important role in the uterus during parturition and milk ejection during lactation. The system is also suggested to play a role in regulation and development of social behavior (Takayanagi et al. 2005). The *OXTR* gene comprises four exons, with the majority of the coding region of the gene located on exon 3 (Fig. 12.2a). The gene span is approximately 20 kb and is located on the short arm of chromosome 3 (3p25) in the human genome.

The topologies of phylogenetic trees differ depending on regions (Fig. 12.2b). The phylogenetic tree of region II showed the same topology as the species tree; that is, human and chimpanzee form a cluster, with a high bootstrap value. In contrast, the phylogenetic tree of region I showed a different topology from the species tree with a high bootstrap value. The results clearly showed that the boundary for different gene trees was located on intron 3 of the *OXTR* gene.

A phylogenetic network can embed different topologies. The phylogenetic network (Fig. 12.3) of *OXTR* genes for human, chimpanzee, gorilla, and orangutan, which was constructed using the character state-based method, embedded three topologies: [(human, chimpanzee), (gorilla, orangutan)], [(human, gorilla),

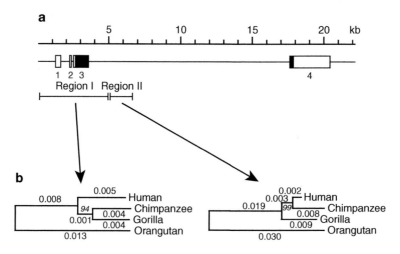

Fig. 12.2 (**a**) Genome structure of the human oxytocin receptor (*OXTR*) gene. *White and black boxes* indicate noncoding and coding exons, respectively. *Numbers below boxes* represent exon numbers. Two analyzed regions (*I* and *II*) are shown. (**b**) Phylogenetic trees constructed by the neighbor-joining method using each region. The evolutionary distances were calculated using Kimura's two-parameter method. Branch length (*normal font*) and bootstrap (*italic*) values are shown for each branch

12 Application of Phylogenetic Network

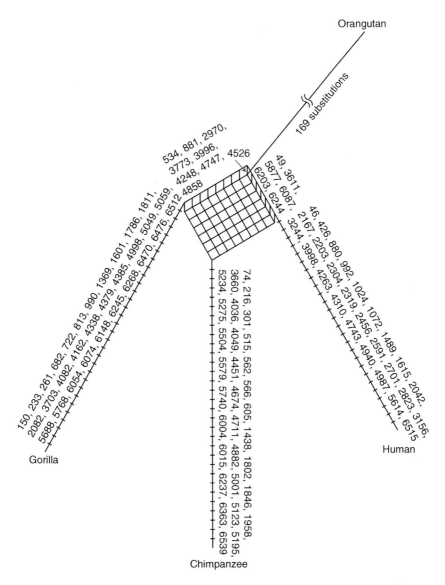

Fig. 12.3 A phylogenetic network of the regions I and II of *OXTR* genes for human, chimpanzee, gorilla, and orangutan. The *edge length* corresponds to a nucleotide difference that is denoted by the *position number* beside it, excluding the orangutan branch. Sites with three character states were eliminated from the analysis

(chimpanzee, orangutan)], and [(chimpanzee, gorilla), (human, orangutan)]. Six sites showed the same topology, that is, [(human, chimpanzee), (gorilla, orangutan)], as the species tree, and four of them were located on region II (Fig. 12.2a). In contrast, eight sites that showed the topology [(chimpanzee, gorilla), (human, orangutan)] were located on region I (Fig. 12.2a). Only one site showed the topology [(human, gorilla), (chimpanzee, orangutan)] on region I. The phylogenetic network (Fig. 12.3) clearly showed that a recombination event of the *OXTR* gene occurred in the ancestral species of human, chimpanzee, and gorilla. The human and gorilla species probably possessed each parental allele for the ancestral recombination event, and the chimpanzee inherited the recombinant allele. How a recombination event is illustrated in a phylogenetic network is explained in the next section.

12.4 Recombination Event and Phylogenetic Network

Figure 12.4 explains how to infer a recombination event from a phylogenetic network using hypothetical data. An ancestral allele (X) produced two descendant alleles (P1 and P2) by accumulating five and four substitutions in each lineage (Fig. 12.4a). The relationship among three alleles (X, P1, and P2) is shown in Fig. 12.4b. Let us assume that a recombination event occurred between the P1 and P2 alleles on the site between 7 and 8, producing two recombinant alleles, R1 and R2 (Fig. 12.4c).

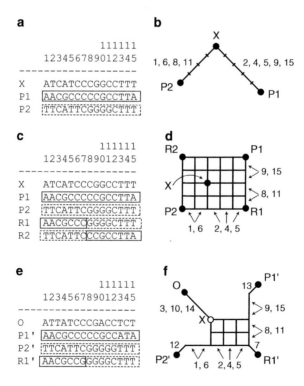

Fig. 12.4 Explanation of recombination in a phylogenetic network using hypothetical data (**a–f**). (**a**), (**c**), and (**e**) are nucleotide sequences; (**b**), (**d**), and (**f**) are corresponding phylogenetic networks. See text for details. (Modified from Kitano et al. 2009)

The phylogenetic network for these five alleles (X, P1, P2, R1, and R2) is shown in Fig. 12.4d. The horizontal edge of the rectangle in Fig. 12.4d corresponds to upstream sequence differences (sites 1, 2, 4, 5, and 6), and the vertical edge corresponds to downstream differences (sites 8, 9, 11, and 15) accumulated from the ancestral allele (X) to the two parental ones. Because five substitutions occurred at sites 2, 4, 5, 9, and 15 in the lineage from the ancestral (X) to the parental allele P1 and four substitutions occurred at sites 1, 6, 8, and 11 in the lineage from the ancestral (X) to the parental allele P2 (Fig. 12.4a,b), the ancestral sequence is located in the middle of the rectangle (Fig. 12.4d). After the recombination, six nucleotide substitutions (sites 3, 7, 10, 12, 13, and 14) accumulated to produce the present-day sequences P1', P2', R1', and O (outgroup) from P1, P2, R1, and X (ancestor), respectively (Fig. 12.4e). Because survival of both recombinant alleles seems distant, it is assumed that the R2 lineage became extinct. The phylogenetic network shown in Fig. 12.4f represents the relationship of the three present-day alleles (P1', P2', and R1') and an outgroup allele (O). Two parental alleles (P1' and P2') are located on the diagonal with long external branches, while the recombinant allele (R1') has a short external branch and is located on another diagonal with the outgroup allele (O). Nucleotide substitutions on the short external branch of the recombinant allele (R1') were regarded as those that occurred after the recombination event. In contrast, long external branches of two parental alleles (P1' and P2') contain nucleotide substitutions both before and after the recombination event. This work is based on the 2009 study of Kitano et al.

12.5 Relic of Ancient Recombinations in Gibbon ABO Blood Group Genes

Human A and B alleles of the *ABO* blood group gene code for glycosyltransferases, which transfer *N*-acetylgalactosamine and galactose, respectively, to a common precursor (Yamamoto et al. 1990, 1995). Two critical sites for the distinction between A and B activities of the glycosyltransferase have been identified in exon 7 (Yamamoto and Hakomori 1990), and these nucleotide differences are concordant with serological studies (Moor-Jankowski et al. 1964). Kitano et al. (2009) determined the nucleotide sequences of approximately 2.2-kb regions from a part of exon 5 to a part of exon 7 of the *ABO* blood group genes for five individuals of agile gibbon (*Hylobates agilis*), 12 individuals of white-handed gibbon (*Hylobates lar*), and six individuals of siamang (*Symphalangus syndactylus*). They constructed the phylogenetic network of these gibbon haplotypes and a human A type allele (DDBJ/EMBL/GenBank accession no. AJ536122) as an outgroup, and classified 24 haplotypes into seven clusters (Fig. 12.5). Clusters 1, 2, 3, and 4 were A type, and clusters 5, 6, and 7 were B type. There were some large reticulations, suggesting recombination events, in this network. They found relics of five ancient intragenic recombinations that occurred about 2–7 million years ago, as summarized in Fig. 12.6 (Kitano et al. 2009). This result established the coexistence of divergent allelic lineages of the

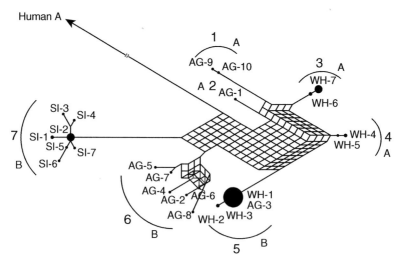

Fig. 12.5 A phylogenetic network of *ABO* blood group genes for three gibbon (*AG* Agile gibbon, *WH* White-handed gibbon, *SI* Siamang) species and one human allele (DDBJ/EMBL/GenBank accession no. AJ536122). A *full circle* shows one haplotype, and its size indicates the number of haplotype copies. The *edge length* corresponds to the number of nucleotide differences, excluding the human A branch. The clusters of gibbons are numbered from *1* to *7*, with *A* or *B* type. (Modified from Kitano et al. 2009)

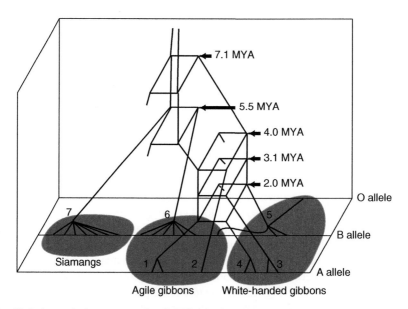

Fig. 12.6 An evolutionary scenario of *ABO* blood group genes for three lesser ape species. Numbers *1* to *7* in the current plane designate seven clusters of Fig. 12.5. Five recombination events that occurred about 2–7 million years ago (MYA) are shown by *parallelograms*. (Modified from Kitano et al. 2009)

ABO blood group gene for a long period in the ancestral gibbon species, and this study on lesser apes further strengthened positive selection as a common feature of primate *ABO* blood group genes.

Because evolution is complex, sometimes it cannot be represented by a phylogenetic tree. In such a case, a phylogenetic network would be helpful to resolve problems in a complex evolutionary process.

References

Bandelt HJ, Dress AWM (1992) A canonical decomposition theory for metrics on a finite set. Adv Math 92:47–105

Bandelt HJ, Forster P, Sykes C et al (1995) Mitochondrial portraits of human populations using median networks. Genetics 141:743–753

Bryant D, Moulton V (2004) Neighbor-Net: an agglomerative method for the construction of phylogenetic networks. Mol Biol Evol 21:255–265

Gimpl G, Fahrenholz F (2001) The oxytocin receptor system: structure, function, and regulation. Physiol Rev 81:629–683

Grünewald S, Forslund K, Dress A et al (2007) QNet: an agglomerative method for the construction of phylogenetic networks from weighted quartets. Mol Biol Evol 24:532–538

Hendy MD, Penny D (1993) Spectral analysis of phylogenetic data. J Classif 10:5–24

Huson DH, Bryant D (2006) Application of phylogenetic networks in evolutionary studies. Mol Biol Evol 23:254–267

Inoue T, Kimura T, Azuma C et al (1994) Structural organization of the human oxytocin receptor gene. J Biol Chem 269:32451–32456

Kimura M (1980) A simple method for estimating evolutionary rate of base substitutions through comparative studies of nucleotide sequences. J Mol Evol 16:111–120

Kimura T, Tanizawa O, Mori K et al (1992) Structure and expression of a human oxytocin receptor. Nature (Lond) 356:526–529

Kitano T, Saitou N (1999) Evolution of Rh blood group genes have experienced gene conversions and positive selection. J Mol Evol 49:615–626

Kitano T, Noda R, Sumiyama K et al (2000) Gene diversity of chimpanzee ABO blood group genes elucidated from intron 6 sequences. J Hered 91:211–214

Kitano T, Liu YH, Ueda S et al (2004) Human-specific amino acid changes found in 103 protein-coding genes. Mol Biol Evol 21:936–944

Kitano T, Noda R, Takenaka O et al (2009) Relic of ancient recombinations in gibbon ABO blood group genes deciphered through phylogenetic network analysis. Mol Phylogenet Evol 51:465–471

Liu YH, Takahashi A, Kitano T et al (2008) Mosaic genealogy of the *Mus musculus* genome revealed by 21 nuclear genes from its three subspecies. Genes Genet Syst 83:77–88

Makarenkov V (2001) T-REX: reconstructing and visualizing phylogenetic trees and reticulation networks. Bioinformatics 17:664–668

Moor-Jankowski J, Wiener AS, Rogers CM (1964) Human blood group factors in non-human primates. Nature (Lond) 202:663–665

Noda R, Kitano T, Takenaka O et al (2000) Evolution of the ABO blood group gene in Japanese macaque. Genes Genet Syst 75:141–147

Saitou N, Nei M (1987) The neighbor-joining method: a new method for reconstructing phylogenetic trees. Mol Biol Evol 4:406–425

Saitou N, Yamamoto F (1997) Evolution of primate ABO blood group genes and their homologous genes. Mol Biol Evol 14:399–411

Sumiyama K, Kitano T, Noda R et al (2000) Gene diversity of chimpanzee ABO blood group genes elucidated from exon 7 sequences. Gene (Amst) 259:75–79

Takayanagi Y, Yoshida M, Bielsky IF et al (2005) Pervasive social deficits, but normal parturition, in oxytocin receptor-deficient mice. Proc Natl Acad Sci USA 102:16096–16101

Tamura K, Dudley J, Nei M et al (2007) MEGA4: Molecular Evolutionary Genetics Analysis (MEGA) software version 4.0. Mol Biol Evol 24:1596–1599

Willson SJ (2006) Unique reconstruction of tree-like phylogenetic networks from distances between leaves. Bull Math Biol 68:919–944

Yamamoto F, Hakomori S (1990) Sugar-nucleotide donor specificity of histo-blood group A and B transferases is based on amino acid substitutions. J Biol Chem 265:19257–19262

Yamamoto F, Clausen H, White T et al (1990) Molecular genetic basis of the histo-blood group ABO system. Nature (Lond) 345:229–233

Yamamoto F, McNeill PD, Hakomori S (1995) Genomic organization of human histo-blood group ABO genes. Glycobiology 5:51–58

Part III
Chromosome Genomics

Chapter 13
Comparative Primate Molecular Cytogenetics: Revealing Ancestral Genomes, Marker Order, and Evolutionary New Centromeres

Roscoe Stanyon, Nicoletta Archidiacono, and Mariano Rocchi

Abbreviations

BACs	Bacterial artificial chromosomes
BES	Bacterial artificial chromosome end sequences
BLAST	Basic local alignment search tool
CAR	Contiguous ancestral regions
ChIP	Chromatin immunoprecipitation
DOP-PCR	Degenerate oligonucleotide primed-PCR
ENC	Evolutionary new centromere
FACS	Fluorescence-activated cell sorter
FISH	Fluorescent in situ hybridization DNA
PAC	P1 artificial chromosomes
SB	Synteny block
WCP	Whole chromosome paints
YAC	Yeast artificial chromosomes

13.1 Introduction

The reconstruction of the history of life on our planet is one of the fundamental goals of evolutionary biology. Darwin (1859) first showed that evolution operated through natural selection and descent from common ancestors. Darwin's theory that

---

R. Stanyon (✉)
Laboratory of Anthropology, Department of Evolutionary Biology,
University of Florence, Via del Proconsolo 12, 50122, Florence, Italy
e-mail: roscoe.stanyon@unifi.it

N. Archidiacono • M. Rocchi
Department of Genetics and Microbiology, University of Bari, Via Amendola 165/A, Bari, Italy

H. Hirai et al. (eds.), *Post-Genome Biology of Primates*, Primatology Monographs,
DOI 10.1007/978-4-431-54011-3_13, © Springer 2012

all forms of life on this planet are connected by descent from common ancestors is now an undisputed fact. This fact means that all forms of life on our planet are related and all species have common ancestors.

Huxley (1863) showed how the morphological comparison between living species using Darwinian theory could illuminate aspects of evolution. Until recently, the evidence and knowledge of evolution were based almost exclusively on comparative anatomy and paleontology, but over the past half century the study of evolution has ever more become the study of the evolution of genomes, even if the fundamental questions (when, where, and how) remain the same.

Evolutionary biologists often depict the history of life as a branching tree. Some of our ancestors are located on branches close by and others are found on very distant branches. Humans and chimpanzees, our closest living relatives, diverged from a common ancestor about 6 million years ago. Cats and humans probably had a common ancestor around 90 million years ago, whereas humans and chickens had a common ancestor more than 350 million years ago, and so forth.

Without doubt, paleontology provides indispensable, direct material evidence of ancient life forms. Reconstructing evolutionary history through phylogenomics may appear extremely limited because genomics, after all, are restricted to the study of living species (even if the study of ancient DNA can provide a window on the past, but only up to about 50,000 years ago). However, one advantage of reconstructing the history of life with phylogenomics compared to paleontology is that although we cannot be sure any particular fossil left decedents, we know without doubt that all living species had ancestors.

One revelation of comparative molecular cytogenetics is that genomes of living species are very similar and are therefore highly conserved over evolutionary time. The high conservation of chromosomes and genes makes it possible to reconstruct ancestral genomes using molecular cytogenetic methods such as chromosome painting and, more recently, fluorescent in situ hybridization (FISH) with cloned DNA (bacterial artificial chromosomes, BACs).

Comparative molecular cytogenetics represents a type of archeology or paleontology of the genome. We can now develop hypotheses about the content of the genome of long-extinct ancestral species for many branching points on the tree of life, and nowhere is this better known than in the primates. Reconstructing the genome of these ancestors is an obligatory goal of comparative cytogenetics. These reconstructions provide insight into evolutionary forces that have sculpted the genomes of extant species. We are beginning not only to trace the phylogenomic pathways that our ancestors took but also to understand how the genomes of modern primates came to be organized the way they are.

Certainly, one difference between cytogenetics and other genomic levels is that we need living cells, which are generally obtained by in vitro cell culture, because chromosomes are seen only during cell division. This limitation makes sampling more difficult because a blood draw or biopsy is required whereas DNA can be extracted from bodily excretions, hairs, and other noninvasive methods. This disadvantage is offset by the fact that comparative molecular cytogenetics is a much more economical and rapid method of surveying the genome. Other phylogenomic methods, especially complete genome assembly, are still very expensive and very time consuming. In practice, that means that currently we have a much more extensive and taxonomically rich array of species studied at the cytogenetic level. We should add, however, that it is becoming increasingly evident that comparative molecular cytogenetics and sequencing have much to offer each other, and we believe that these two fields will become increasingly integrated (Rocchi et al. 2006). In this review, we focus on the cytogenetic level of the genome organization, chromosomes and karyotypes, and discuss sequencing data when it interfaces with molecular cytogenetics.

Comparative molecular cytogenetic of primates is now leading the way to showing how evolutionary perspectives can provide compelling underlying explicative grounds for contemporary genomic phenomena (Capozzi et al. 2009). In particular, molecular cytogenetic provides a pictorial illustration of Dobzhansky's well-known statement that "nothing in biology makes sense except in the light of evolution."

13.2 Landmarks in Primate Cytogenetics

13.2.1 Classical Cytogenetics

Cytogenetic comparisons between species has a long history, going back to the early decades of the last century. However, data began to be accumulated quickly only after 1956 when the correct chromosome number of humans was determined.

It is a curious fact that knowledge of the correct chromosome number (2n) of some primates predated that of humans. The correct diploid number of the chimpanzee was published 16 years earlier (Yeager et al. 1940). The rhesus macaque diploid number was known almost 20 years earlier (Shiwago 1939), and Makino correctly described the Taiwanese macaque diploid number in 1952.

Comparative cytogenetics can be divided into three historical periods, closely linked to technical advances. The first period of classical cytogenetics, a result of the widespread application of tissue culture methods and hypotonic treatment, corresponds to the period from 1956 to about 1970. In this period the chromosomes were grossly stained and only the number of chromosomes, their relative size, and the position of the centromere could be determined. On the basis of classical cytogenetics it was easily shown that the various primates had different numbers and forms of chromosomes; that is, they had different karyotypes. Karyotypes could vary greatly even between related species, and hypotheses about ancestral genomes were framed in terms of the diploid number (2n) and the number of chromosome arms (fundamental number). Early primate cytogeneticists provide these basic data on a great many primate species (Chu and Bender 1961). We now know primate chromosome numbers range from 16 to 80 in primates.

Some researchers proposed that the ancestral primate karyotypes had a high diploid number and that evolution proceeded by chromosomal fusions. Others proposed that fission was the main evolutionary mechanism and proposed low

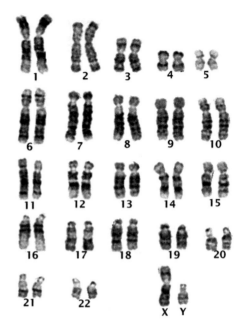

Fig. 13.1 A G-banded karyotype of the cotton-top tamarin (*Sanguinus oedipus*), a South American monkey. G-banding is produced by digesting the chromosomes with trypsin, a protease, and then staining with Giemsa stain. Note that the karyotype is arranged according to length and centromere (*primary constriction along the chromosome*) position

ancestral diploid numbers, whereas others took the middle ground, proposing that both mechanisms were important and the ancestral diploid number was near to the modal number of living mammals. We now know that the ancestral placental diploid number of primates was near the modal number, and that both fissions and fusions were important, but not the sole mechanism in the evolution of genomes.

13.2.2 Chromosome Banding

Chromosome banding was introduced in the 1970s. Banding consists of different intensity of staining along the length of chromosomes (Fig. 13.1). The first banding technique was known as Q-banding, for quinacrine, the fluorescent stain used to produce the banding (Caspersson et al. 1970). Later, trypsin digestion and staining with Giemsa stain led to G-banding (Seabright 1971). A minority of workers used R-banding, which produced a pattern mostly the reverse of that seen with G-banding (Dutrillaux et al. 1972). With chromosome banding, each chromosome in the karyotype could be easily identified and better comparisons of chromosome morphology between species could be made.

Primate species used in biomedical research from which biological samples were more easily obtained made up the majority of reports. For instance, among Papionini (comprising macaques and baboons), by 1983 there were more than 50 banding reports. By the 1990s, a good percentage of primate species belonging to all major taxonomic groups were eventually studied using banding (cf. Dutrillaux 1979).

One of the important insights using chromosome banding was the proposed high chromosome homology and conservation among humans, apes, and monkeys (Stock and Hsu 1973). These observations were even extended to distantly related species, such as between humans and cats (O'Brien and Nash 1982) or humans and rabbits (Dutrillaux et al. 1980). These workers were right about high levels of chromosome conservation even if many chromosomal homologies they proposed were wrong. Establishing chromosomal homology between distantly related species or species characterized by rapid chromosomal evolution remained speculative until the advent of molecular cytogenetics, and especially chromosome painting, in the late 1980s and early 1990s.

13.2.3 Molecular Cytogenetics

Distinguishing between homology and convergence in cytogenetics became possible during the past 20 years with the introduction of molecular methods in cytogenetics. With molecular methods it was possible to compare the underlying DNA by FISH. This method takes advantage of the conservative, precise replication and binding characteristics of the double helix DNA.

The most commonly employed probes were whole chromosome paints (WCP). As the name implies, these probes contain DNA from a single, entire chromosome. A hybridization cocktail is made that usually contains, in addition to the chromosome-specific DNA, cot-1 DNA (highly repeated fraction of genome) and salmon sperm DNA (a DNA carrier). The cot-1 DNA serves as a competitor DNA to drop out the repeat part of the genome that may not be chromosome specific. After denaturation, the probe is applied directly (in situ) to metaphase of the target species, which has also been denatured. A coverslip is mounted and sealed. After incubation from one to several days, the slides are washed and detected. The probe is then assigned or mapped to the corresponding chromosome or chromosome segments in the target species by observation and imaging on a fluorescent microscope. Thus, the name FISH (fluorescent in situ hybridization). The strength of the technique is that interspecies chromosomal homology is established not on the basis of banding similarities but on the basis of DNA content. Chromosome painting was rapidly shown to be superior to chromosome banding for mapping the chromosomal homology between species.

Initially, chromosome paints were provided by phage or plasmid DNA libraries that were labeled by nick translation. The advent of chromosome sorting by flow cytometry and degenerate oligonucleotide primed-polymerase chain reaction (DOP-PCR) brought a significant simplification to chromosome painting (Ferguson-Smith et al. 2005). Chromosome sorting is now the principal source of chromosome-specific DNA for chromosome painting (Stanyon and Stone 2008). A single chromosome suspension is produced from a rapidly growing cell culture. The chromosomes are then stained with GC- and AT-specific fluorochromes and passed through a FACS (fluorescence-activated cell sorter). A bivariate plot of chromosome fluorescence

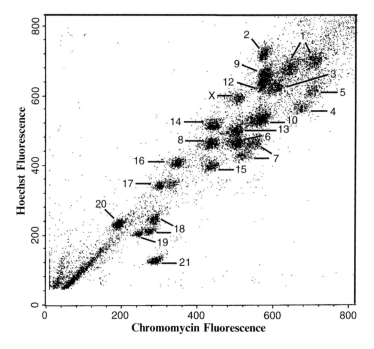

Fig. 13.2 Flow karyotype of *Cebuella pygmaeus* (CPY, pygmy marmoset), a New World primate. A single chromosome suspension is prepared from rapidly growing cell cultures. The chromosomes are stained with AT- and GC-specific fluorochromes. They are then passed through a flow cytometer with two lasers, each tuned to the specific fluorochrome. The bivariate plot that results allows the operator to gate on a particular chromosome, which is deflected into an test tube by a electromagnetic field. Pure samples of individual chromosomes can then be manipulated in the laboratory by polymerase chain reaction (PCR) to produce whole chromosome-specific probes, commonly known as chromosome paints. [The authors gratefully acknowledge late Gary Stone (NCI-Frederick) for assistance in preparing the flow karyotype]

allows the operator to gate on specific chromosomes and sort them directly into PCR reaction tubes (Fig. 13.2). A primary round of DOP-PCR is used to directly amplify the sorts, and then a secondary PCR reaction is used to label the primary products. The labeled chromosome paints are then hybridized in situ, singularly or in combination, to standard chromosome metaphase preparations, and later detected and analyzed with a fluorescence microscope (Fig. 13.3a).

With chromosome sorting, probes can be made from any species, and reciprocal chromosome painting is possible between species. Paints from one species are used to hybridize chromosomes of another species and vice versa. More than two paint sets allow multidirectional chromosome painting. Such reciprocal and multidirectional chromosome painting permits the delineation of subchromosomal homology and a much more precise inference of breakpoints involved in chromosome rearrangements.

Spectral karyotyping (SKY) and M-FISH are extensions of chromosome painting in which all chromosomes (usually human) are hybridized to a metaphase to produce

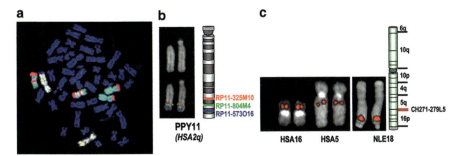

Fig. 13.3 (a) A metaphase chromosome spread of the proboscis monkey (*Nasalis larvatus*) hybridized with various whole human chromosome paints. Human chromosome 1 is hybridized in *green*, chromosome 3 in *yellow*, and 19 in *red*. Note that only two chromosome are hybridized by human chromosome 3 (*yellow*). A single pair of hybridization signals indicates that the synteny is maintained between humans. However, both chromosomes 1 and 19 show multiple pairs of signals, indicating that the synteny was not conserved. These chromosomes are found associated (*red and green signals* together on the same monkey chromosomes) thus forming new, derived syntenies. (b) Example of cohybridization fluorescent in situ hybridization (FISH) experiment of three bacterial artificial chromosomes (BAC) clones, performed to establish their reciprocal marker order in the orangutan chromosome 11 (HSA2q). (c) The bacterial artificial chromosome end sequences (BES) of BAC CH271-279L5 (*Nomascus leucogenys* gibbon) map to distinct chromosomes in humans (HSA5q and HSA16p), indicating a gibbon-specific rearrangement. In a FISH experiment, indeed, the BAC yielded a single signal in gibbon (*right*) and a split signal, to 16p and 5q in humans, supporting, therefore, the bioinformatic findings

a multicolored image of the complete hybridization pattern. Although these methods have found application in tumor cytogenetics, they have proved to be of limited utility in comparative cytogenetics because of difficulties imposed by diminishing hybridization efficiencies encountered with increasing phylogenetic distance.

13.3 Comparative Primate Molecular Cytogenetics Using Chromosome Paints

Chromosome painting allows us to determine how many chromosomes or chromosome segments are conserved between the genomes of two or any number of species. We can also determine how many chromosome rearrangements are needed to transform the genome of one species into another. Parsimony analysis is then commonly used to determine ancestral chromosomal syntenies (the largest linkage group known) and associations. Commonality, combined with the inclusion of appropriate "outgroups," helps determine if a chromosome is syntenic and what associations were present in an ancestral genome. An outgroup is one or more species from the next phylogenetic branch out from the group under discussion. For instance, the appropriate outgroup for great apes and humans would be the Old

World monkeys. In the reconstruction of ancestral genomes, human chromosome numbering is normally used as the reference.

Several simple observations derive from the application of these analyses (Wienberg and Stanyon 1997). A chromosomal synteny may often be disrupted *independently* by chromosomal rearrangements, but it is highly unlikely that the same syntenic group was brought together independently in different lineages, especially with the same marker order. When a chromosomal synteny is found intact between various species, this condition is likely to be ancestral. For instance, a chromosomal synteny found in widely divergent mammalian orders would suggest that it was present in the ancestor of all mammals. In the early stages of reconstructing the ancestral genome of living placental mammals, it became clear that all chromosomes homologous to the human chromosome 3 and human chromosome 21 were syntenic in the ancestor (Müller et al. 2000). This syntenic chromosome association, written as 3/21, was later disrupted in the evolutionary line leading to humans. Today it is not found in Old World monkeys, apes, or humans, but it is conserved in New World monkeys and prosimians, and is universal in almost all other mammals, including marsupials, and even in chickens.

The origin of the human chromosome 1 synteny illustrates how molecular cytogenetics proceeds. There were two competing hypotheses of the origin of human chromosome 1. In the first hypothesis this synteny was recent and derived from two independent chromosomes (corresponding to 1q21.2-pter and 1q21.2-qter). Indeed, two or more segments were found in more than 40 species from 11 orders of placental mammals. The alternative hypothesis viewed the intact synteny, at 285 million base pairs (Mbp), as the largest physical unit in the ancestral placental genome. This synteny was initially found intact only in catarrhine primates (Old World monkeys, great apes, and humans) and dolphins. Later, the intact synteny was found in all four placental mammalian superorders: Afrotheria (elephants, manatees, hyraxes, tenrecs, aardvark, and elephant shrews), Xenarthra (sloths, anteaters, and armadillos), Euarchontoglires (rodents and lagomorphs as a sister taxon to primates, flying lemurs, and tree shrews), and Laurasiatheria (cetaceans, artiodactyls, perissodactyls, carnivores, pangolins, bats, and insectivores). Further, multidirectional chromosome painting and gene mapping showed that the breakpoints in mammalian species in which the chromosome 1 synteny were fragmented segments were not the same. These data integrating both molecular cytogenetics and other levels of phylogenomic analysis led to the conclusion that the second hypothesis was correct, and it is now accepted that human chromosome 1 was syntenic in the ancestral placental mammalian karyotype and in the ancestral karyotype of primates (Murphy et al. 2003).

We should keep in mind, however, that it is the derived chromosome rearrangements, not the conserved chromosome syntenies, that are phylogenetically informative. In chromosome painting a particular type of rearrangement is revealed: translocations. Chromosomal rearrangements are known to be rare, and therefore it seems logical to assume that fragmentation of syntenies and especially associations of various syntenic groups, which are seen in a limited array of lineages, most probably indicate common derived traits that phylogenetically link species.

13.3.1 Phylogenetic Relationship Between Humans and Other Primates

One difficult problem until the mid-1990s was the exact phylogenetic relationship (branching order) between humans and great apes. Paleontological phylogenies at this time placed all great apes, both Asian and African, together in the Pongidae, with a very early divergence of the human line between 15 and 30 million years ago. However, biomolecular data beginning with immunological comparisons that became predominant by the mid-1990s viewed humans as more closely related to the African apes with the human line originating about 5 million years ago. It is interesting to note that by the 1970s chromosome comparisons had already shown that humans were more closely related to African apes and by the early 1980s had already begun to indicate that there was a closer phylogenetic relationship between humans and chimpanzees (Yunis and Prakash 1982), a view which became even more secure by the mid-1990s and that today no one would challenge.

There are at least three major rearrangements that unite chimpanzees and humans after the divergence of the gorilla and various independent rearrangements (mostly inversions) in all African ape and human lines (Fig. 13.4). At each point in this recent human tree, we can reconstruct the ancestral karyotype – for the great ape/human ancestor, for the African ape/human ancestor, and for the chimpanzee/human ancestor – with great confidence even just on the banding data. Chromosome painting confirmed most, but not all, of the conclusions from chromosome banding. As expected, the synteny of most human chromosomes (each human chromosome had an equivalent in the great apes) was found intact in all great apes with only the exception of chromosome 2 and a reciprocal translocation between the human homologues 5 and 17 in the gorilla (Stanyon et al. 1992). However, the chromosome painting data were of no utility to view the remaining rearrangements known from banding, which are intrachromosomal (inversions).

Although the banding data had proved accurate for comparison between humans and their closest relative, the great apes, comparisons between more distant species or species that do not have conserved genomes banding are difficult and sometimes impossible. As mentioned previously, with banding only morphology is compared, and it can be difficult to separate homology from convergence. Chromosome painting soon resolved that problem in all the major taxonomic divisions of primates. Next, we review the data accumulated on chromosome painting first in more than 50 species of primates (Stanyon et al. 2008).

13.3.2 Chromosome Painting in the Primate Order

The first species fully mapped with human chromosome paints was the Japanese macaque (2n=42) in 1992 (Wienberg et al. 1992). An incredible conservation of synteny over 25 million years of evolution was found for all human chromosomes,

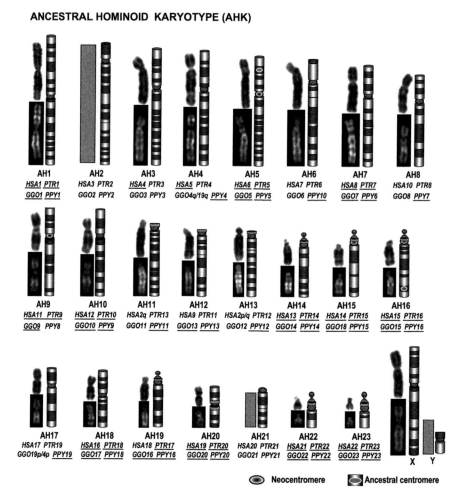

Fig. 13.4 Diagrams of hypothetical ancestral hominoid (AH) karyotype (*AHK*), as described in Stanyon et al. (2008). Corresponding chromosomes (same group synteny) in each hominoid species are reported *below* the diagram. Species that have conserved also the same marker order arrangement with respect to the AH are *underscored*. Examples of G-banded (*upper*) and Q-banded (*below*) chromosomes are on the *left* of each diagram. No living species retained the AH2 (human 3), AH21 (human 20), and AHY marker order organization

but with three different associations: 7/21, 14/15, and 20/22. This pattern of high conservation is generally followed by the majority of, but not all, primate and mammalian species.

However, the lesser apes (gibbons and siamang), species more closely related to humans than the Old World monkeys (OWM), have chromosomes in which the human syntenies are often highly fragmented and shuffled (see Chap. 14). It was known from chromosome banding that comparisons between humans and lesser apes or among lesser apes which differed in diploid number (2n = 38, 44, 50, and 52) were quite difficult. Some attempts were made, and although the correct hypothesis of multiple reciprocal translocations was advanced, the homologies proposed were often incorrect. Indeed, comparative chromosome painting was initiated precisely to resolve the problem of identifying homologies between humans and lesser apes. The utility of this technique was immediately apparent. Chromosome painting allowed, in a series of papers, the mapping of chromosomal between human and the four gibbon karyomorphs (Jauch et al. 1992; Koehler et al. 1995a; Koehler et al. 1995b; Yu et al. 1997; Nie et al. 2001).

The dozen or more species of lesser apes can be divided into four karyomorphs: 2n = 38, 44, 50, and 52.The painting data showed that human chromosomes were highly fragmented and reshuffled in each of these karyomorphs. Translocation, often reciprocal, played a major role in transforming these genomes. For instance, *Hylobates lar* chromosomes (2n = 44) were composed of up to five segments of different human chromosomes. The 24 siamang autosomes (2n = 50), composed of 60 segments homologous to human chromosomes, were the result of at least 33 translocations. In the concolor gibbon, the autosomes were composed of 63–67 segments as the result of at least 31 translocations.

13.3.2.1 Colobines

As in the case of the Japanese macaque, colobine karyotypes are also highly conserved. Both Asian and African colobines have the same diploid number, and all share the syntenic association of homologues to human chromosome 21/22 (Bigoni et al. 1997a, b; Nie et al. 1998). However, a series of independent rearrangements marks the lineages leading to African and Asian colobines. There is no equivalent to chromosome 1 in any colobine species. In African colobines, there are at least two reciprocal translocations of this chromosome, yielding two derived syntenic associations, 1/10 and 1/17. In Asian colobines, a 1/19 reciprocal translocation is found, whereas a reciprocal translocation between 3 and 19 is found in African colobines. A reciprocal translocation of 6 and 16 appears to be a distinguishing characteristic of the genus *Trachypithecus*. This rearrangement was found in all *Trachypithecus* species published so far with the exception of the purple-faced langur (*T. vetulus*).

13.3.2.2 Guenons

Diploid numbers in the Cercopithecini vary from 48 to 72. Chromosome painting results show that fissions are driving these species toward higher diploid numbers. The karyotype of the ancestral Cercopithecini (guenons) is derived for two fissions (chromosomes 3 and 5) and a fusion (20/21) and had a diploid number of 2n = 48. *Allenopithecus nigroviridis* is the only living guenon that maintains this karyotype. Thereafter there are two independent lines leading to increased diploid number (Stanyon et al. 2005).

13.3.2.3 New World Primates

It has only become recently appreciated that these platyrrhines represent a closely knit phylogenetic array. In support of this perspective, the chromosome painting data convincingly illustrate that most of the chromosome rearrangements that distinguish these primates from catarrhines (or better, from the ancestral primate karyotype) occurred in the last common ancestor of all these species. Such data allow the reconstruction of the ancestral karyotype of New World monkeys (Dumas et al. 2007) and support their monophyletic origin from Africa.

Traditionally, New World monkeys (Platyrrhini) were classified into two families, Cebidae and Callitrichidae, but interpretations of molecular data do not support this division. There is now ample support for a division into three groups called Cebidae, Pitheciidae, and Atelidae, and again chromosomal rearrangements lend support to this phylogenetic arrangement (Dumas et al. 2007). Modern taxonomies of primates basically follow this arrangement (Groves 2001). We can note that chromosome painting supports grouping *Callimico goeldii* within marmosets with *Saimiri* as the sister taxon to marmosets and tamarins (Neusser et al. 2001).

There is a tremendous amount of chromosomal variability in these monkeys. *Callicebus lugens*, for example, has the lowest diploid number found in primates (2n = 16) (Stanyon et al. 2003). Cytogenetics confirms that New World monkey biodiversity is still not well known and that species numbers are probably underestimated in traditional taxonomy. Molecular cytogenetic data indicate that many taxa including Atelidae (genera *Lagothrix*, *Brachyteles*, *Alouatta*, and *Ateles*), Cebidae (genus *Aotus*), and Pitheciidae (genus *Callicebus*) are karyologically derived, and there are probably additional species hidden within many of these taxa.

13.3.2.4 Strepsirrhine Chromosome Painting

Chromosome painting in the strepsirrhines (prosimians) shows that both lemuriform and lorisiform primates have highly derived, rearranged karyotypes (Müller et al. 1997, 1999; Stanyon et al. 2006; Nie et al. 2006). None of these primates has a karyotype sufficiently conserved to be considered ancestral or primitive for the primates. It is informative that only fissions of homologues to human chromosomes 1 and 15 provide significant evidence of a cytogenetic link between Lemuriformes and Lorisiformes. The association of human chromosomes 7/16 in both strepsirrhines, however, strongly suggests that this association was present in the ancestral primate genome.

13.3.3 Lack of Molecular Clock

One conclusion that came from the early painting comparisons of humans and other primates, especially lesser apes, was that clearly there is no molecular clock for chromosome evolution. In other words, evolutionary rates in chromosome rearrangements vary greatly from one evolutionary line to another (Wienberg and Stanyon 1997; Trifonov et al. 2008). As we have seen, high chromosome conservation was the rule for most catarrhines, including humans, with the exception of lesser apes, which showed extremely rapid rates of chromosome evolution, and to a lesser extent cercopthecines. Many platyrrhines and strepsirrhines have highly rearranged karyotypes.

13.3.4 The Ancestral Karyotype of Placental Mammals and the Origin of the Primate Genome

The events leading to the establishment of the primate karyotype can only be known if we are able to develop reasonable hypotheses about the ancestral eutherian karyotype, which has been one of the aims of interspecies chromosome painting studies. During the past few years various hypotheses were advanced on the content of the ancestral karyotype of all living placental mammals. Most authors appeal to the concept of parsimony and consider that likely ancestral chromosomes are commonly present in species of a number of divergent eutherian orders (Svartman et al. 2004). Although some questions remain, the general picture that has emerged is one of conservation, reflected in the general uniformity of proposals.

Discussion has centered on the presence or absence of various syntenic associations: 1/19, 4/8, 7/16, 10/12/22 (Ferguson-Smith and Trifonov 2007). The presence or absence of these associations is important to determine the composition of the probable ancestral primate genome. There is now general agreement that most of these associations, with the exception of 1/19, were certainly present.

Given the foregoing consideration, the current chromosome painting hypothesis (buttressed by data from the opossum genome assembly) is that the ancestral eutherian karyotype would contain the following 46 chromosomes: 1, 2p-q, 2q, 3/21, 4/8p, 5, 6, 7a, 7b/16p, 8q, 9, 10q, 10p/12a/22a, 11, 12b/22b, 13, 14/15, 16q/19q, 17, 18, 19p, 20, X, and Y.

The transition to the primate genome therefore involved three fissions: 10p from 12a/22a, the disruption of the association 16q/19q, and the fission of association 4/8p. A fusion of 8p and 8q then gave origin to the synteny of chromosome 8. Given these results, we can propose that the last common ancestor of all living primates had a diploid number of 50 with the following chromosomes: 1, 2p-q, 2q, 3/21, 4, 5, 6, 7, 7/16p, 8, 9, 10p, 10q, 11, 12a/22a, 12b/22b, 13, 14/15, 16q, 17, 18, 19p, 19q, 20, X and Y.

We can note that there is no cytogenetic reason to consider Scandentia (Tupaia) as the closest mammalian order to primates. Chromosome painting between Primates, Scandentia, and Dermoptera show that these latter two are equidistant sister orders to Primates (Nie et al. 2008). Our unpublished data indicate that an earlier report on tree shrew chromosome painting missed the 4/8 association (Dumas et al. 2012; Müller et al. 1999).

13.4 FISH of Cloned DNA probes

Painting probes have the advantage of spotting, in a few experiments, all the translocation differences between the species under study. This technology, however, does not provide information on intrachromosomal rearrangements, inversions in particular. The latter cases can be disclosed only by establishing a precise marker order along the chromosomes. Marker order has been historically accomplished using the classical linkage studies or by radiation hybrid experiments. Both methods are laborious and are usually performed in the frame of preparative work toward the full sequencing of a genome. Specific problems of marker order along a chromosome can be solved by cohybridization multicolor FISH experiments using probes that yield a dot signal (Fig. 13.3b). Their specific use in this context is illustrated next. Here we show the main features of different kind of probes suitable for FISH.

13.4.1 YAC Clones

A good FISH signal can be obtained using a DNA clone of appropriate length. The minimum length required to produce a reproducible FISH signal depends on many variables. The length of the probe is the first variable, but only the actual size of its single-copy portion has to be considered. Repetitive sequences [Alu, LINE (long interspersed elements), etc.] have to be blocked by adding cold COT-1 DNA to the hybridization mixture to avoid their hybridization to the overall genome.

Historically, the first clones efficiently and widely used in FISH experiments were the yeast artificial chromosomes (YAC) clones. The length of the insert was usually very large, up to 2 Mb. They had, however, two main disadvantages: chimerism (two or more DNA segments, from different chromosomal location, co-cloned in the same YAC), and difficulties in obtaining a pure human DNA, separated from the yeast DNA. The YAC, indeed, is a minor part of the yeast genome, and its separation from the yeast bulk DNA through pulsed-field gel electrophoresis is unpractical for routine use.

13.4.2 BAC, PAC, and Fosmids

In early 1990, a new type of clone was introduced: BAC and PAC (P1 artificial chromosomes). These clones increased up to about 300 kb the potential length of the insert, which was limited to 40 kb in the already available fosmid/cosmid clones. The insert length of the fosmid/cosmid clones is quite precise (40 kb) because of the constriction of the packaging process used for their generation. All these kinds of probes share the feature that the cloned DNA fragment is introduced in appropriate vectors and then inserted into bacteria. The length of these new types of clones, even if much shorter than YAC clones, is usually more than enough to produce a good FISH signal. Their significant advantage over YACs is because (1) their DNA is easily separated from bacterial DNA during DNA extraction (made very straightforward by specific extraction kits), and (2) their ends can be easily sequenced. These features made BAC clones very valuable for genome sequencing projects based on the shotgun approach because their ends can unequivocally link two contigs. Subsequently, the precise position of the clones on the assembled sequence can be easily derived from the two end sequences of the clone [bacterial artificial chromosome end sequences (BES) for BACs]. Some genome browsers do this mapping systematically, and the position of each clone is graphically displayed (see specific tracks of human BAC/PAC and fosmid end-pairs at the University California Santa Cruz genome browser (http://genome.ucsc.edu). In this way, specific clones can be selected on the basis of their precise position on the human sequence.

It is important to note that human clones usually work well in FISH experiments in apes and Old World monkeys. In New World monkeys, or in more distant primates, some clones may fail to produce detectable FISH signals because of sequence divergence related to the phylogenetic distance (the rule of thumb is 1% sequence divergence/5 million years). In these cases, species-specific BACs can be traced in different ways. BES of entire libraries have been produced for some primates: chimpanzee, orangutan, *Nomascus leucogenys*, macaque, and marmoset, for instance. The macaque BES are displayed at this site: http://brl.bcm.tmc.edu/pgi/rhesus/dataAccess.rhtml. A sequence can be BLASTed against a specific genome to reveal BES that identify appropriate clones. If the library, but not the BES, is available, then, using an appropriate probe, the high-density filters in which the library has been arrayed can be screened in search of specific BACs. A consistent number of primate BAC libraries along with their high-density filters are available from P. de Jong (Oakland; http://bacpac.chori.org/).

13.4.3 *Cautions in Interpreting Comparative FISH Results*

The breakpoint definition in a species is technically quite similar to the definition of breakpoints in a constitutional rearrangements, with one major difference: the constitutional rearrangements are always derivative. In comparative studies, one of the two species is not always a good reference species. Both species could be derived with respect to their last common ancestor. The best example, in this respect, is provided by the human chromosome 3. If human chromosome 3 and the corresponding chromosome in the Sumatra orangutan are compared, it has to be kept in mind that two distinct pericentric inversions occurred, since the separation of the two species, in each of the two lineages (Fig. 13.5). The assumption that the human is the reference is an anthropocentric scenario.

13.5 Neocentromeres

Human clinical neocentromeres are perfectly functioning, analphoid centromeres that occasionally emerge in ectopic chromosomal regions (for a review, see Marshall et al. 2008). Most neocentromeres were seeded in acentric, supernumerary chromosomal fragments, allowing, therefore, their mitotic survival. These findings added further oddities to the black hole of chromosomes: the centromere. Neocentromeres, indeed, completely lack the complexity of a normal centromere, which consists of a large array of satellite DNA (alpha satellite in primates) flanked by clusters of segmental duplications (She et al. 2004). The presence of the supernumerary chromosome negatively affects the phenotype, and the clinical problems bring these patients

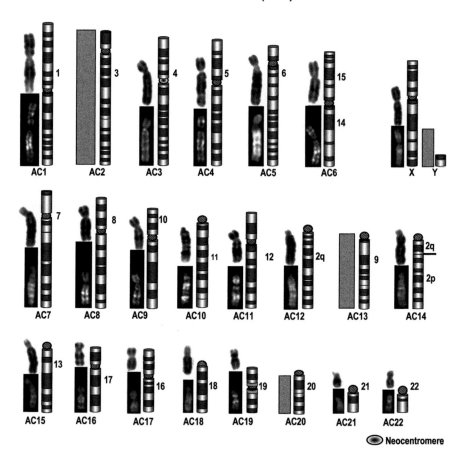

Fig. 13.5 Diagrams of hypothetical ancestral catarrhine (AC) karyotype (ACK), as described in Stanyon et al. (2008). Corresponding human chromosomes are reported on the *right*. No living species retained the AC2, AC13, AC20, and AHY marker order organization

to observation. In few interesting instances, however, the neocentromere just repositioned along the chromosome, apparently leaving unchanged both the old centromere and the sequences harboring the new centromere (Amor et al. 2004; Capozzi et al. 2009; Ventura et al. 2004). The ChIP-on-chip analysis (see below), performed in one case, confirmed that a single-copy sequence had acquired the centromere function. These centromere movements do not generate any cytogenetic imbalance and the individuals are therefore perfectly normal. Indeed, they were discovered by chance, usually because of a prenatal cytogenetic analysis that revealed the centromere repositioning in the fetus. Cytogenetic analysis extended to family members showed that the neocentromere was present in at least two generations, confirming its perfectly normal mitotic and meiotic behavior (Capozzi et al. 2009).

13.6 Evolutionary New Centromeres

Evolutionary studies performed by detailed marker order comparison of homologous chromosomes between different species have disclosed a phenomenon that intimately resembles the centromere movement in these human clinical cases. The evolutionary new centromere (ENC) was obviously seeded in a single individual, then spread in the population, and was finally fixed. ENC have been documented in a variety of eukaryotes, in particular, primates. The human cases, therefore, have been considered as ENC "in progress."

The first clear case of ENC in primates was reported by Montefalcone et al. (1999). Since then, a large number of ENCs has been documented, in a variety of taxa including plants (Han et al. 2009; Carbone et al. 2006; Lomiento et al. 2008). Most of them have been described in primates, but very likely this depends on the fact that primates are better studied from a molecular cytogenetic point of view. In some primate species they are very frequent. If human and macaque are compared, 14 centromeres appear evolutionarily new: 9 occurred in the macaque lineage and 5 occurred in the human lineage after their divergence from the common ancestor. It is worth noting that no ENC was identified following the completion of the sequence of a genome. Centromeres, because of their highly repetitive nature, are usually annotated just as a sequence gap, and gaps are relatively frequent in many shotgun assembled genomes.

13.6.1 Features of Evolutionary New Centromeres

The analysis of a relative large number of ENCs allowed the identification of some common features. The satellite block of the inactivated centromere is usually lost quite rapidly. In sequenced primates, so far as we know, no remains of alpha satellite DNA, typical of primate centromeres, were found in the domain where an ancestral centromere inactivated, with one exception. Human chromosome 2, as mentioned, is the result of a telomere–telomere fusion of two ancestral chromosomes. Remains of alpha satellite DNA can be found at 2q21.1, where the second centromere of the initially dicentric chromosome inactivated. We do not know when the fusion occurred, but surely after the human–chimpanzee divergence, which dates back to 5–6 million years ago. Primate centromeres, and more generally mammalian centromeres, are also characterized by the presence of large clusters of segmental duplications flanking the heterochromatic block (She et al. 2004). In contrast to the loss of satellite DNA, occasionally clusters of segmental duplications have been found in regions encompassing domains where ancestral centromeres inactivated. The best example is found at human 15q24-26. It is now well known that chromosomes 14 and 15 were joined in a single ancestral synteny. Ape chromosomes corresponding to human 14 and 15 are the results of a chromosomal fission of this ancestral chromosome (chromosome 7 in macaque). The fission point corresponded to the short arm of chromosome 15 and to the telomeric region of chromosome 14q (Ventura et al. 2003). The ancestral centromere, at 15q25, inactivated and lost any alpha satellite sequence, whereas a large cluster of segmental duplications persisted. This region is also evolutionarily interesting for an additional reason (see following).

The parallelism of ENCs with human clinical neocentromeres and, more specifically, with the few examples of human repositioned centromeres reported in normal individuals (see foregoing), suggests that ENC seeding points were initially devoid of alpha satellite DNA. The progression of these ENs, however, appears quite obligate: they rapidly acquire the normal complexity, in term of satellite DNA and pericentromeric segmental duplications. On this basis, the nine evolutionary new macaque centromeres are indistinguishable from the old ones. Similar to the "old" centromeres, they harbor large blocks of alphoid DNA (Ventura et al. 2007). The organization of the pericentromeric region of the macaque chromosome 4 (human 6) was studied in detail, using the corresponding human chromosomal region as a reference, because it represents the ancestral organization of the region before the neocentromere repositioning event in the Old World monkey ancestor. A 250-kb single-copy segment was extensively duplicated around this centromere. The acquisition of satellite DNA and pericentromeric duplications appears, therefore, as an ineluctable fate of neocentromeres that were fixed in the population.

13.6.2 Genes in Neocentromeric Domains

What about the fate of genes occasionally located in a neocentromeric domain? The neocentromere, per se, does not affect gene expression (Saffery et al. 2003). The structural reshuffle of the sequence where the neocentromere was seeded, however, can be hypothesized to affect gene integrity. These two considerations may predict that only neocentromeres that were seeded in a gene desert have an evolutionary perspective. This hypothesis was tested on 14 ENCs and the correlation was statistically highly significant (Lomiento et al. 2008).

CENP-A and CENP-C are two of the centromeric proteins exclusively located at functional centromeres. DNA immunoprecipitation using antibodies raised against these two proteins, followed by the hybridization of the immunoprecipitated

DNA-to-DNA oligo arrays, identifies with certainty the single-copy sequences responsible for neocentromere function. Comparison of the main features of these DNA sequence domains did not reveal clear sequence features linking the sequence to the centromere function.

13.6.3 Reuse of Centromere Domains for Neocentromere Seeding

Although no genes or sequences are linked to a neocentromere, it does appear that some centromere domains have been used repeatedly for neocentromere seeding. Indeed, examples are accumulating that the same domain can be used as a neocentromere seeding point more than one time in evolution or in clinical cases. The first example came from the 15q24-26 domain, which is the region harboring an inactivated centromere (see foregoing). Interestingly, a cluster of human clinical neocentromeres maps to this region (Ventura et al. 2003). The centromere of macaque chromosome 17 (human 13) is an ENC and maps to the middle of this chromosome (13q21). This same domain was used as an ENC seeding point in the pig (Cardone et al. 2006). The divergence between these two species dates back to approximately 95 million years ago. One of the three human cases of centromere repositioning in a normal individual is found at 3q26. This region corresponds to the domain where an ENC emerged in the common ancestor of the Old World monkeys, 25–40 million years ago. The clinical neocentromere on chromosome 9 reported by Capozzi et al. (2008) maps to 9q33.1. Again, this was the domain where an ENC was seeded in the OWM ancestor. Interestingly, the human neocentromere arose to rescue a small ring chromosome, 12 Mb in size, generated by an excision event in one chromosome 9 of the mother of the propositus. The neocentromere came under observation because the propositus inherited, from his mother, the deleted (excised) chromosome 9 but not the ring chromosome. Additionally, a "normal" repositioned centromere (see earlier), segregating in a three-generation family, occurred on chromosome 6. The centromere moved to 6p22.1. A careful analysis of the evolutionary history of this chromosome showed that this was the normal location of the centromere about 17 million years ago (Capozzi et al. 2009). These examples point to these domains as regions where neocentromeric latency lasted for millions of years.

13.7 Marker Order of Ancestral Karyotypes

The use of FISH with cloned DNA and especially BACs during the past 10 years has allowed remarkable progress in determining marker order along the primate chromosome. As we have seen, these results allowed the important phenomena of the neocentromere to emerge. These methods also allowed hypotheses about the ancestral marker order and centromere position in ancestral karyotypes at five major branching points on the primate evolutionary tree (cf. Stanyon et al. 2008): ancestral

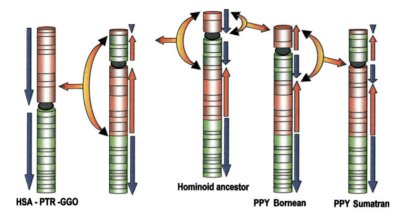

Fig. 13.6 Evolution of chromosome 3 in hominoids, according to Ventura et al. (2004). For details, see text

primate, ancestral anthropoid, ancestral platyrrhine, ancestral catarrhine (Fig. 13.6), and ancestral hominoid (Fig. 13.4). The application of BAC FISH to additional primates will help improve the hypothesis of the content of ancestral genomes during the next few years.

13.8 Molecular Cytogenetics and Sequencing Projects

The human genome sequencing project has utilized a dedicated "hierarchical" approach, in which a "golden path" of overlapping clones, essentially BAC/PAC, almost completely covering the entire genome, has been first ordered and then sequenced. The sequencing of all the other genomes has utilized a more simple and less expensive approach, called "shotgun," in which a reasonable amount of small-insert DNA clones are sequenced and then assembled using dedicated software. This kind of assembly is more error-prone because of the complexity of eukaryotic genomes. Segmental duplications and disperse repetitive elements are not easy to handle. A distinct, sequence independent validation and testing of these assemblies is represented by the FISH technology, that can prove or disprove assembly hypotheses. Useful information can be gained by studies on the evolutionary history of chromosomes. The rationale of this approach has been illustrated in detail (Rocchi et al. 2006). The assembly of chromosomes 12 and 13 in the macaque, which correspond to human chromosome 2, is a paradigmatic example. Human chromosome 2 is the result of a telomere–telomere fusion of two ancestral acrocentric chromosomes corresponding to macaque chromosomes 12 (2q) and 13 (2p). A stretch of head-to-head telomeric sequences precisely identifies the fusion point (chr 2: 114,076,792–114,077,148; March 2006 release). In humans, no inversion of this region occurred after the chromosomal fusion event. The fusion point, therefore, precisely delimits sequences belonging to macaque chromosome 12 from those of 13. As a consequence, the 24.7-Mb segment (chr 2: 114,076,736–138,830,121) of 2q, mapping below the fusion, and placed, on the macaque assembly, on macaque 13 (2p), is erroneous (Roberto et al. 2008).

Some primate genomes show very rearranged karyotypes with respect to their common ancestor. Gibbons are the best known example in this respect (Müller et al. 2003; and Chap. 9). Utilizing a large panel of human BAC clones, employed in cohybridization experiments on *H. lar* and on *N. leucogenys* gibbons, we were able to pinpoint the many rearrangements that differentiated these two species with respect to the hypothetical gibbon ancestor, with respect to man, and with respect to their common ancestor (Misceo et al. 2008; Roberto et al. 2007). The arrangements of the synteny blocks in these two gibbons are graphically summarized at http://www.biologia.uniba.it/lar/ and http://www.biologia.uniba.it/gibbon/.

The BES for the entire *Nomascus* BAC library has been produced. This genus represented a good example of how molecular cytogenetics and limited sequencing (BES) can provide an extreme detailed synteny arrangement. The systematic mapping, by BLAST, of all the *Nomascus* BES, identified those BACs encompassing breakpoints in the lineages leading to humans. Their BES, indeed, map to completely different location in humans, in case of translocations, or to different positions along the chromosome in case of inversion. Conversely, we experimentally identified, using human BAC clones in reiterative FISH experiments on *Nomascus* chromosomes, the human BACs spanning breakpoints of rearrangements that occurred in the lineages leading to *Nomascus* after the separation of their ancestral lineages. In this way we constructed a very detailed map of synteny conservation that will be very helpful when the *Nomascus* is fully sequenced. FISH experiments using these peculiar BAC clones are depicted in Fig. 13.3b.

13.8.1 Future of Cytogenetics in the Massive Sequencing Age

What place does cytogenetics have in the age of massive sequencing? This is a question frequently asked and often pondered by cytogeneticists. Certainly the resolution of sequencing cannot be approached by cytogenetics. Although there is talk of $1,000 genomes and the sequencing of thousands of vertebrate species, the current reality is that there are still only a very limited number of complete genome assemblies, not anywhere near the phylogenetic richness of species studied by molecular cytogenetics. So for the moment there appears to be work for cytogeneticists and their global descriptions of species genomes. What about the future in 5 or 10 years? As illustrated here, cytogenetics can serve as a test for any genome assembly. Inconsistencies in results between cytogenetics and bioinformatics point out opportunities to reciprocally test hypotheses and improve ancestral genome reconstructions. For instance, appropriate cohybridization FISH experiments of cloned DNA, essentially BACs (BAC-FISH), can independently test contiguous ancestral regions (CAR) orientation, adjacencies, and chromosomal breakpoints suggested by bioinformatics. This point represents common ground where sequencing and cytogenetics can meet to generate high-quality synteny block (SB) analyses: the end sequencing of appropriate BAC libraries. A BAC can be allocated to a SB by FISH, or, much more precisely, by placing its end sequences (BES) on the human sequence, which is usually used as a reference. This method establishes an extremely precise connection between cytogenetic and bioinformatic data sets. Appropriate BACs can easily verify inconsistencies among different data sets. Another limitation already mentioned the fidelity of sequence assembly as a consequence of the "shotgun" sequence methodology. These problems are particularly evident around centromere and pericentromeric regions, which are often sequencing black holes.

Therefore, our prediction is that cross-talk between molecular cytogenetics and sequencers will increase until these two fields become integrated (Rocchi et al. 2006). This integrated science of phylogenomics will make notable contributions to reconstructing ancestral genomes, tracing the origins of human chromosomes and interpreting current genomic phenomenon within a holistic, evolutionary framework.

References

Amor DJ, Bentley K, Ryan J et al (2004) Human centromere repositioning "in progress". Proc Natl Acad Sci USA 101:6542–6547

Bigoni F, Koehler U, Stanyon R et al (1997a) Fluorescence in situ hybridization establishes homology between human and silvered leaf monkey chromosomes, reveals reciprocal translocations between chromosomes homologous to human Y/5, 1/9, and 6/16, and delineates an X1X2Y1Y2/X1X1X2X2 sex-chromosome system. Am J Phys Anthropol 102:315–327

Bigoni F, Stanyon R, Koehler U et al (1997b) Mapping homology between human and black and white colobine monkey chromosomes by fluorescent in situ hybridization. Am J Primatol 42:289–298

Capozzi O, Purgato S, Verdun di Cantogno L et al (2008) Evolutionary and clinical neocentromeres: two faces of the same coin? Chromosoma (Berl) 117:339–344

Capozzi O, Purgato S, D'Addabbo P et al (2009) Evolutionary descent of a human chromosome 6 neocentromere: a jump back to 17 million years ago. Genome Res 19:778–784

Carbone L, Nergadze SG, Magnani E et al (2006) Evolutionary movement of centromeres in horse, donkey, and zebra. Genomics 87:777–782

Cardone MF, Alonso A, Pazienza M et al (2006) Independent centromere formation in a capricious, gene-free domain of chromosome 13q21 in Old World monkeys and pigs. Genome Biol 7:R91

Caspersson T, Zech L, Johansson C et al (1970) Identification of human chromosomes by DNA-binding fluorescent agents. Chromosoma (Berl) 30:215–227

Chu EHY, Bender MA (1961) Chromosome cytology and evolution in primates. Science 133:1399–1405

Darwin C (1859) Origin of species by means of natural selection, or the preservation of favoured races in the struggle for life. John Murray, London

Dumas F, Stanyon R, Sineo L et al (2007) Phylogenomics of species from four genera of New World monkeys by flow sorting and reciprocal chromosome painting. BMC Evol Biol 7(suppl 2):S11

Dumas F, Houck ML, Bigoni F et al (2012) Chromosome painting of the pygmy tree shrew shows that no derived cytogenetic traits link Primates and Scandentia. Cytogenet Genome Res (in press)

Dutrillaux B (1979) Chromosomal evolution in primates: tentative phylogeny from *Microcebus murinus* (Prosimian) to man. Hum Genet 48:251–314

Dutrillaux B, Finaz C, de Grouchy J et al (1972) Comparison of banding patterns of human chromosomes obtained with heating, fluorescence, and proteolytic digestion. Cytogenetics 11:113–116

Dutrillaux B, Viegas-Pequignot E, Couturier J (1980) Great homology of chromosome banding of the rabbit (*Oryctolagus cuniculus*) and primates, including man (author's translation. Ann Genet 23:22–25

Ferguson-Smith MA, Trifonov V (2007) Mammalian karyotype evolution. Nat Rev Genet 8:950–962

Ferguson-Smith MA, Yang F, Rens W et al (2005) The impact of chromosome sorting and painting on the comparative analysis of primate genomes. Cytogenet Genome Res 108:112–121

Groves CP (2001) Primate taxonomy. Smithsonian Institution Press, Washington, DC

Han Y, Zhang Z, Liu C et al (2009) Centromere repositioning in cucurbit species: implication of the genomic impact from centromere activation and inactivation. Proc Natl Acad Sci USA 106:14937–14941

Huxley TH (1863) Evidence as to man's place in nature. Williams & Norwood, London

Jauch A, Wienberg J, Stanyon R et al (1992) Reconstruction of genomic rearrangements in great apes and gibbons by chromosome painting. Proc Natl Acad Sci USA 89:8611–8615

Koehler U, Arnold N, Wienberg J et al (1995a) Genomic reorganization and disrupted chromosomal synteny in the siamang (*Hylobates syndactylus*) revealed by fluorescence in situ hybridization. Am J Phys Anthropol 97:37–47

Koehler U, Bigoni F, Wienberg J et al (1995b) Genomic reorganization in the concolor gibbon (*Hylobates concolor*) revealed by chromosome painting. Genomics 30:287–292

Lomiento M, Jiang Z, D'Addabbo P et al (2008) Evolutionary-new centromeres preferentially emerge within gene deserts. Genome Biol 9:R173

Makino S (1952) A contribution to the study of the chromosomes in some Asiatic mammals. Cytologia 16:288–301

Marshall OJ, Chueh AC, Wong LH et al (2008) Neocentromeres: new insights into centromere structure, disease development, and karyotype evolution. Am J Hum Genet 82:261–282

Misceo D, Capozzi O, Roberto R et al (2008) Tracking the complex flow of chromosome rearrangements from the Hominoidea ancestor to extant *Hylobates* and *Nomascus* gibbons by high-resolution synteny mapping. Genome Res 18:1530–1537

Montefalcone G, Tempesta S, Rocchi M et al (1999) Centromere repositioning. Genome Res 9:1184–1188

Müller S, O'Brien PC, Ferguson-Smith MA et al (1997) Reciprocal chromosome painting between human and prosimians (*Eulemur macaco macaco* and *E. fulvus mayottensis*). Cytogenet Cell Genet 78:260–271

Müller S, Stanyon R, O'Brien PC et al (1999) Defining the ancestral karyotype of all primates by multidirectional chromosome painting between tree shrews, lemurs and humans. Chromosoma (Berl) 108:393–400

Müller S, Stanyon R, Finelli P et al (2000) Molecular cytogenetic dissection of human chromosomes 3 and 21 evolution. Proc Natl Acad Sci USA 97:206–211

Müller S, Hollatz M, Wienberg J (2003) Chromosomal phylogeny and evolution of gibbons (Hylobatidae). Hum Genet 113:493–501

Murphy WJ, Fronicke L, O'Brien SJ et al (2003) The origin of human chromosome 1 and its homologs in placental mammals. Genome Res 13:1880–1888

Neusser M, Stanyon R, Bigoni F et al (2001) Molecular cytotaxonomy of New World monkeys (Platyrrhini): comparative analysis of five species by multi-color chromosome painting gives evidence for a classification of *Callimico goeldii* within the family of Callitrichidae. Cytogenet Cell Genet 94:206–215

Nie W, Liu R, Chen Y et al (1998) Mapping chromosomal homologies between humans and two langurs (*Semnopithecus francoisi* and *S. phayrei*) by chromosome painting. Chromosome Res 6:447–453

Nie W, Rens W, Wang J et al (2001) Conserved chromosome segments in *Hylobates hoolock* revealed by human and *H. leucogenys* paint probes. Cytogenet Cell Genet 92:248–253

Nie W, O'Brien PC, Fu B et al (2006) Chromosome painting between human and lorisiform prosimians: evidence for the HSA 7/16 synteny in the primate ancestral karyotype. Am J Phys Anthropol 129:250–259

Nie W, Fu B, O'Brien PC et al (2008) Flying lemurs – the 'flying tree shrews'? Molecular cytogenetic evidence for a Scandentia-Dermoptera sister clade. BMC Biol 6:18

O'Brien SJ, Nash WG (1982) Genetic mapping in mammals: chromosome map of domestic cat. Science 216:257–265

Roberto R, Capozzi O, Wilson RK et al (2007) Molecular refinement of gibbon genome rearrangement. Genome Res 17:249–257

Roberto R, Misceo D, D'Addabbo P et al (2008) Refinement of macaque synteny arrangement with respect to the official rheMac2 macaque sequence assembly. Chromosome Res 16:977–985

Rocchi M, Archidiacono N, Stanyon R (2006) Ancestral genomes reconstruction: an integrated, multi-disciplinary approach is needed. Genome Res 16:1441–1444

Saffery R, Sumer H, Hassan S et al (2003) Transcription within a functional human centromere. Mol Cell 12:509–516

Seabright M (1971) A rapid banding technique for human chromosomes. Lancet 2:971–972

She X, Horvath JE, Jiang Z et al (2004) The structure and evolution of centromeric transition regions within the human genome. Nature (Lond) 430:857–864

Shiwago P (1939) Recherches sur le caryotype du *Rhesus macacus*. Bull Biol Med Exp (USSR) 9:3–8

Stanyon R, Stone G (2008) Phylogenomic analysis by chromosome sorting and painting. Methods Mol Biol 422:13–29

Stanyon R, Wienberg J, Romagno D et al (1992) Molecular and classical cytogenetic analyses demonstrate an apomorphic reciprocal chromosomal translocation in *Gorilla gorilla*. Am J Phys Anthropol 88:245–250

Stanyon R, Bonvicino CR, Svartman M et al (2003) Chromosome painting in *Callicebus lugens*, the species with the lowest diploid number (2n = 16) known in primates. Chromosoma (Berl) 112:201–206

Stanyon R, Bruening R, Stone G et al (2005) Reciprocal painting between humans, De Brazza's and patas monkeys reveals a major bifurcation in the Cercopithecini phylogenetic tree. Cytogenet Genome Res 108:175–182

Stanyon R, Dumas F, Stone G et al (2006) Multidirectional chromosome painting reveals a remarkable syntenic homology between the greater galagos and the slow loris. Am J Primatol 68:349–359

Stanyon R, Rocchi M, Capozzi O et al (2008) Primate chromosome evolution: ancestral karyotypes, marker order and neocentromeres. Chromosome Res 16:17–39

Stock AD, Hsu TC (1973) Evolutionary conservatism in arrangement of genetic material. A comparative analysis of chromosome banding between the rhesus macaque (2n equals 42, 84 arms) and the African green monkey (2n equals 60, 120 arms). Chromosoma (Berl) 43:211–224

Svartman M, Stone G, Page JE et al (2004) A chromosome painting test of the basal eutherian karyotype. Chromosome Res 12:45–53

Trifonov VA, Stanyon R, Nesterenko AI et al (2008) Multidirectional cross-species painting illuminates the history of karyotypic evolution in Perissodactyla. Chromosome Res 16:89–107

Ventura M, Mudge JM, Palumbo V et al (2003) Neocentromeres in 15q24-26 map to duplicons which flanked an ancestral centromere in 15q25. Genome Res 13:2059–2068

Ventura M, Weigl S, Carbone L et al (2004) Recurrent sites for new centromere seeding. Genome Res 14:1696–1703

Ventura M, Antonacci F, Cardone MF et al (2007) Evolutionary formation of new centromeres in macaque. Science 316:243–246

Wienberg J, Stanyon R (1997) Comparative painting of mammalian chromosomes. Curr Opin Genet Dev 7:784–791

Wienberg J, Stanyon R, Jauch A et al (1992) Homologies in human and *Macaca fuscata* chromosomes revealed by in situ suppression hybridization with human chromosome specific DNA libraries. Chromosoma (Berl) 101:265–270

Yeager CH, Painter TS, Yerkes RM (1940) The chromosomes of the chimpanzee. Science 91:74–75

Yu D, Yang F, Liu R (1997) A comparative chromosome map between human and *Hylobates hoolock* built by chromosome painting. Yi Chuan Xue Bao 24:417–423

Yunis JJ, Prakash O (1982) The origin of man: a chromosomal pictorial legacy. Science 215:1525–1530

Chapter 14
Chromosomal Evolution of Gibbons (Hylobatidae)

Stefan Müller and Johannes Wienberg

Abbreviations

BAC   Bacterial artificial chromosome
FISH   Fluorescence in situ hybridization
WAT   Whole-arm translocation

14.1 Introduction

Together with the great apes, chimpanzee, bonobo, gorilla, orangutan, and human, gibbons (Hylobatidae, the lesser or small apes) are included in the superfamily Hominoidea. Humans and gibbons are estimated to have separated from their common hominoid ancestor between 15 and 20 million years ago (Goodman 1999). Gibbons are highly endangered species that inhabit the forests of South and Southeast Asia and represent the most variable members of this taxon. Although gibbons are our closest relatives besides great apes, there has been considerable debate about their taxonomy, phylogeny, and evolution. Early systematics divided gibbons into two distinct subgenera that included the siamang (*Symphalangus*, one species) and *Hylobates* (all other lesser apes) (Napier and Napier 1967). More recently, from studies on various biological traits it became evident that four distinct clades of lesser apes should be recognized. The most recent taxonomic classification of gibbons includes 15 species from four genera: *Hoolock* (hoolock gibbon), with 2

S. Müller (✉)
Institut für Humangenetik, Klinikum der Ludwig-Maximilians-Universität, Munich, Germany
e-mail: s.mueller@lrz.uni-muenchen.de

J. Wienberg
Anthropology and Human Genetics, Department of Biology II, Ludwig-Maximilians-Universität, Munich, Germany

Chrombios GmbH, Raubling, Germany

H. Hirai et al. (eds.), *Post-Genome Biology of Primates*, Primatology Monographs,
DOI 10.1007/978-4-431-54011-3_14, © Springer 2012

species, *Symphalangus* (siamang), consisting of a single species, *Hylobates*, with 7 species, and *Nomascus* (crested gibbons), with 5 species (Brandon-Jones et al. 2004; Mootnick 2006; Mootnick and Groves 2005; Roos et al. 2007). This division also reflects the different karyomorphs found in these four clades, with representatives such as the hoolock gibbon (*Hoolock hoolock*, formerly *Bunopithecus*), 2n = 38, the white-handed gibbon (*Hylobates lar*), 2n = 44, the siamang (*Symphalangus syndactylus*), 2n = 50, and the white-cheeked gibbon (*Nomascus concolor*), 2n = 52.

On the basis of various molecular, behavioral, and morphological analyses, different phylogenetic trees have been published for gibbons. For example, Roos and Geissmann (2001) suggested that the evolutionary branching sequence was *Nomascus* {*Symphalangus* {*Hoolock* and *Hylobates*}}, whereas Garza and Woodruff (1992) identified *Symphalangus* as the most basal group of the Hylobatidae, followed by *Nomascus*, with *Hylobates* and *Bunopithecus* as the last to diverge. Another molecular analysis suggested either *Hoolock* or *Nomascus* as the basal clade of hylobatids, whereas *Symphalangus* and *Hylobates* were consistently found to be the last genera to diverge (Takacs et al. 2005). These competing interpretations of gibbon phylogeny suggest that extant gibbons may have diverged in a very short evolutionary time, which renders it difficult to obtain enough informative markers to establish a conclusive phylogenetic tree. Alternatively, it has been speculated that gibbon phylogenetic relationships may instead represent a true polytomy, or that gibbon speciation was further complicated by hybridization events (Arnold and Meyer 2006). This possibility, however, seems quite unlikely, because many gibbon species are geographically isolated and there are only two sympatric species, with no indication of hybridization except for artificial breeds in zoos. For example, Myers and Shafer (1979) described a "Siabon," an intergeneric hybrid between *Hylobates* and *Symphalangus*. More recently, a "Larcon" resulted from an accidental cross between a female white-handed gibbon (*Hylobates lar*) and a male concolor-group gibbon (*Nomascus* sp.) (Hirai et al. 2007). Taking into consideration the pronounced differences in the chromosomal genome organization of the parental species, however, fertility of the intergeneric hybrid offspring would not be expected.

14.2 Chromosome Reshuffling in Gibbons

Although gibbons are very closely related to human and great apes, comparison of classical G- or R-banding patterns of gibbon chromosomes could rarely identify homologies between lesser and great apes. Except for the well-conserved X chromosome, only a few gibbon chromosomes showed a banding pattern similar to a putative homologue in the human or great apes (de Grouchy et al. 1978). In contrast, Old World monkeys such as macaques and baboons, which are known to be more distantly related to humans than gibbons, were found to have conserved most chromosomal syntenies with great apes (Wienberg et al. 1992; see also Chap. 13, by Stanyon, this volume). Comparative analyses of chromosome morphology between gibbon species revealed additional extensive differences in chromosome banding

14 Chromosomal Evolution of Gibbons (Hylobatidae)

patterns, suggesting various translocations and other rearrangements (Dutrillaux et al. 1975; Marks 1982; Stanyon 1983; Van Tuinen and Ledbetter 1983). Despite the technical limitation of these early studies, it had already become clear that gibbons experienced a dramatic lineage-specific change in chromosome morphology not found in any other primates.

14.3 Large-Scale Rearrangements in Gibbons Revealed by Chromosome Painting

Cross-species fluorescence in situ hybridization (FISH) analysis of gibbon chromosomes, and in particular the use of human chromosome-specific painting probes, for the first time allowed a detailed description of these complex evolutionary chromosome changes (Wienberg et al. 1990). During the following years, the four gibbon karyomorphs were analyzed in detail by this method: three species of the genus *Hylobates* (*H. lar*, *H. agilis*, *H. klossii*) (Jauch et al. 1992), the white-cheeked gibbon (Koehler et al. 1995b; Schröck et al. 1996), the siamang (Koehler et al. 1995a), and the hoolock (Yu et al. 1997). In their analysis of chromosome rearrangements in three of the four genera that had been studied at that time, Koehler et al. (1995a) suggested that the majority of the rearrangements should have occurred in their common ancestor, whereas shared derived rearrangements of putative sister genera could not be identified with certainty. Thus, because of the complexity of many of the rearrangements, it was not possible to propose a firm phylogenetic interpretation of gibbon chromosome evolution using human chromosome-specific painting probes alone.

To overcome these limitations, a "reciprocal" or "multidirectional" chromosome painting strategy was introduced to obtain more detailed information about chromosome homologies between species (Arnold et al. 1996). In addition to human paint probes, paint probes from at least one representative of the species group to be investigated was applied in cross-species FISH. This method also allowed the identification of homologous chromosome subregions and therefore a more precise interpretation of the origin of complex chromosome rearrangements. This strategy was helpful for the establishment of detailed homology maps between human and gibbon species from all four genera (*N. leucogenys*, *S. syndactylus*, *H. hoolock*, and *H. lar*) (Müller et al. 1998, 2003; Nie et al. 2001).

14.4 Chromosomal Polymorphisms in Gibbons

Examinations within distinct gibbon species using classical banding information further revealed polymorphisms for inversions and translocations (Couturier et al. 1982; Couturier and Lernould 1991; Stanyon et al. 1987; Van Tuinen and Ledbetter

1983). Because only a few individuals had been analyzed in these early studies, it was not clear whether these rearrangements are true polymorphisms or defined karyological differences of subspecies or of species not yet recognized. Furthermore, some of these "polymorphisms" may have been the result of artificial matings between individuals from different subspecies kept in captivity.

More recently van Tuinen et al. (1999) analyzed more than 60 individuals of the genus *Hylobates* by chromosome banding and described extensive inversion/translocation chromosome polymorphisms. At least some of these rearrangements, namely, the different inversions of *Hylobates* chromosome 8 and the derivative from one of these inversion morphs, the whole-arm translocation (WAT 8/9), was subsequently described as a marker for geographically distinct populations from the genus *Hylobates*, which today are classified as separate species. The WAT8/9 was found to be present in the agile gibbons *Hylobates agilis unko* and *H. agilis agilis* from Sumatra, but it was absent in Bornean *H. albibarbis* and *H. muelleri*.

Concerning the *Nomascus* group, banding analysis revealed four distinct karyotypes resulting from a combination of a translocation between chromosomes 1 and 22 and an inversion of chromosome 7 (Couturier and Lernould 1991). The respective subtypes 1a/1b, 22a/22b, and 7a/7b were initially also described as polymorphisms present in the different (sub)species of *Nomascus*. Individuals from *Nomascus concolor leucogenys* were shown to have chromosome forms 1a, 7a, and 22a; *N. c. siki* was described with chromosomes 1b, 7a, and 22b; and *N. c. gabriellae* with 1b, 7b, and 22b (Couturier and Lernould 1991; Koehler et al. 1995b). A more recent study demonstrated population-specific differences, because southern white-cheeked gibbons (*N. leucogenys siki*) were described having chromosome types 1b, 7b, and 22b, and northern white-cheeked gibbons (*N. leucogenys leucogenys*) showed chromosome types 1a, 7a (or 7b), and 22a (Roberto et al. 2007). The distribution of the putatively polymorphic inversion of *Nomascus* chromosome 7 was further investigated in more than 50 individuals from different (sub)species of gibbon, using a polymerase chain reaction (PCR)-based screening approach, and taking advantage of the known breakpoint sequence (Carbone et al. 2006, 2009b): notably, this inversion was shown to be taxon specific for *N. leucogenys leucogenys* (NLE). All tested NLE individuals were homozygous for this inversion, in disagreement with the interpretation by Couturier and Lernould (1991). The study by Carbone et al. (2009b) emphasizes that chromosomal characters may represent informative markers that can be employed in conservation and breeding efforts, and also that careful sampling and the analysis of individuals with proven origin is essential when trying to differentiate between polymorphisms and taxon-specific chromosome forms.

14.5 The Ancestral Gibbon Karyotype

In an attempt to reconstruct the inferred ancestral gibbon karyotype based on multi-directional chromosome painting data from the four gibbon genera (Müller et al. 2003), the orangutan was included as an "outgroup" for lesser apes because it has a karyotype that is supposed to be most similar to the ancestral form of human and apes (Müller and Wienberg 2001). Thus, comparing gibbon with orangutan instead of human chromosomes helped to reduce "noise" from additional derived chromosome rearrangements, mostly inversions (reviewed by Stanyon et al. 2008; see Chap. 13 by Stanyon), that occurred in the African ape–human lineage but not during the early stages of ape–orangutan genome evolution.

According to the cross-species chromosome painting analyses, in the common ancestor of hominoids and before the split of gibbons and great apes, a fission of the syntenic association 14/15 took place, leading to separate chromosomes 14 and 15. Thus, an association of human homologues to 14/15 is not found in the ancestral gibbon karyotype. The inferred ancestral gibbon karyotype included homologues to human chromosomes 7, 9, 13, 14, 15, 20, 21, 22, X, and Y with conserved synteny, because these chromosomes were found entirely conserved in at least one gibbon species. It further incorporated chromosome segments with conserved synteny between the orangutan and at least one extant gibbon, which were either found as separate chromosomes in one or more gibbon karyotypes or were involved in different rearrangements in each phylogenetic lineage. These findings included segments homologous to human chromosome 1 (three segments), 2 (two segments), 3, 4, 5, and 6 (two segments), and 8, 11, 12, and 17, respectively. Finally, syntenic associations of segments homologous to different human chromosomes were included: 3/8, 3/12/19, 5/16, 5/16/5/16, 19/12/19, 10/4 (twice), 12/3/8, 17/2/17/2, and 18/11. By conclusion, the inferred ancestral karyotype of all extant gibbons would have had 2n = 66 chromosomes and already differed from the putative ancestral hominoid by at least 24 rearrangements: 5 reciprocal translocations, 8 inversions, 10 fissions, and 1 fusion (Müller et al. 2003) (see Fig. 14.1).

14.6 Gibbon Chromosomal Phylogeny

A cladistic reconstruction based on the identification of ancestral versus derived chromosome forms placed the genus *Hoolock* as the most basal group of the Hylobatidae, followed by *Hylobates* and with a last common ancestor of *Symphalangus* and *Nomascus* (Müller et al. 2003). The hoolock was not found to share any further derived chromosome changes with other gibbons, but 16 of its 18 autosomes were shown to be the product of species-specific rearrangements, including 4 reciprocal translocations, 19 fusions, and 5 fissions. The inferred last common ancestor of white-cheeked gibbon, siamang, and white-handed gibbon would differ from the ancestral gibbon karyotype by fusion of human syntenic segments 2/7, 8/3/11/18, 4/5, and 22/5/16 and the inversion in the chromosome 6 homologue. In the last common ancestor of the white-cheeked gibbon and the siamang, 1 fusion and 4 fissions may have occurred. Again, both species showed a large number of species-specific rearrangements: a minimum of 6 fusions, 7 reciprocal translocations, and 1 inversion in the white-cheeked gibbon and 7 fusions, 3 reciprocal translocations, and 1 inversion in the siamang.

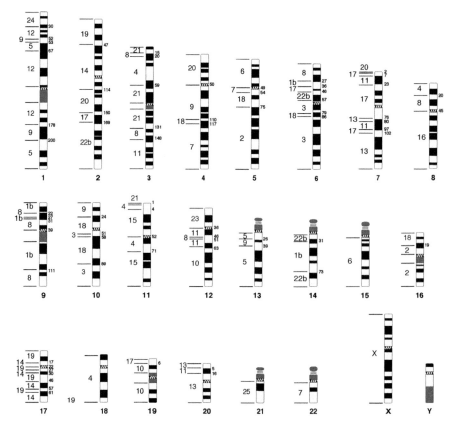

Fig. 14.1 Localization of evolutionary chromosome breakpoints in the northern white-cheeked gibbon (*Nomascus leucogenys leucogenys*) with respect to human chromosomes (according to Müller et al 1998; Nie et al. 2001; Carbone et al. 2006; Roberto et al. 2007). To the *left* of the ideograms of human G-banded chromosomes, *N. L. leucogenys* homologous chromosome segments are displayed. To the *right*, the approximate localization of the breakpoint with respect to the human reference sequence in Mbp is given. (Note: Not all these breakpoints correspond to chromosome rearrangements that occurred in the gibbon lineage. Instead, some breakpoints are hallmarks of more recent evolutionary changes that took place in the great ape/human lineage, for example, the fusion of human chromosome 2. See also Chap. 13 by Stanyon, this volume)

14.7 High-Resolution Molecular Analysis of Gibbon Chromosome Breakpoints

Within primates, the molecular structure of chromosome rearrangements have been most thoroughly analyzed in the human and chimpanzee genome (Kehrer-Sawatzki and Cooper 2007 for review). Until recently, DNA sequencing data were not available in this quality for other species. Close to or within breakpoints, DNA sequence data revealed mostly segmental duplications (SDs) and all sorts of repetitive elements such as SINEs, LINEs, LTRs, α-satellites, and $(AT)_n$ repeats. Further, various disrupted genes were observed that may have lost functions. Hardly any "fusion genes" and dramatic "position effects" have yet been identified in primates; however, various genes at or close to chromosome breakpoints show different expression patterns in human and chimpanzee.

Recent high-resolution genomic analysis of gibbon chromosome breakpoints indicate that the molecular composition of their breakpoints may not be too much different from that found in other primates (Carbone et al. 2006, 2009a; Girirajan et al. 2009). Gibbon chromosome rearrangements have been analyzed in more detail taking advantage of a bacterial artificial clone (BAC) library (CHORI-271) of the northern white-cheeked gibbon (*Nomascus leucogenys leucogenys*). To identify clones that span breakpoints, a combination of various techniques were used. Hybridizing fluorescently labeled DNA from flow-sorted gibbon chromosomes on a array of 32,000 human BAC DNAs spotted on glass slides ("array painting") (Fiegler et al. 2003) identified various human BAC clones that would span breakpoints in the gibbon genome (Carbone et al. 2006). These findings were verified by FISH mapping of the respective BAC probes to gibbon chromosomes. As expected for chromosome rearrangements, most of these BACs produced hybridization signals at two different locations. A different approach for identifying BACs spanning breakpoints was a computer-based analysis of gibbon BAC end sequences by aligning these with the human reference sequence (Carbone et al. 2006; Roberto et al. 2007). Reciprocal FISH experiments then identified BACs from the gibbon library that were covering the breakpoints in a resolution of 80–200 kbp.

Subsequent DNA sequence analysis of numerous BACs spanning breakpoints identified no gibbon-specific sequence element to be common among independent rearrangements (Carbone et al. 2006, 2009a; Girirajan et al. 2009). Instead, as for other rearrangements in primate genomes, at or in close proximity with the exact breakpoints mostly segmental duplications and interspersed repetitive elements were located. *Alu* elements were most abundant among the repeats found around gibbon breakpoints but, interestingly, these were shown to have a higher CpG content compared to other *Alu*s. Furthermore, bisulfite allelic sequencing revealed that these gibbon *Alu* elements have a lower average density of methylated cytosines than their human orthologues (Carbone et al. 2009a). The epigenetic alteration of genomic sequences by DNA methylation is known as a potent regulator of mobile DNA activity and may explain a higher susceptibility of gibbons for genomic instability.

In addition to segmental duplications and repeats, various disrupted genes were observed that frequently were members of gene families (for example, the ABCC family on human chromosome 16). Presence of gene family members at breakpoint regions suggests that these rearrangements may have been occurred through nonallelic homologous recombination. Further, the presence of genes with similar function close to the disrupted genes may have relaxed natural selection against disruptions of chromosomes. Another interesting observation was the detection of various "micro-rearrangements" in 8 of 23 sequenced BACs in the report by Carbone et al. (2009a) that fell below the resolution of BAC end sequencing. Segments derived from other chromosomes were found in the respective BACs, indicating that gibbon chromosome reshuffling may even more pronounced than anticipated so far.

14.8 Nuclear Genome Organization and Patterns of Chromosome Evolution in Gibbons

The interplay between the chromosomal organization of the human genome and its three-dimensional arrangement in the interphase nucleus with respect to the local genomic landscape has been the subject of numerous investigations. Initial studies on the radial arrangement of chromatin revealed a remarkable evolutionary conservation of a gene density-correlated distribution pattern over a period of at least 30 million years (Neusser et al. 2007; Tanabe et al. 2002). In human and as in various non-human primates, gene-dense chromatin is preferentially located in the nuclear center, whereas gene-poor chromatin shows an orientation toward the nuclear periphery. This probabilistic but highly nonrandom three-dimensional (3D) genome organization is strictly maintained in gibbons, despite their extremely reshuffled genomes. Gene density-correlated patterns of the nonrandom higher-order nuclear architecture may further have important consequences for the predominance of certain evolutionary genomic changes, especially on the predisposition for nonrandom breakage and rejoining of centrally located, gene-dense chromosomes or chromosome regions. Interestingly, available molecular cytogenetic and sequence data on the genomic environment surrounding the more than 100 evolutionary chromosome breakpoints in gibbons indicate that the majority of breaks occurred in gene-dense chromosome segments located in the interior of the nucleus (Neusser et al. 2007; Roberto et al. 2007). From an evolutionary perspective, this hypothesis is appealing, because by reshuffling the genomic landscape its adaptive potential could be enhanced while a functionally competent internal nuclear compartment would be evolutionarily preserved. Notably, this pattern seems to be preserved throughout mammals, in whom evolutionary breaks occurred at significantly higher frequency compared to gene-poor chromosomes (Kemkemer et al. 2009).

References

Arnold ML, Meyer A (2006) Natural hybridization in primates: one evolutionary mechanism. Zoology (Jena) 109:261–276

Arnold N, Stanyon R, Jauch A et al (1996) Identification of complex chromosome rearrangements in the gibbon by fluorescent in situ hybridization (FISH) of a human chromosome 2q specific microlibrary, yeast artificial chromosomes, and reciprocal chromosome painting. Cytogenet Cell Genet 74:80–85

Brandon-Jones D, Eudey AA, Geissmann T et al (2004) Asian primate classification. Int J Primatol 25:97–164

Carbone L, Vessere GM, ten Hallers BF et al (2006) A high-resolution map of synteny disruptions in gibbon and human genomes. PLoS Genet 2:e223

Carbone L, Harris RA, Vessere GM et al (2009a) Evolutionary breakpoints in the gibbon suggest association between cytosine methylation and karyotype evolution. PLoS Genet 5:e1000538

Carbone L, Mootnick AR, Nadler T et al (2009b) A chromosomal inversion unique to the northern white-cheeked gibbon. PLoS One 4:e4999

Couturier J, Lernould JM (1991) Karyotypic study of four gibbon forms provisionally considered as subspecies of *Hylobates* (*Nomascus*) *concolor* (Primates, Hylobatidae). Folia Primatol (Basel) 56:95–104

Couturier J, Dutrillaux B, Turleau C et al (1982) Comparative karyotyping of our gibbon species or subspecies (author's translaation). Ann Genet 25:5–10

de Grouchy J, Turleau C, Finaz C (1978) Chromosomal phylogeny of the primates. Annu Rev Genet 12:289–328

Dutrillaux B, Rethore MO, Aurias A et al (1975) Karyotype analysis of 2 species of gibbons (*Hylobates lar* and *H. concolor*) with different banding species. Cytogenet Cell Genet 15:81–91

Fiegler H, Gribble SM, Burford DC et al (2003) Array painting: a method for the rapid analysis of aberrant chromosomes using DNA microarrays. J Med Genet 40:664–670

Garza JC, Woodruff DS (1992) A phylogenetic study of the gibbons (*Hylobates*) using DNA obtained noninvasively from hair. Mol Phylogenet Evol 1:202–210

Girirajan S, Chen L, Graves T et al (2009) Sequencing human-gibbon breakpoints of synteny reveals mosaic new insertions at rearrangement sites. Genome Res 19:178–190

Goodman M (1999) The genomic record of humankind's evolutionary roots. Am J Hum Genet 64:31–39

Hirai H, Hirai Y, Domae H et al (2007) A most distant intergeneric hybrid offspring (Larcon) of lesser apes, *Nomascus leucogenys* and *Hylobates lar*. Hum Genet 122:477–483

Jauch A, Wienberg J, Stanyon R et al (1992) Reconstruction of genomic rearrangements in great apes and gibbons by chromosome painting. Proc Natl Acad Sci USA 89:8611–8615

Kehrer-Sawatzki H, Cooper DN (2007) Structural divergence between the human and chimpanzee genomes. Hum Genet 120:759–778

Kemkemer C, Kohn M, Cooper DN et al (2009) Gene synteny comparisons between different vertebrates provide new insights into breakage and fusion events during mammalian karyotype evolution. BMC Evol Biol 9:84

Koehler U, Arnold N, Wienberg J et al (1995a) Genomic reorganization and disrupted chromosomal synteny in the siamang (*Hylobates syndactylus*) revealed by fluorescence in situ hybridization. Am J Phys Anthropol 97:37–47

Koehler U, Bigoni F, Wienberg J et al (1995b) Genomic reorganization in the concolor gibbon (*Hylobates concolor*) revealed by chromosome painting. Genomics 30:287–292

Marks J (1982) Evolutionary tempo and phylogenetic inference based on primate karyotypes. Cytogenet Cell Genet 34:261–264

Mootnick A (2006) Gibbon (Hylobatidae) species identification recommended for rescue or breeding centers. Primate Conserv 21:103–138

Mootnick A, Groves C (2005) A new generic name for the hoolock gibbon (Hylobatidae). Int J Primatol 26:971–975

Müller S, Wienberg J (2001) "Bar-coding" primate chromosomes: molecular cytogenetic screening for the ancestral hominoid karyotype. Hum Genet 109:85–94

Müller S, O'Brien PC, Ferguson-Smith MA et al (1998) Cross-species colour segmenting: a novel tool in human karyotype analysis. Cytometry 33:445–452

Müller S, Hollatz M, Wienberg J (2003) Chromosomal phylogeny and evolution of gibbons (Hylobatidae). Hum Genet 113:493–501

Myers RH, Shafer DA (1979) Hybrid ape offspring of a mating of gibbon and siamang. Science 205:308–310

Napier J, Napier P (1967) A handbook of living primates. Academic Press, London

Neusser M, Schubel V, Koch A et al (2007) Evolutionarily conserved, cell type and species-specific higher order chromatin arrangements in interphase nuclei of primates. Chromosoma (Berl) 116:307–320

Nie W, Rens W, Wang J et al (2001) Conserved chromosome segments in *Hylobates hoolock* revealed by human and *H. leucogenys* paint probes. Cytogenet Cell Genet 92:248–253

Roberto R, Capozzi O, Wilson RK et al (2007) Molecular refinement of gibbon genome rearrangements. Genome Res 17:249–257

Roos C, Geissmann T (2001) Molecular phylogeny of the major hylobatid divisions. Mol Phylogenet Evol 19(3):486–494

Roos C, Thanh VN, Walker L, Nadler T (2007) Molecular systematics of Indochinese primates. Vietn J Primatol 1:41–53

Schröck E, du Manoir S, Veldman T et al (1996) Multicolor spectral karyotyping of human chromosomes. Science 273:494–497

Stanyon R (1983) A test of the karyotypic fissioning theory of primate evolution. Biosystems 16:57–63

Stanyon R, Sineo L, Chiarelli B et al (1987) Banded karyotypes of the 44-chromosome gibbons. Folia Primatol (Basel) 48:56–64

Stanyon R, Rocchi M, Capozzi O et al (2008) Primate chromosome evolution: ancestral karyotypes, marker order and neocentromeres. Chromosome Res 16:17–39

Takacs Z, Morales JC, Geissmann T et al (2005) A complete species-level phylogeny of the Hylobatidae based on mitochondrial ND3-ND4 gene sequences. Mol Phylogenet Evol 36:456–467

Tanabe H, Müller S, Neusser M et al (2002) Evolutionary conservation of chromosome territory arrangements in cell nuclei from higher primates. Proc Natl Acad Sci USA 99:4424–4429

Van Tuinen P, Ledbetter DH (1983) Cytogenetic comparison and phylogeny of three species of Hylobatidae. Am J Phys Anthropol 61:453–466

Van Tuinen P, Mootnick AR, Kingswood SC et al (1999) Complex, compound inversion/translocation polymorphism in an ape: presumptive intermediate stage in the karyotypic evolution of the agile gibbon *Hylobates agilis*. Am J Phys Anthropol 110:129–142

Wienberg J, Jauch A, Stanyon R et al (1990) Molecular cytotaxonomy of primates by chromosomal in situ suppression hybridization. Genomics 8:347–350

Wienberg J, Stanyon R, Jauch A et al (1992) Homologies in human and *Macaca fuscata* chromosomes revealed by in situ suppression hybridization with human chromosome specific DNA libraries. Chromosoma (Berl) 101:265–270

Yu D, Yang F, Liu R (1997) A comparative chromosome map between human and *Hylobates hoolock* built by chromosome painting. Yi Chuan Xue Bao 24:417–423

Chapter 15
Evolution and Biological Meaning of Genomic Wastelands (RCRO): Proposal of Hypothesis

Hirohisa Hirai

Abbreviations

FISH   Fluorescence in situ hybridization
GW   Genomic wasteland
HERV   Human endogenous retrovirus
PRINS   Primed in situ
RCRO   Retrotransposable compound repeated DNA organization
StSat   Subterminal satellite

15.1 Introduction

A major question in the life sciences is what makes our species human. Chromosomes might have been a part of the pivotal machineries in human evolution (see Chap. 13 by Stanyon, this volume). Chromosomes of higher eukaryotes contain many repetitive DNA sequences, which may have played important roles in forming the chromosome shape and may have affected chromosomal rearrangements and/or genetic diversification. Humans and great apes share similar cytogenetic karyotypes, with a few significant chromosomal structural differences, including several pericentric and paracentric inversions, and the fusion of two chromosomes to form human chromosome 2 (Yunis and Prakash 1982). However, the subtelomeric regions of most chromosomes are different between humans and African apes. The African apes (chimpanzees, bonobos, and gorillas) have large heterochromatin blocks at the terminal regions, but humans and orangutans do not have such structures (Marks 1985; Stanyon et al. 1986; Haaf and Schmid 1987; Hirai 2001).

H. Hirai (✉)
Primate Research Institute, Kyoto University, 41-2 Kanrin, Inuyama, Aichi 484-8506, Japan
e-mail: hhirai@pri.kyoto-u.ac.jp

H. Hirai et al. (eds.), *Post-Genome Biology of Primates*, Primatology Monographs,
DOI 10.1007/978-4-431-54011-3_15, © Springer 2012

Genome sequence analyses have revealed both similarities and distinctions between closely related organisms as well as between phylogenetically remote organisms. These analyses also revealed that more than 45% of the human genome is composed of repeat elements or their fossils (International Human Genome Sequencing Consortium 2001). Segment duplication also seems to have played a significant role in biological evolution (Bailey and Eichler 2006). In addition, a human–chimpanzee comparative genome analysis revealed that the difference between the chimpanzee and human genomes at the nucleotide level was 1.23% (Fujiyama et al. 2002). Nevertheless, the phenotypes of these two species are largely distinct from each other. What, besides gene differences, makes us human? Such questions are not easily answered, but need to be investigated in the post-genome era from novel points of view.

More intensive genome analyses revealed that the difference between the chimpanzee and human genomes including the presence/absence of repetitive sequences is about 2.7% (Cheng et al. 2005). These differences affect gene regulation and several modifications related to methylation, histone deacetylation, chromatin remodeling, etc., so-called epigenetics. Histones modified by these reactions act to dynamically alter chromatin structure and gene expression regulation. These genomic reactions are very important and are therefore discussed at the molecular level in other chapters of this volume.

Chromosome subtelomeric regions appear to be important for maintaining the biological diversity of organisms (Mefford and Trask 2002). Subtelomeric regions possess highly variable genetic structures produced by ectopic recombination between nonhomologous chromosomes and consequently play important roles in bringing about variability, plasticity, and diversity among genomes. Then, how is this subtelomere genetic diversity influenced by the structural subtelomere differences between humans and African apes already mentioned? I am interested in the heterochromatin (C-band) blocks of the subtelomeric regions that are characteristically observed in African apes because they are different from the corresponding regions in humans and specific to each species. Such differences probably produce specific biological features in humans and African apes, although the block itself has no transcriptional activity and is genetically inert. Here I infer the chromosomal and biological evolution based on retrotransposable compound repeated DNA organization [RCRO: so-called genomic wastelands (GWs)], and I discuss differences of cytological and genetic functions resulting from the differences of GWs composed of RCROs in chimpanzees (representative of possessors of terminal RCROs) and humans (representative of non-possessors thereof).

15.2 Definition of Technical Terms

In this chapter, there are some different technical terms with similar but subtly different meanings. I need to explain the nuance of their meanings to make my logic understandable and clear.

1. Genomic wasteland (GW) is defined in a literal sense, that is, a chromosome region with genetically inert segments, a so-called aggregate of junk DNA consisting of repetitive sequences, transposable element ruins, DNA wreckage, and so on.
2. Constitutive heterochromatin is cytologically observed as a C-band that can be detected by a series of reactions: alkaline hydrolysis of DNA, DNA extraction with warm saline, and Giemsa staining (Sumner 1972, 2003). The regions are composed of satellite or highly repeated DNA and are transcriptionally inactive (John 1988). The minimum quantity of DNA that is a necessary threshold for C-band-positive staining is between 10.5 and 17.5 Mb of DNA (Kunze et al. 1996). To characterize C-bands at the molecular level, each C-band should thus be checked for the DNA sequence configuring the constitutive heterochromatic block.
3. Retrotransposable compound repeated DNA organization (RCRO) is designated as a component containing clusters of transposable elements in GWs. RCROs were first found in terminal C-band blocks in the chimpanzee (Hirai et al. 2005). Chimpanzee RCROs were composed of at least four DNA components [telomere, subterminal satellite, human endogenous retrovirus (HERV)-K, and HERV-W].
4. Subterminal satellite (StSat) was first found in chimpanzee and gorilla as a repeat sequence located at the proximal part adjacent to the telomere sequence (Royle et al. 1994).

The range of biological meaning that is expressed by these terms is as follows. The first is GW, which has the widest meaning as a general term for clusters of heterochromatin parts or genetically inert regions. The second is C-band, which is detected as a block with repetitive DNA sequences of more than 10 Mb length. The third is RCRO, which exists as a specific DNA component in C-bands, although not in all C-bands. The fourth is StSat, which is a DNA component of RCRO.

C-band patterns are comparatively specific for each species of primates as well as for all other groups of organisms. Figure 15.1 shows the C-band characteristics of humans, chimpanzees, bonobos, gorillas, siamangs, and rhesus macaques. C-band patterns are specific for each species, as has long been observed. In particular, although the human and rhesus macaque lack remarkable C-bands in the subterminal regions of most chromosomes, the other four species possess such bands. I am interested in the significance of such large differences occurring even between closely related species, for example, humans and chimpanzees, and what effects the chromosomal properties have on genetic or genomic outcomes. As the first step to answer these questions, I tried to dissect the regions of interest using a molecular cytogenetic technique, PRINS (primed in situ) labeling.

15.3 PRINS Labeling Technique

I used a PRINS labeling technique to localize several repetitive DNA sequences. This technique is simple and useful in terms of the rapid reaction and high specificity for detecting repetitive sequence blocks. The arrayed blocks of repetitive sequences

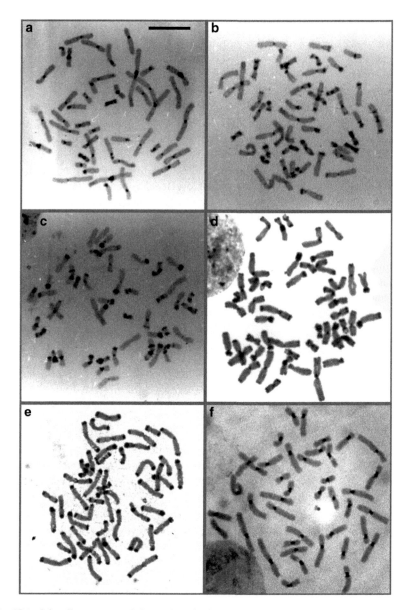

Fig. 15.1 C-banding patterns of six species of primates: human (**a**), chimpanzee (**b**), bonobo (**c**), gorilla (**d**), siamang (**e**), and rhesus macaque (**f**). *Black regions* are constitutive heterochromatin (C-bands); *gray* regions are euchromatin. *Bar* 10 μm

specific for the primers can be precisely determined as fluorescent blocks. This technique is therefore very useful for performing molecular dissection of regions consisting of arrays combining several different repetitive sequences. Usually such blocks of genetically inert components composed of repetitive sequences such as

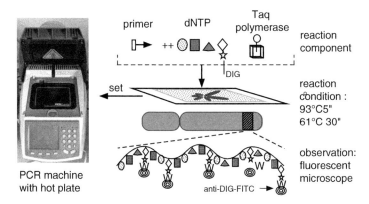

Fig. 15.2 Technique of primed in situ (PRINS) labeling

GWs can hardly be detected even by techniques of bacterial artificial chromosome (BAC) cloning or whole genome sequencing. The PRINS labeling technique is probably the best way to examine assemblies of GWs. The PRINS reaction is performed as follows. For one example, telomere sequences can be localized using a primer of $(CCCTAA)_7$ on the hotplate of a polymerase chain reaction (PCR) machine with incubation at 93°C for 5 min (denaturation of template DNA) and then at 61°C for 30 min (primer annealing and strand extension). Detection of signals is carried out with a fluorescence microscope attached to a CCD camera utilizing IPLab software (Fig. 15.2; see also Reiter et al. 1999; Hirai 2001; Hirai et al. 2005).

15.4 What Is an RCRO?

To molecularly dissect the characteristics of C-band blocks existing in African apes, several candidate primers for examining repetitive sequences were tried with the PRINS technique to detect the locations of the blocks. Thus far, four primers to detect four elements [subterminal satellite (StSat), telomeres, and two families of HERV-K and HERV-W] have been shown to label the C-band blocks. The C-band blocks thus consist of at least these four repetitive sequences, and accumulation of these sequences must have led to heterochromatinization. I named such regions RCRO (Hirai et al. 2005).

RCRO is composed of so-called genetically inert sequences, resulting in C-band blocks, although RCRO components are not always localized in all the C-bands in the genome, and some of the members may have transposition activity. Such an activity has not yet been proved at the molecular level. However, because intercalary insertions of RCRO exist in some homologous chromosomes (colinear chromosomes) that do not have any structural changes among different species, for example, chromosome 7 of human, chimpanzee, and bonobo (Fig. 15.3), such insertions may

Fig. 15.3 Locations of RCROs (retrotransposable compound repeated DNA organization) detected by PRINS labeling in human (*H*), chimpanzee (*C*), and bonobo (*B*). *Black regions* are zones positive for RCRO. Humans have no RCROs. Chimpanzees and bonobos showed patterns of RCROs differing from each other. Chromosome bands were drawn using the schematic illustration of Yunis and Prakash (1982)

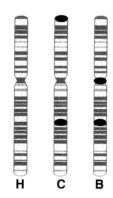

be induced by transposon activity. Although we need much more extensive experimentation to examine this, the speculation is derived from observation data that some chromosomes homologous between humans and chimpanzees/bonobos do not have structural changes except for differences of RCRO localization. As already mentioned, I dissected the subterminal heterochromatin blocks that are specific characteristics of the chromosomes of African apes (Fig. 15.1) and proved that the characteristic heterochromatin blocks (RCROs) consisted of at least four different repetitive sequences (for details, see Hirai et al. 2005).

15.5 Characteristics of StSat

As the first step for starting molecular dissection analysis of RCRO, I focused on StSat, which is a member array of the RCRO (Hirai et al. 2005). StSat was previously found in chimpanzees and gorillas but not detected in orangutans or humans (Royle et al. 1994). StSat is composed of repeat units of a 32-bp-long consensus motif sequence (GATATTTCCATGTTTATACAGATAGCGGTGTA). It was suggested, therefore, that StSat was already in the ancestor of African apes and humans and was lost in the human lineage. Studies of this array are expected to provide a useful window into primate chromosomal architecture, because the StSat blocks generate quite different structures among the chromosomes of hominoids with and without StSat repeats (Hirai et al. 2005). We surveyed the chromosomes of 22 species of a wide range of primates for the presence of StSat using the PRINS labeling technique with a primer specific for the motif sequence. The species examined were human, chimpanzee, bonobo, gorilla, orangutan, concolor gibbon, siamang, lar gibbon, agile gibbon, rhesus macaque, green monkey, patas monkey, tufted capuchin, owl monkey, common marmoset, spider monkey, common squirrel monkey, cotton-top tamarin, aye-aye, brown lemur, ring-tailed lemur, and Verreaux's sifaka.

15 Evolution and Biological Meaning of Genomic Wastelands... 233

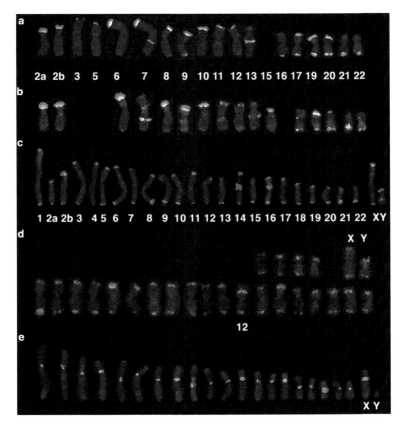

Fig. 15.4 PRINS reaction with subterminal satellite (StSat) primer on primate chromosomes: chimpanzee (**a**), bonobo (**b**), gorilla (**c**), siamang (**d**), and rhesus macaque (**e**). *Whitish regions* are positive reaction areas; *numbers* are chromosome numbers; *X*, X chromosome; *Y*, Y chromosome

The primer for StSat labeled most chromosomes of chimpanzee, bonobo, gorilla, siamang, and rhesus macaque (Fig. 15.4). The three species of African apes had the sequence mostly in subterminal regions. Siamangs had the sequence at subterminal regions, similarly to African apes, but rhesus macaque had it in the centromeric regions, not the subterminal regions. In an additional fluorescent in situ hybridization (FISH) experiment to detect localization with an StSat clone, however, the plasmid clone did not hybridize with siamang or rhesus macaque chromosomes (unpublished data). To clarify the difference between the data obtained by PRINS and FISH, we need to verify the DNA sequence of signals detected by the PRINS reaction in siamang and rhesus macaque using more effective techniques. Nevertheless, the PRINS technique with StSat primer specifically reacted with heterochromatic blocks localized at subterminal and centromeric regions.

A previous study using four species of hominids (human, chimpanzee, gorilla, and orangutan) suggested that the StSat sequences were contained in the genome of the African progenitor, which has been expanded into chimpanzees and gorillas, and was subsequently lost from the human genome (Royle et al. 1994). Our PRINS analyses newly found that bonobo, siamang, and rhesus macaque also have the component reacting with the StSat primer at subterminal and centromeric regions, respectively, in addition to gorilla and chimpanzee. The remaining 17 species examined did not have them. Our BLAST search against the genome sequences of the gorilla, bonobo, and chimpanzee with a query sequence (the consensus sequence for the 32-bp-long repeat units) revealed that these species have sequences almost identical to the StSat repeats. The results of BLAST search and molecular analyses suggested that the StSat repeats had already formed multiple arrays in the common ancestor of African apes. Taken together, our findings demonstrated that humans lost the repeats after their divergence from chimpanzees (Koga et al. 2011). The existence of the StSat sequence in the subterminal and centromeric C-band blocks of siamang and rhesus macaque thus cannot be explained in the evolutionary pathway at present.

15.6 Intragenomic Dispersion of Terminal RCROs

The StSat sequence was found to be inserted adjacent to the telomere sequence and therefore localized at the subtelomeric region (Royle et al. 1994). Signals observed in the species positive for the StSat were very strong in most chromosomes. In this chapter, StSat is used as a representative of RCRO, and I discuss the mechanisms of dispersion of terminal RCROs using StSat data. Before the start of dispersion, insertion (possibly by retrotransposition) and subsequent amplification (heterochromatinization) of tandem repeats at a single StSat site are the first essential requirements. The insertion mechanism has yet to be clarified, but one likely candidate is retrotransposition, because RCRO appears to include retrotransposable elements. We will try to elucidate the mechanism with appropriate molecular biological techniques. These insertions and amplifications could be a trigger for intrachromosomal dispersion of StSat.

The telomeric regions of all chromosomes form a cluster at the initiation of chromosome pairing in meiosis (the bouquet stage: Scherthan et al. 1996; Scherthan 2007; Egel 2008). The amplified DNA blocks at the subtelomeric regions are prone to form firmer clusters. The species with terminal RCRO blocks thus prolong the bouquet stage up to pachytene (provisionally, 32% bouquet frequency), as seen in the meiotic prophase of the chimpanzee (Hirai et al. 2005; Hirai et al., unpublished data). Species without the RCRO blocks in the subtelomeric regions do not seem to do this, as seen in the meiotic prophase of the rhesus macaque (provisionally, 0% bouquet frequency) (Hirai et al. 2005; Hirai et al., unpublished data). The wild-type mouse without the terminal heterochromatin blocks (or RCROs) also shows lower bouquet frequency (0.8%) (Liebe et al. 2004). In addition, the prolongation probably increases the opportunities for homologous or nonhomologous chromosomes to associate. Consequently, the association increases the chance of recombination between terminal

15 Evolution and Biological Meaning of Genomic Wastelands... 235

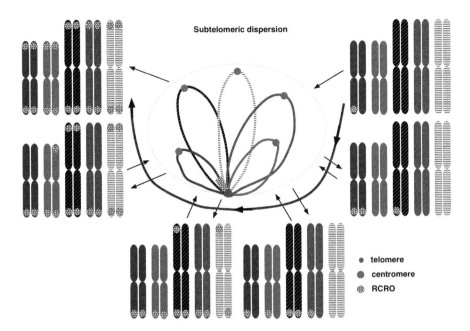

Fig. 15.5 Intragenomic dispersion mechanisms of RCROs in subtelomeric regions that were inferred from observations of the chromosome configuration in male meiotic prophase (leptotene-pachytene) of chimpanzees. The *large oval* in the center of the figure shows the chromosomal configuration in meiotic prophase (leptotene-pachytene). The length and locations of RCROs may be gradually accumulated and increased, respectively, through every event of amplification and transfer in meiotic cell division and in early-embryonic mitotic cell division. Finally, all chromosomes may have acquired RCROs in each region, likely in gorilla and siamang, although these postulates have to be confirmed by more extensive experiments in the future

RCRO blocks and nonterminal RCRO regions of different chromosomes, ectopic recombination, resulting in intragenomic dispersion by transfer of the terminal RCRO sequence. The phenomenon of ectopic recombination has been pointed out in some other organisms as well (Pfeiffer et al. 2000; Rothkamm et al. 2003). Further amplification and recombination will produce nested dispersion. More repeats of the cycle may result in transposition and accumulation of the terminal RCRO repeats to and in more subtelomeric regions (Fig. 15.5; unpublished data).

Besides strong signals at subtelomeric regions, signals at interstitial regions were also found in chimpanzee, bonobo, gorilla, and siamang (Fig. 15.4). We need to consider a different mechanism for the formation of interstitial RCRO blocks, because the interstitial blocks seem to have been also made in chromosomes without structural changes such as inversion and translocation, especially in chimpanzee and bonobo (Fig. 15.3). A possible mechanism would be the insertion of the array into interstitial regions of chromosomes by transposition. We showed at the chromosome level that the StSat array is colocalized with telomere and HERV sequences in some interstitial regions (Hirai et al. 2005). We should thus elucidate whether RCRO components have a transposable function.

15.7 Biological Effects Brought About by Accumulation of Terminal RCRO

The events of intragenomic dispersion and accumulation of such genetically inert arrays must be tightly linked to the genesis of genetic diversity and biological evolution. To explain my inferences, I propose here a hypothesis. As already mentioned, the bouquet stage of meiosis provides an opportunity to assemble the terminal regions of all chromosomes and consequently gives a chance to exchange chromosome segments by recombination between interlocked chromosomes in the cluster, which are subsequently resolved by both nonhomologous end joining and homologous recombination. This mechanism is also suggested by molecular genetic studies of human subtelomeric regions (Mefford and Trask 2002). Subtelomeres of humans are extraordinarily dynamic and variable regions near the ends of chromosomes. Extrapolating from the rhesus macaque chromosomes without terminal heterochromatin blocks (RCROs), as observed cytologically, human chromosomes do not seem likely to have a prolonged bouquet stage either (unpublished data). However, human chromosome ends are prone to ectopic recombination, as deduced from the high degree of gene diversity of the subtelomeric regions (Mefford and Trask 2002).

If so, what biological differences do heterochromatic and nonheterochromatic subtelomeres have in cell division? These differences can probably be investigated by comparing human and chimpanzee subtelomeric regions, because these two species have very similar gene orders in many euchromatic regions, as revealed by genome sequence projects (Fujiyama et al. 2002). I would therefore like to propose here a chromosomal landscape related to the foregoing question. The chimpanzee is a representative of a genome with large subtelomeric RCROs, whereas the human and rhesus macaque are representatives of a genome with nonsubtelomeric RCROs. According to reviews of human subtelomeres (Mefford and Trask 2002; Reithman et al. 2005), these regions are unusually dynamic and are variable mosaics of multichromosomal blocks of sequences. Such continued shuffling of sequences among extant subtelomeric duplicons has been observed in yeast and malaria as well as human (Freitas-Junior et al. 2000; van Overveld et al. 2000; Mefford et al. 2001; Teng and Zakian 1999; Linardopoulou et al. 2001). On the other hand, chimpanzees and gorillas did not show as much variability as humans (Mefford and Trask 2002). Although the data are not yet sufficient for proof, here I propose a preliminary hypothesis related to the effects of subtelomeric structures on genetic variability (unpublished data). The human chromosome type can produce different gene orders and duplications of the subtelomeres in several chromosomes by ectopic recombination, resulting in variability and plasticity. However, the chimpanzee chromosome type may not lead to such gene shuffling at the subtelomeric regions, for the following reason.

Assuming that recombination occurs at a similar distance from the telomere end in both human and chimpanzee, the recombination site in the chimpanzee subtelomere would be within the blocks of RCRO consisting of a repetitive sequence array (Fig. 15.6). In a previous study, most chimpanzee chromosomes showed frequent meiotic crossing-over (chiasmata) in subterminal regions with RCROs (Hirai et al.

Fig. 15.6 Different results of ectopic recombination (×) between chromosomes with (*H*, human) and without (*C*, chimpanzee) RCROs (*black regions*). RCRO possessors recombine only between RCROs, but RCRO non-possessors recombine gene loci between nonhomologous chromosomes. This difference causes differences of genetic diversity in subtelomeric regions between these two closest species

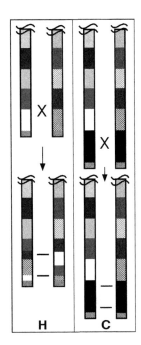

2005). This observation suggests that the subtelomeric RCRO regions are probably prone to recombine with RCROs of nonhomologous chromosomes as well. Thus, recombination between such RCROs (GWs) themselves does not provide meaningful genetic variability because of their inherent, overall sequence homogeneity, even though this recombination is more frequent than that between other euchromatic regions, and instead prevents diversification of genes surrounding the region. From this it can be inferred that "an increase of genomic wasteland brings about loss of genetic variability." That is, such genomic wastelands influence the diversity of the surrounding genes. This hypothesis will be tested using genome analyses or bioinformatics in future post-genomic research on primates.

15.8 Perspective Remarks

Humans and chimpanzees have somewhat different chromosomes in terms of fusion, inversions, and heterochromatinization, although they have basically homologous gene orders. The hypothesis, so-called increase is loss, as already mentioned is one plausible postulate that can be inferred from the biological and genetic characteristics of humans and chimpanzees. Differences of RCROs may result in phenotypic differentiation as well. In addition, there are some interstitial RCROs as well as terminal RCROs. Our investigation using meiotic cells of a male chimpanzee revealed that such interstitial RCROs appear to prevent crossing-over in the vicinity regions; for example, chimpanzee 7q31 with a large interstitial RCRO (unpublished data). This conclusion is strongly supported by the fact that fine-scale analysis of genetic recombination indicated that the same region, 7q31, formed a hotspot and a cold-spot of genetic recombination, respectively, in humans and chimpanzees (Winkler et al. 2005). The regions with RCROs probably suppress chiasma formation by a position effect of the reiterated structures, which suggests to us that humans and chimpanzees may have different genetic organization from each other even between orthologous genes in chromosome 7.

Such variations in recombination rates would be a very important factor in genetic differentiation between, for example, humans and chimpanzees. Because such differentiation promotes genetic differences in limited regions, reduction of genetic recombination affects genomic diversity by disrupting ancestral linkage disequilibrium and producing new combinations of alleles in the affected region. Genomic changes, including gene silencing, around such specialized regions will be clarified in the future by analyzing genome information and gene expression.

If these speculations are correct, position effects of and gene silencing by RCROs might have acted to produce divergence of these two species with similar genes during the past 5 million years. Accordingly, these differentiations probably did and will continue to cause large biological differences between chimpanzees and humans. The two species, which have very similar genome sequences in euchromatic regions, show different susceptibility to some viral diseases, cancers, degenerative disorders, etc. (Olson and Varki 2002). In addition, a previous study suggested that humans and chimpanzees were still hybridizing for about 1 million years after the species split (Patterson et al. 2006), which might imply that the early human lineage still had the same karyotype as the great apes (Disotell 2006). RCROs may also have some relationship to such karyotypic differentiation, because at present humans and chimpanzees are quite different from one another with respect to RCROs as well as the formation of human chromosome 2.

Chromosome differentiations, including insertion and accumulation changes such as RCROs, could thus be related to genetically differentiating and evolutionary events. In the case of chromosome 7, for example, there is some evidence that duplication and deletion in 7q11 and 7q31 originating from interchromosomal recombination and unequal meiotic recombination can lead to disorders related to speech and language (Someville et al. 2005; Fisher 2005). Gene conversion events resulting from homologous recombination have been shown to cause human genetic diseases (Chen et al. 2007). In contrast, chimpanzees are deduced to have fewer such chromosome rearrangements, as a consequence of inhibition of crossing-over around the RCRO regions, so that chimpanzees may be much less susceptible to such genetic disorders than humans, on the assumption that chimpanzees have the same genes in these regions as humans. Moreover, in humans and chimpanzees, proteins encoded by genes on chromosomes that have undergone structural rearrangements have evolved faster than those on colinear chromosomes without changes (Navarro and Barton 2003a), and chromosome change may accelerate speciation (Rieseberg 2001; Navarro and Barton 2003b). These possibilities will be examined in post-genomic analyses.

15 Evolution and Biological Meaning of Genomic Wastelands...

Acknowledgments Primate biomaterials were supplied from KUPRI, Japan; JMC, Japan; and Ouji Zoo, Japan, through the GAIN project; PSSP, IPB, Indonesia; and PBZT, Madagascar. I thank Dr. A Koga for his critical reading of the manuscript and valuable comments, and Dr. Elizabeth Nakajima for revision of the English. This research was supported in part by the Global COE Program (A06 to Kyoto University) of the Ministry of Education, Culture, Sports, Science and Technology-Japan and a grant of the Japan Society for the Promotion of Science (20405016, 22247037).

References

Bailey JA, Eichler EE (2006) Primate segmental duplications: crucibles of evolution, diversity and disease. Nat Rev Genet 7:552–564

Chen J-M, Cooper DN, Chuzhanova N et al (2007) Gene conversion: mechanisms, evolution and human disease. Nat Rev Genet 8:762–775

Cheng Z, Ventura M, She X et al (2005) A genome-wide comparison of recent chimpanzee and human segmental duplications. Nature (Lond) 437:88–93

Disotell TR (2006) 'Chumanzee' evolution: the urge to diverge and merge. Genome Biol 7:240

Egel R (2008) Meiotic crossing-over and disjunction: overt and hidden layers of description and control. In: Egel R, Lankenau D-H (eds) Recombination and meiosis: crossing-over and disjunction. Springer, Berlin, pp 1–30

Fisher SM (2005) On gene, speech, and language. N Engl J Med 353:1655–1657

Freitas-Junior LH, Bottius E, Pirrit LA et al (2000) Frequent ectopic recombination of virulence factor genes in telomeric chromosome cluster of *P. falciparum*. Nature (Lond) 407:1018–1022

Fujiyama A, Watanabe H, Toyoda A et al (2002) Construction and analysis of a human–chimpanzee comparative clone map. Science 295:131–134

Haaf T, Schmid M (1987) Chromosome heteromorphisms in the gorilla karyotype. J Hered 78:287–292

Hirai H (2001) Relationship of telomere sequence and constitutive heterochromatin in the human and apes as detected by PRINS. Methods Cell Sci 23:29–35

Hirai H, Matsubayashi K, Kumazaki K et al (2005) Chimpanzee chromosomes: retrotransposable compound repeat DNA organization (RCRO) and its influence on meiotic prophase and crossing-over. Cytogenet Genome Res 108:248–254

International Human Genome Sequencing Consortium (2001) Initial sequencing and analysis of the human genome. Nature (Lond) 409:860–921

John B (1988) The biology of heterochromatin. In: Verma RS (ed) Heterochromatin: molecular and structural aspects. Cambridge University Press, Cambridge, NY, pp 1–128

Koga A, Notohara M, Hirai H (2011) Evolution of subterminal satellite (StSat) repeats in hominids. Genetica 139:167–175

Kunze B, Weichenhan D, Virks P et al (1996) Copy numbers of a clustered long-range repeat determine C-band staining. Cytogenet Cell Genet 73:86–91

Liebe B, Alsheimer M, Hoog C, Benavente R, Scherthan H (2004) Telomere attachment, meiotic chromosome condensation, pairing, and bouquet stage duration are modified in spermatocytes lacking axial elements. Mol Biol Cell 15:827–837

Linardopoulou E, Mefford HC, Nguyen O et al (2001) Transcriptional activity of multiple copies of a subtelomerically located olfactory receptor gene that is polymorphic in number and location. Hum Mol Genet 10:2373–2383

Marks J (1985) C-band variability in the common chimpanzee, *Pan troglodytes*. J Hum Evol 14:669–675

Mefford HC, Trask BJ (2002) The complex structure and dynamic evolution of human subtelomeres. Nat Rev Genet 3:91–102

Mefford HC, Linardopoulou E, Coil D et al (2001) Comparative sequencing of a multicopy subtelomeric region containing olfactory receptor genes reveals multiple interactions between non-homologous chromosomes. Hum Mol Genet 21:2363–2372

Navarro A, Barton NH (2003a) Accumulating postzygotic isolation genes in parapatry: a new twist on chromosomal speciation. Evolution 57:447–459

Navarro A, Barton NH (2003b) Chromosomal speciation and molecular divergence-accelerated evolution in rearranged chromosomes. Science 300:321–324

Olson MV, Varki A (2002) Sequencing the chimpanzee genome: insights into human evolution and disease. Nat Rev Genet 4:20–28

Patterson N, Richter DJ, Gnerre S et al (2006) Genetic evidence for complex speciation of humans and chimpanzees. Nature (Lond) 441:1103–1108

Pfeiffer P, Goedecke W, Obe G (2000) Mechanisms of DNA double-strand break repair and their potential to induce chromosomal aberrations. Mutagenesis 15:289–302

Reiter LT, Liehr T, Rautenstrauss B et al (1999) Localization of mariner DNA transposons in the human genome by PRINS. Genome Res 9:839–843

Reithman H, Ambrosini A, Paul S (2005) Human subtelomere structure and variation. Chromosome Res 13:505–515

Rieseberg LH (2001) Chromosomal rearrangements and speciation. Trends Ecol Evol 16:351–358

Rothkamm K, Kruger I, Thompson LH, Lobrich M (2003) Pathways of DNA double-strand break repair during the mammalian. Mol Cell Biol 23:5706–5715

Royle NJ, Barid DM, Jeffereys AJ (1994) A subterminal satellite located adjacent to telomeres in chimpanzees is absent from the human genome. Nat Genet 6:52–56

Scherthan H (2007) Telomere attachment and clustering during meiosis. Cell Mol Life Sci 64:117–124

Scherthan H, Weich S, Schwegler H et al (1996) Centromere and telomere movements during early meiotic prophase of mouse and man are associated with the onset of chromosome pairing. J Cell Biol 134:1109–1125

Someville MJ, Mervis CB, Young EJ et al (2005) Severe expressive-language delay related to duplication of the Williams–Beuren locus. N Engl J Med 353:1694–1701

Stanyon R, Chiarelli B, Gottlieb K, Patton W (1986) The phylogenetic and taxonomic status of *Pan paniscus*: a chromosomal perspective. Am J Phys Anthropol 69:489–498

Sumner AT (1972) A simple technique for demonstrating centromeric heterochromatin. Exp Cell Res 75:304–306

Sumner AT (2003) Chromosomes: organization and function. Blackwell, Oxford

Teng S-C, Zakian VA (1999) Telomere-telomere recombination is an efficient bypass pathway for telomere maintenance in *Saccharomyces cervisiae*. Mol Cell Biol 19:8083–8093

van Overveld PGM, Lemmers RJFL, Deidda G et al (2000) Interchromosomal repeat array interactions between chromosomes 4 and 10: a model for subtelomeric plasticity. Hum Mol Genet 19:2879–2884

Winkler W, Myers SR, Richter DJ et al (2005) Comparison of fine-scale recombination rates in humans and chimpanzees. Science 308:107–111

Yunis JJ, Prakash O (1982) The origin of man: a chromosomal pictorial legacy. Science 215:1525–1530

Part IV
Evolution of Humans and Non-Human Primates

Chapter 16
Molecular Phylogeny and Evolution in Primates

Atsushi Matsui and Masami Hasegawa

Abbreviations

K/T	Cretaceous/tertiary
Mya	Million years ago
NWM	New World monkey
OWM	Old World monkey
R	Purines (adenine and guanine)
rRNA	Ribosomal RNA
SINE	Short interspersed element
Y	Pyrimidines (cytosine and thymine)

16.1 Tree of Primates and Molecular Phylogenetic Problems in Primates

There are more than 350 species of extant primates in the world today (Groves 2005). Traditionally, the order Primates was classified into two suborders, the Prosimii (prosimians), which includes lemurs, lorises, and tarsiers, and the Anthropoidea (anthropoids), which includes New World monkeys (NWMs), Old World monkeys

A. Matsui (✉)
Department of Cellular and Molecular Biology, Primate Research Institute,
Kyoto University, 41-2 Kanrin, Inuyama, Aichi 484-8506, Japan
e-mail: amatsui@pri.kyoto-u.ac.jp

M. Hasegawa
School of Life Sciences, Fudan University, 220 Handan Road, Shanghai 200433, China

The Institute of Statistical Mathematics, Tachikawa, Tokyo 190-8562, Japan
e-mail: masamihase@gmail.com

H. Hirai et al. (eds.), *Post-Genome Biology of Primates*, Primatology Monographs,
DOI 10.1007/978-4-431-54011-3_16, © Springer 2012

(OWMs), and hominoids (humans and apes). However, there is another classification of Primates into two suborders, Strepsirrhini (lemurs and lorises), meaning "curved nose," and the Haplorhini (tarsiers, NWMs, OWMs, and hominoids), meaning "simple nose," based on the shape of the nose. Recent molecular studies revealed the monophyly of haplorhines (Schmitz et al. 2001; Matsui et al. 2009), supporting the latter classification. Moreover, NWMs and OWMs/hominoids are exactly equivalent to Platyrrhini, meaning "flat nose," and Catarrhini, meaning "narrow nose," respectively, in the latter classification.

First, we show the simplified tree topology, which is a synthesized tree discussed in the following subsections, concerning the relationships among the main groups of living primates (Fig. 16.1).

16.1.1 Lemurs (Lemuriformes)

Lemurs are endemic to Madagascar and account for more than 15% of all extant primates in the world. In the lemuriforme clade, five extant families (Lemuridae, Cheirogaleidae, Daubentoniidae, Indriidae, and Lepilemuridae) are well defined. Nevertheless, the relationship among the four families, except for Daubentoniidae (containing only a single extant species, the aye-ayes), remains highly controversial and can be represented only as a multifurcation (Fig. 16.2a).

Previously, because of its morphological specializations, Daubentoniidae has sometimes been placed at the basal position of the strepsirrhine clade, thereby suggesting a diphyletic Lemuriformes (Groves 1989; Adkins and Honeycutt 1994) and even at the basal position of primates (Oxnard 1981). However, regardless of the data set used in molecular phylogenetics (Yoder 1994; Yoder et al. 1996; Porter et al. 1995; Pastorini et al. 2002; Poux and Douzery 2004) and by karyotype comparisons (Rumpler et al. 1988), the position of Daubentoniidae is well resolved as the sister group of all other Lemuriformes Karanth et al. (2005) also suggested that Daubentoniidae represents the earliest offshoot in lemuriformes with additional analyses of ancient DNA from two subfossil lemur species.

As already mentioned, the phylogenetic relationships among the four families other than Daubentoniidae has been highly controversial. Pastorini et al. (2003) suggested that the Indriidae is the basal lineage among the problematic four families by using partial mitochondrial data, but they did not sufficiently resolve the relationships among the other three families (Lemuridae, Cheirogaleidae, and Lepilemuridae). Roos et al. (2004) suggested that the Lemuridae and Indriidae are sister relationships, but the positions of the other two families (Cheirogaleidae and Lepilemuridae) remain unresolved by analyses of SINE (short interspersed element) integrations. Using nuclear DNA data, Poux et al. (2005) proposed the following relationships: the Lemuridae is a sister lineage to the other three families, the Cheirogaleidae and Lepilemuridae form a sister clade, and the position of the Indriidae is sister to the Cheirogaleidae/Lepilemuridae. Horvath et al. (2008) strongly suggested the same relationships as those of Poux et al. (2005) by using a

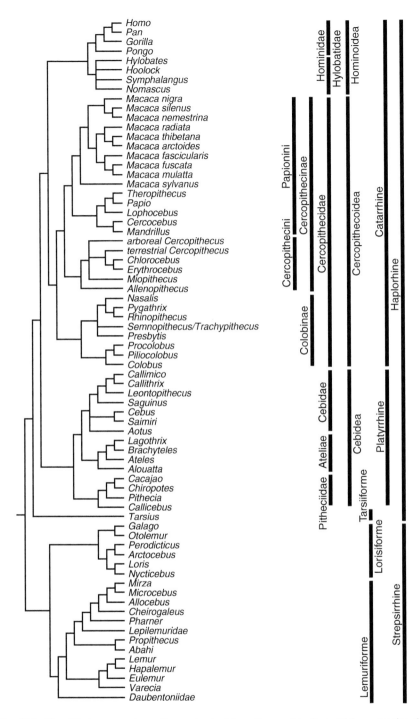

Fig. 16.1 Simplified tree topology: the synthesized tree discussed in Sect. 16.1, concerning the relationships among the main groups of living primates

Fig. 16.2 (a) Multifurcation relationships among five families within the Lemuriformes. (b) Branching order within the Lemuriformes proposed by recent molecular phylogenetic studies

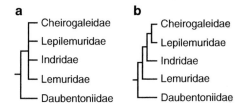

larger data set composed of mitochondrial and nuclear DNA (Fig. 16.2b). Furthermore, concerning the relationships among genera in the Lemuridae, *Varecia* is placed with a strong support at the most basal position in the family, *Lemur* and *Hapalemur* form a sister clade, and *Eulemur* is sister to the *Lemur/Hapalemur* clade (Pastorini et al. 2003; Horvath et al. 2008) (Fig. 16.1).

16.1.2 Lorises (Lorisiformes)

The Lorisiformes is classified into the African Galagidae and the African–Asian Lorisidae (Groves 2005). In the Lorisiformes, Galagonidae is monophyletic, whereas the status of Lorisidae is controversial between morphological and molecular studies.

General morphological data (Schwartz 1992) support a monophyletic origin of Lorisidae. In contrast, Porter et al. (1997a, b) and Goodman et al. (1998) suggested Galagidae (*Otolemur*) and Asian Lorisidae (*Nycticebus*) grouped into one clade excluding African Lorisidae (*Perodicticus*) as an outgroup by using nuclear DNA data. Yoder et al. (2001) analyzed partial mitochondrial DNA and nuclear DNA data from three Lorisidae genera (*Nycticebus*, *Loris*, and *Perodicticus*) and found that the Asian Lorisidae (*Nycticebus* and *Loris*) formed a clade that was sister to the Galagidae, with African Lorisidae (*Perodicticus*) as the most basal taxon. Furthermore, by analyses of the molecular and morphological data, Masters et al. (2005) could not fully resolve the relationships among the Lorisiformes. Poux and Douzery (2004) discussed that the instability of the position of the potto (*Perodicticus*; African Lorisidae) might be caused by the slowest nucleotide substitution rate of the potto among Lorisiformes, and accordingly by the long-branch attraction (Felsenstein 1978) between Galagidae and Asian Lorisidae, by analyses of the nuclear gene encoding the interstitial retinoid-binding protein. These analyses suggest that Lorisidae is paraphyletic.

As already mentioned, the systematic relationships among the African–Asian lorises, particularly the position of the pottos, have been contentious in molecular studies. We also could not fully resolve the relationships among Lorisiformes, even with the whole mitochondrial genomes, because of the inconsistency of the position of pottos (Matsui et al. 2009). We detected three distinct lineages, African Lorisidae (potto, *Perodicticus*), Asian Lorisidae (*Loris* and *Nycticebus*), and monophyletic Galagidae (*Galago* and *Otolemur*), within the Lorisiformes (Fig. 16.3a).

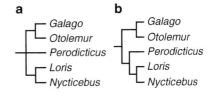

Fig. 16.3 (a) Trifurcation relationship composed of Galagidae, Asian Lorisidae, and African Lorisidae among the Lorisiformes. (b) Monophyly of the Lorisidae and Galagidae shown by molecular phylogenetic studies

Recently, Roos et al. (2004) suggested monophyly of the Lorisidae and of the Galagidae with three and six SINE integrations, respectively (Fig. 16.3b). Furthermore, several insertions suggested a common ancestry geologically of the African and Asian lorises. Masters et al. (2007) also suggested that the monophyly of the Lorisidae and of Galagidae using molecular and morphological data.

These results imply that the living Lorisidae is a monophyletic group and that two widespread geographic subclades in Asia and Africa have evolved in the fragmented area.

16.1.3 Tarsiers (Tarsiiformes)

One of the most controversial problems in primate phylogenetics at the intraordinal level is the position of tarsiers among primates. Tarsiers are the only surviving genus of formerly diverse ancestors in Tarsiiformes, which shares morphological characters of both Strepsirrhini and Anthropoidea. Practically, primates are divided into two suborders, of which the taxon content differs according to the phylogenetic position of tarsiers.

Several molecular studies suggested that tarsiers have a close relationship with Anthropoidea to form the Haplorhini (Schmitz et al. 2001; Poux and Douzery 2004; Gibson et al. 2005) or with Strepsirrhini to form the Prosimii (Eizirik et al. 2001; Murphy et al. 2001; Jow et al. 2002; Schmitz et al. 2002; Hudelot et al. 2003). Some data even suggested tarsiers to be a basal group of primates (Arnason et al. 2002). This incongruence is partly derived from the different data used for phylogenetic analyses. Nuclear DNA sequence comparisons tend to point toward a sister relationship of tarsiers and Anthropoidea, whereas mitochondrial DNA do not consistently support this affiliation or only marginally support it.

The base composition bias of mitochondrial DNA from several mammal species has been reported. Schmitz et al. (2002) compared the base composition of 26 mammalian mitochondrial genomes including the mitochondrial genome of *Tarsius bancanus*. They suggested that the overall nucleotide composition changed dramatically; decrease of T and A composition and increase of C composition on the lineage lead to higher primates at both silent and non-silent sites, and these changes of nucleotide composition have caused a change of amino acid composition.

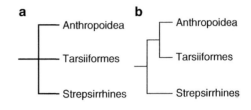

Fig. 16.4 (a) Trifurcation relationship among primates at intraordinal level concerning the position of tarsiers. (b) Monophyly of the Haplorhini (Anthropoidea and Tarsiiformes) and Strepsirrhini (*strepsirrhines*) shown by molecular phylogenetic studies

Furthermore, Gibson et al. (2005) carried out a comprehensive analysis of base composition in 69 mammalian mitochondrial genomes and examined whether the variation in base composition across genes and species affects the phylogenetic analysis. They found significant variation in T and C among these data, and they then tried to incorporate the effects in the phylogenetic reconstruction. As a result, they obtained the monophyly of Haplorhini.

Phillips and Penny (2003) mentioned the incongruence in the deep divergences of the mammalian tree obtained from mitochondrial genomes was caused by the difference of T and C frequencies among different species. The RY-coding [analyses using two nucleotide categories of purines (adenine and guanine: R) and pyrimidines (cytosine and thymine: Y)] was recently used to alleviate the bias caused by compositional differences in mammalian mitochondrial DNA sequences, and it was found to be effective in resolving some of the earliest branchings of the mammalian tree (Phillips and Penny 2003).

Therefore, we examined carefully the problematic characteristics of the mitochondrial genes and their effects on phylogenetic resolution of primates (Matsui et al. 2009). The position of tarsiers among primates could not be resolved by the maximum likelihood and neighbor-joining analyses with several data sets. Concerning the position of tarsiers, any of the three alternative topologies (monophyly of Haplorhini, monophyly of prosimians, and tarsiers being basal in primates) (Fig. 16.4a) could not be rejected at the significance level of 5%, neither at the nucleotide nor at the amino acid level. In addition, significant variations of C and T composition were observed across primate species. Furthermore, we used AGY data sets for phylogenetic analyses to remove the effect of different C/T (pyrimidines: Y) composition bias across species and to retain information from A/G (purines). These analyses provided a medium support for the monophyly of haplorhines, which might have been screened by the variation in base composition of mitochondrial DNA across species.

Recent studies of nuclear DNA seem to have established a consensus with respect to the phylogenetic position of tarsiers (Schmitz et al. 2001; Poux and Douzery 2004). By analyses of SINEs, Schmitz et al. (2001) found three Alu insertions at orthologous loci, suggesting the monophyly of Haplorhini (Anthropoidea and Tarsiiformes), and supported the monophyly of Strepsirrhini (Fig. 16.4b).

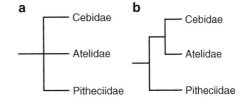

Fig. 16.5 (a) Trifurcation relationship among the Ceboidea at family level. (b) Relationships of three monophyletic families among the Ceboidea by recent molecular phylogenetic studies

16.1.4 New World Monkeys (Ceboidea)

The extant NWMs constitute a monophyletic group, which is exactly equivalent to the infraorder Platyrrhini and the superfamily Ceboidea. They inhabit only South and Central America. Today, the Platyrrhini is divided, according to molecular studies, into three monophyletic families: the Atelidae, the Cebidae (or Callitrichidae), and the Pitheciidae (Goodman et al. 1998; Schneider 2000). Furthermore, six subgroups were recognized among Platyrrhini (Groves 1989; Schneider et al. 1993; Barroso et al. 1997): (1) the spider monkeys (*Ateles*), howler monkeys (*Alouatta*), muriqui (*Brachyteles*), and woolly monkeys (*Lagothrix*) subgroup; (2) the marmosets (*Callithrix, Cebuella, Callimico*) and tamarins (*Saguinus, Leontopithecus*) subgroup; (3) the capuchines (*Cebus*) and squirrel monkeys (*Saimiri*) subgroup; (4) the owl monkeys (*Aotus*) subgroup; (5) the saki monkeys (*Pithecia, Chiropotes*) and uakaris (*Cacajao*); and (6) the titi monkeys (*Callicebus*) subgroup. They are distributed into three monophyletic families as follows: the Atelidae [(1) *Ateles, Alouatta,* and *Brachyteles*], the Cebidae [(2) *Callithrix, Cebuella, Callimico, Saguinus,* and *Leontopithecus*; (3) *Cebus* and *Saimiri*; (4) *Aotus*], and the Pitheciidae [(5) *Pithecia, Chiropotes,* and *Cacajao,* and (6) *Callicebus*]. The following three phylogenetic problems still remain unresolved in Platyrrhini.

First, the relationships among three monophyletic families have not been sufficiently resolved (Fig. 16.5a). This issue has been extensively investigated using various molecular data. Several studies suggested that the Cebidae is basal and has a sister-group relationship with the Atelidae/Pitheciidae clade (Schneider et al. 1993; Harada et al. 1995; Porter et al. 1997a; Canavez et al. 1999a). Poux et al. (2006) suggested that Cebidae and Pitheciidae form a sister-group relationship. Conversely, some studies produced different topologies. Horovitz and Meyer (1995) proposed that the Pitheciidae is the first divergence from the Atelidae and Cebidae clade using mitochondrial sequences. From mitochondrial genome analysis, Hodgson et al. (2009) also found some support for a Cebidae/Atelidae clade with the Pitheciidae as a basal lineage. Recent studies using nuclear DNA data also suggested the same topology (Fig. 16.5b), although the support for the position of the Pitheciidae was not as strong and depends on the methods used in the analyses (von Dornum and Ruvolo 1999; Steiper and Ruvolo 2003; Prychitko et al. 2005; Opazo et al. 2006). Furthermore, Ray et al. (2005) provided robust support for the latter topology, which has a sister relationship between the Atelidae and Cebidae, by SINE analyses.

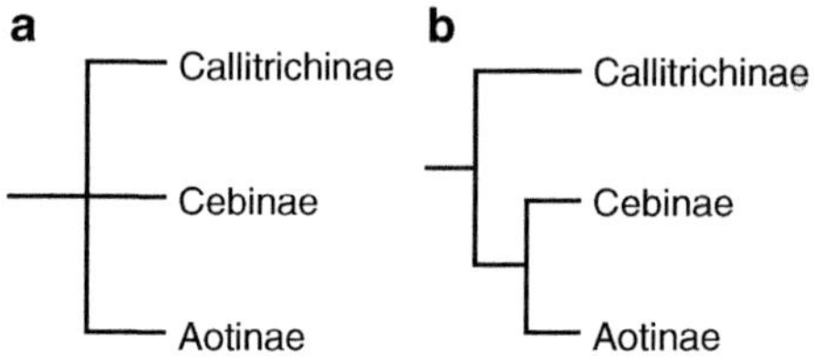

Fig. 16.6 (a) Trifurcation relationship within the Cebidae at subfamily level. (b) Relationships of three monophyletic subfamilies among the Cebidae by recent molecular phylogenetic studies

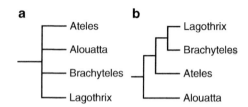

Fig. 16.7 (a) Unresolved relationships of four genera within the Atelidae. (b) Relationships of four genera among the Atelidae by recent molecular phylogenetic studies

Second, there are problematic relationships among Callitrichinae (the marmosets and tamarins), Cebinae (the capuchines and squirrel monkeys), and Aotinae (the owl monkeys) at subfamily level within the Cebidae (Fig. 16.6a). In some studies, the position of Aotinae could not be fully resolved within the Cebidae (Schneider et al. 1993; von Dornum and Ruvolo 1999; Singer et al. 2003). Previous molecular studies suggested that Aotinae forms a sister clade with Callitrichinae (Harada et al. 1995; Porter et al. 1997a,b, 1999; Goodman et al. 1998) or that Aotinae is a basal linage in Cebidae (Horovitz et al. 1998). Recent studies proposed a sister relationship between Aotinae and Cebinae, although with low support (Steiper and Ruvolo 2003; Opazo et al. 2006) (Fig. 16.6b). With SINE analyses, Ray et al. (2005) also provided a low support for this sister relationships between Aotinae and Cebinae by one insertion.

Third, a similar problem exists within Atelidae (*Ateles*, *Alouatta*, *Brachyteles*, and *Lagothrix*) (Fig. 16.7a). Early studies have suggested that *Brachyteles* have a close relationship to *Lagothrix* (Harada et al. 1995; Horovitz and Meyer 1995; Schneider et al. 1993, 1996; Porter et al. 1997a,b; Goodman et al. 1998; Canavez et al. 1999a; Meireles et al. 1999; von Dornum and Ruvolo 1999) (Fig. 16.7b). Based on mitochondrial and nuclear data, Collins (2004) proposed trichotomy among *Brachyteles*, *Lagothrix*, and *Ateles*. According to SINE analyses, Ray et al. (2005) presented a support for a closer relationship between *Lagothrix* and *Ateles* than either is to *Alouatta*. The relationships among these four genera in Atelidae, however, have remained unresolved.

At genus-level relationships among Pitheciidae, the branching order is well characterized as follows: *Callicebus* is a first-diverged species, and *Cacajao* and *Chiropotes* form a sister clade (Schneider et al. 1993; Harada et al. 1995; Porter et al. 1997a,b; Goodman et al. 1998; Horovitz et al. 1998; Canavez et al. 1999a; Meireles et al. 1999; Porter et al. 1999; von Dornum and Ruvolo 1999) (Fig. 16.8a).

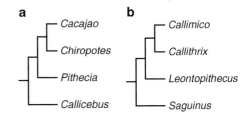

Fig. 16.8 (a) Phylogenetic relationships among the Pitheciidae at genus level. (b) Phylogenetic relationships among the subfamily Callitrichinae at genus level

The relationships among genera of the subfamily Callitrichinae have been agreed upon by most molecular studies (Canavez et al. 1999a, b; Chaves et al. 1999; von Dornum and Ruvolo 1999). *Saguinus* is a first-diverged species, *Leontopithecus* diverged next, and *Callimico* and *Callithrix* form a clade (Fig. 16.8b). Some molecular studies, however, suggested an alternative relationship, in which *Leontopithecus* and *Saguinus* form a clade (Schneider et al. 1993; Harada et al. 1995; Porter et al. 1997b, 1999; Goodman et al. 1998).

16.1.5 Old World Monkeys (Cercopithecoidea)

The extant OWMs constitute a monophyletic group, which is exactly equivalent to the superfamily Cercopithecoidea in the infraorder Catarrhini. Today, they widely inhabit Africa and Asia and are the most diverse group of primates. The superfamily Cercopithecoidea consists of the family Cercopithecidae. The Cercopithecidae is divided into two monophyletic subfamilies: the Cercopithecinae (cheek-pouched monkeys) and the Colobinae (leaf-eating monkeys) (Delson 1994; Raaum et al. 2005; Sterner et al. 2006; Groves 2005).

The subfamily Cercopithecinae consists of two tribes: the Papionini [mangabeys (*Cercocebus*, *Lophocebus*, and *Rungwecebus*), macaques (*Macaca*), baboons (*Papio*), drills and mandrills (*Mandrillus*), and geladas (*Theropithecus*)] and the Cercopithecini [patas monkeys (*Erythrocebus*), Allen's swamp monkeys (*Allenopithecus*), talapoins (*Miopithecus*), guenons (*Cercopithecus*), and vervet monkeys (*Chlorocebus*)] (Strasser and Delson 1987; Delson 1992, 1994; Disotell 2003; Groves 2005; Xing et al. 2005). The highland mangabey (kipunjis, *Rungwecebus*) was independently discovered by two research teams in 2003 and 2004, and recent molecular study assigned it to a new genus (Davenport et al. 2006). Most Cercopithecinaes are distributed in sub-Saharan Africa. However, macaques exceptionally have wide distribution from eastern Asia to northern Africa (Fa 1989), and some small populations of baboons inhabit Arabia. Interestingly, it is well known that the Cercopithecini have colorful pelage patterns, although most Papionini have the grey or brown pelage.

Four issues in phylogeny of the subfamily Cercopithecinae remain unresolved at the tribe and subtribe level. The first question is whether the genus *Macaca* is a monophyletic group in the Papionini. Previous studies suggested the possibility that *Macaca* is a paraphyletic group consisting of two groups [barbary macaques

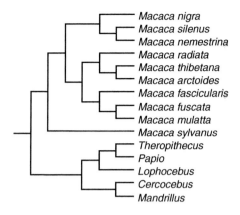

Fig. 16.9 Phylogenetic relationships among the tribe Papionini at genus level (at species level in *Macaca*)

(*Macaca sylvanus*) and other *Macaca* species], based on morphological study (Groves 1989). However, molecular studies supported the monophyly of *Macaca* (Hayasaka et al. 1996; Morales and Melnick 1998; Tosi et al. 2000). Tosi et al. (2003) recently reinforced the monophyly of *Macaca* using gene sequences encoded on the Y chromosome. In addition, Xing et al. (2005) provided strong support for the *Macaca* monophyly by SINE analyses. They also presented monophyly of the subtribe Papionina as a sister clade to *Macaca*; *Papio* and *Theropithecus* form a monophyletic clade, which is a sister group to mangabeys (mentioned below) among Papionina (Fig. 16.9). Furthermore, the relationships within *Macaca* at genus level have not been well characterized in detail. Today, the genus *Macaca* is one of the most successful primates, and it is said that approximately 20–22 species exist in genus *Macaca*, although various definitions of species are proposed by different authors. The *Macaca* is classified into three to six species groups (Fooden 1976; Delson 1980; Groves 2005). Based on SINE analyses, Li et al. (2009) recently proposed that there are four monophyletic species groups within *Macaca*: the silenus group (*M. silenus*, *M. nemestrina*, and *M. nigra*), the sinica group (*M. radiata*, *M. thibetana*, and *M. arctoides*), the fascicularis group (*M. fascicularis*, *M. mulatta*, and *M. fuscata*), and the sylvanus group (*M. sylvanus*) as only one African clade. As for the relationships among these taxa, they suggested that the silenus group is a basal clade in the Asian macaques and that the sinica and fascicularis groups form a sister relationship (Fig. 16.9).

The second question is whether mangabeys (*Cercocebus* and *Lophocebus*) form a monophyletic group in the Papionini. Two genera in mangabeys have often been debated in morphological studies concerning their phylogenetic placement. Several morphological studies showed that the two mangabeys have similarities, suggesting their sister relationship (Szalay and Delson 1979; Strasser and Delson 1987). In contrast, however, others suggested that these mangabeys have diphyletic origins (Groves 1978; Fleagle and McGraw 1999). Molecular studies also suggested the diphyletic origin of mangabeys (Disotell 1994; Harris and Disotell 1998; Page and Goodman 2001) (Fig. 16.9). In addition, Tosi et al. (2003, 2005) recently proposed polyphyly of the mangabeys (*Cercocebus* and *Lophocebus*).

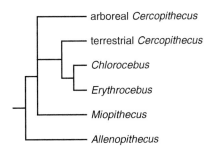

Fig. 16.10 Phylogenetic relationships among the tribe Cercopithecini at genus level

The third issue is whether guenons (*Cercopithecus*) form a monophyletic group in the Cercopithecini. The guenons (*Cercopithecus*) involve the arboreal species (arboreal *Cercopithecus*) and terrestrial species (*Cercopithecus aethiops* and *Cercopithecus lhoesti*). Many molecular studies suggested that the arboreal *Cercopithecus* within Cercopithecini form a monophyletic clade (Dutrillaux et al. 1988; Tosi et al. 2004). In contrast, the monophyly of the terrestrial species or guenons (*Cercopithecus*) has not been supported (Ruvolo 1988; Disotell and Raaum 2002; Tosi et al. 2002). However, Groves (1989) suggested that *C. aethiops* and *Erythrocebus patas* have a close relationship as a terrestrial species among the Cercopithecini. Concerning these problems, Tosi et al. (2003, 2005) proposed the paraphyly of *Cercopithecus* by analyses using a dataset of gene-encoded X- and Y chromosomes: *E. patas*, *C. aethiops*, and *C. lhoesti* form a monophyletic group (Fig. 16.10). Xing et al. (2005) supported the close relationship between terrestrial *Chlorocebus* and *Erythrocebus* by SINE analyses. These findings may imply two distinct clades composed of arboreal and terrestrial species, respectively, within the Cercopithecini.

The fourth issue is the phylogenetic position of *Allenopithecus* and *Miopithecus* in the subfamily Cercopithecinae. It has not been sufficiently resolved whether these species are involved in tribe Papionini or tribe Cercopithecini. Morphological studies by different authors suggested different affiliations of *Allenopithecus* and *Miopithecus* (Szalay and Delson 1979; Strasser and Delson 1987; Groves 1989). Some molecular studies showed that *Allenopithecus* and *Miopithecus* are involved in the tribe Cercopithecini (Disotell and Raaum 2002; Tosi et al. 2002) (Fig. 16.10). Recent molecular study also reported that the positions of *Allenopithecus* and *Miopithecus* fall into the tribe Cercopithecini (Fig. 16.10) (Tosi et al. 2003, 2005).

The subfamily Colobinae is roughly classified into two groups: the African clade [black and white colobuses (*Colobus*), red colobuses (*Procolobus*), and olive colobuses (*Piliocolobus*)] and the Asian clade [proboscis monkeys (*Nasalis*), snub-nosed monkeys (*Rhinopithecus*), douc langurs (*Pygathrix*), pigtailed langurs (*Simias*), surilis (*Presbytis*), gray langurs (*Semnopithecus*), and lutungs (*Trachypithecus*)] (Collura et al. 1996; Messier and Stewart 1997; Page et al. 1999; Disotell 2003; Groves 2005; Xing et al. 2005). Within the Asian clade there are two groups: the odd-nosed group (*Nasalis*, *Simias*, *Pygathrix*, and *Rhinopithecus*) and the langur and leaf monkeys group (*Presbytis*, *Semnopithecus*, and *Trachypithecus*) (Groves 1970; Jablonski

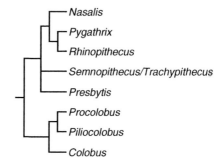

Fig. 16.11 Phylogenetic relationships among the subfamily Colobinae at genus level

1998; Jablonski and Peng 1993; Xing et al. 2005). African colobuses have stub thumbs. The Colobinae species are usually arboreal and mainly eat leaves and hard fruits, have multi-chambered stomachs, and are also called leaf-eating monkeys (Chivers and Hladik 1980; Strasser and Delson 1987). Three phylogenetic issues of the subfamily Colobinae remain unresolved. The evolutionary relationships among Asian Colobinae especially remain contested and have not been fully demonstrated by molecular studies (Messier and Stewart 1997; Zhang and Ryder 1998).

The first question is whether the odd-nosed group is monophyletic within Asian Colobinae. Previously, several studies suggested that the odd-nosed group is not monophyletic (Bigoni et al. 2003, 2004; Jablonski 1998; Wang et al. 1995). Sterner et al. (2006), however, clearly showed their monophyly by using whole mitochondrial sequences (Fig. 16.11). This monophyletic relationship was strongly supported also by SINE analyses (Xing et al. 2005). Concerning the branching order within the odd-nosed group, they proposed that *Nasalis* is a first-diverged species and that *Pygathrix* and *Rhinopithecus* form a sister clade, although morphological studies by different authors suggested different relationships (Peng et al. 1993; Bigoni et al. 2003, 2004).

The second concern is the comprehensive relationships among the langur and leaf monkeys group. Different classifications for three genera (*Presbytis*, *Semnopithecus*, and *Trachypithecus*) have been proposed by different authors (Delson 2000; Groves 2005). The evolutionary relationships in these genera have not yet been resolved. Morphological studies suggested that *Semnopithecus* and *Trachypithecus* form a sister clade (Strasser and Delson 1987; Groves 2001). Recent mitochondrial genome analyses, however, suggested that *Presbytis* and *Trachypithecus* form a sister clade (Sterner et al. 2006). On the other hand, Ting et al. (2008) proposed that *Semnopithecus* and *Trachypithecus* are grouped into one clade, which is concordant with morphological analyses, using the dataset of genes encoded on the X chromosome. They statistically showed that the mitochondrial data set again supported a different topology from the X-chromosome data and that the position of *Presbytis* and odd-nosed groups among Asian Colobinae are quite different using mitochondrial data and X-chromosome data. Osterholz et al. (2008) also suggested the sister-group relationship of *Semnopithecus* and *Trachypithecus* by using the Y-chromosomal gene data, mitochondrial genes, and retroposon integrations. Now, the close relationship of *Semnopithecus* and *Trachypithecus* seems to be supported by recent studies, as already mentioned (Fig. 16.11). Furthermore, there is disagreement about the relationships, particularly within the genus *Trachypithecus*. Gray langurs (*Semnopithecus*) are often called Hanuman langurs, and they inhabit all areas of the Indian subcontinent. *Trachypithecus* inhabit mainly Southeast Asia, and purple-faced langurs (*Trachypithecus vetulus*) and Nilgiri langurs (*Trachypithecus johnii*) live in south India, which is an area disconnected from Southeast Asia. In contrast, *Presbytis* inhabit only Southeast Asia. For this reason, the phylogenetic question is to which group these two species (purple-faced langurs and Nilgiri langurs) are more closely related, whether *Semnopithecus* in India or *Trachypithecus* in Southeast Asia (Fig. 16.11). Some molecular studies suggested that Hanuman langurs (*Semnopithecus*) and purple-faced langurs (*Trachypithecus*) form a sister clade as an Indian species clade and Southeast Asian *Trachypithecus* form another clade (Messier and Stewart 1997; Zhang and Ryder 1998). More recently, Karanth et al. (2008) also suggested these relationships using mitochondrial and nuclear datasets. They and Osterholz et al. (2008) mentioned that hybridization or introgression might have occurred between *Semnopithecus* and *Trachypithecus* clades during their evolutionary history in India. To solve these relationships, further research is required, and the relationship of *Presbytis*, *Semnopithecus/Trachypithecus*, and the odd-nosed group among Asian Colobinae should be treated as a trichotomy at present (Fig. 16.11). It might be necessary to change the current classification of *Semnopithecus* and *Trachypithecus* in the future.

The third issue is that the branching order of the African Colobinae has not been fully resolved by molecular studies. There are three distinct clades among African Colobinae: black-and-white colobuses (*Colobus*), the olive colobuses (*Piliocolobus*), and the red colobuses (*Procolobus*). Ting (2008) showed that olive colobuses (*Piliocolobus*) and the red colobuses (*Procolobus*) are grouped into a sister clade, using mitochondrial data (Fig. 16.11), and analyzed the phylogeny of African Colobinae in detail. This relationship is concordant with previous studies of morphology and vocalizations.

16.1.6   Hominoids (Hominoidea)

The superfamily Hominoidea has two families: Hominidae (great apes) and Hylobatidae (gibbons, small apes).

The family Hominidae involves four extant genera [chimpanzees (*Pan*), gorillas (*Gorilla*), orangutans (*Pongo*), and humans (*Homo*)], and, except for humans, they inhabit only Africa and Asia. The relationships among the Hominidae have been extensively studied because of the direct relevance to human origins. Horai et al. (1995) proposed that humans have a sister relationship with chimpanzees among Hominidae, using whole mitochondrial sequences, and the phylogeny of the Hominidae has been established (Fig. 16.12a).

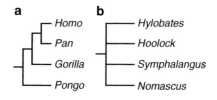

Fig. 16.12 (a) Phylogenetic relationships among the family Hominidae. (b) Phylogenetic relationships among the family Hylobatidae proposed by recent molecular phylogenetic studies

The family Hylobatidae contains four genera (*Hylobates, Hoolock, Nomascus,* and *Symphalangus*) (Groves 2005; Mootnick and Groves 2005; Takacs et al. 2005). Each genus is a monophyletic group within the Hylobatidae. Previously, there was only one genus, *Hylobates*. However, recent molecular studies showed that the present four gibbon genera are more genetically diverged than are humans and chimpanzees and recommended that they should be treated as different genera (Brandon-Jones et al. 2004; Roos and Geissmann 2001). In addition, the genus *Bunopithecus* was recently changed to *Hoolock* for the hoolock gibbons (Mootnick and Groves 2005). Because gibbons (Hylobatidae) are relatively smaller than other apes, they are often called the lesser apes. They inhabit Southeast Asia (northeast India, southern China, Malay Peninsula, Java, Borneo, and Sumatra). Although the phylogeny of the Hominidae has been elucidated, the phylogenetic relationships among the four genera in Hylobatidae remain unresolved. Roos and Geissmann (2001) suggested that *Nomascus* first diverged and *Hoolock* (*Bunopithecus*) and *Hylobates* form a sister clade using control region (D-loop) data of mitochondrial DNA. On the other hand, Takacs et al. (2005) gave some support for either *Hoolock* (*Bunopithecus*) or *Nomascus* as the most basal genus and *Hylobates* to be most recently derived genus using a mitochondrial dataset. Recently, two studies using mitochondrial genome data suggested that *Nomascus* was a first-diverged genus and that *Hylobates* and *Symphalangus* form a sister clade, with high support value among these three genera (Matsudaira and Ishida 2010; Chan et al. 2010). However, the phylogenetic relationships among the four genera of Hylobatidae remain unclear by molecular studies (Thinh et al. 2010; Israfil et al. 2011) and should be further examined in future (Fig. 16.12b).

16.2 Divergence Times of Primates Estimated with Whole Mitochondrial Genome Data

To estimate speciation dates within primates, we used the new complete mitochondrial genome data from 11 primates together with those from 15 primates that contained the mitochondrial genome sequence of sifakas determined by us from their feces samples (Matsui et al. 2007), and 26 non-primate mammals available in

16 Molecular Phylogeny and Evolution in Primates

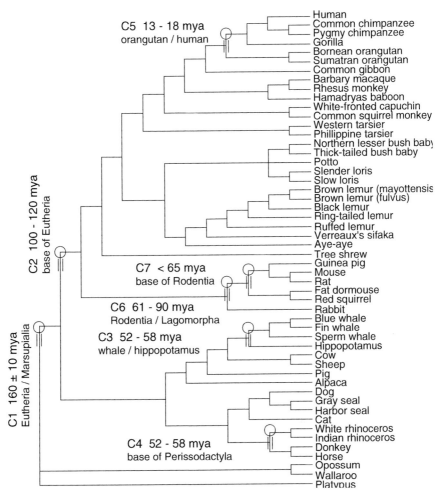

Fig. 16.13 Seven constraint points: *C1*, Eutheria/Marsupilia; *C2*, base of Eutheria (Boreoeutheria/Afrotheria/Xenarthra); *C3*, whale/hippopotamus; *C4*, base of perissodactyla (horse/rhinoceros); *C5*, orangutan/human; *C6*; base of Glires (Rodentia/lagomorph); *C7*, base of Rodentia (cavimorph/myomorph/sciurid)

public databases (Matsui et al. 2009). The significant rate of heterogeneity of mitochondrial DNA among eutherian lineages suggested that a variable rate clock has to be applied to estimate reliable divergence dates (Hasegawa et al. 2003). We analyzed the amino acid sequences of 12 proteins and nucleotide sequences of 2 ribosomal RNAs (rRNAs) encoded in mitochondrial DNA with a Bayesian method, which allows different rates on different branches, of Thorne and Kishino (Thorne et al. 1998; Thorne and Kishino 2002). Seven fossil-based divergence dates were used as constraints to calibrate the relaxed clock (Fig. 16.13). Only one constraint point chosen within the primates as the time of divergence between the orangutan

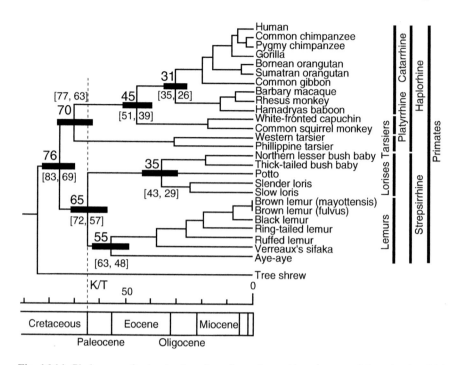

Fig. 16.14 Phylogram of primates based on the amino acid sequences of the mitochondrial protein data set. *Numbers on each node* indicate the divergence dates estimated. *Black bars on nodes* represent 95% credibility interval for the divergence dates

and the human lineages (C5). The orangutan lineage first appeared in the fossil record as *Sivapithecus* about 12–13 million years ago (Mya) (Kelley 2002), but the fossil datings of Sivapithecus were once questioned (Pilbeam et al. 1990). Recent statistical analyses of cranial and postcranial characters (Begun et al. 1997; Ward 1997) suggested that the divergence between the *Sivapithecus*–orangutan and the African apes–human lineages occurred before 13 Mya (Ward 1997; Stewart and Disotell 1998), which is the current consensus among paleoanthropologists. Thus, we gave a constraint to the orangutan/human divergence at 13–18 Mya, in which the older bound of 18 Mya is the date of the connection between Africa and Eurasia (Waddell and Penny 1996). We further gave constraints to six nodes based on the paleontological records and previous research in mammal evolution, and the tree topology of non-primate mammals was adopted into our analyses from the mitochondrial protein analysis by Hasegawa et al. (2003). For the major nodes of primates, Fig. 16.14 depicts a time scale for placental mammal evolution based on the amino acid sequences of the mitochondrial DNA protein data set with the platypus as an outgroup. The estimated divergence dates are summarized in Table 16.1.

The divergence between Haplorhini and Strepsirrhini was estimated to be 76.0±3.3 (69.3–82.5, 95% credibility interval) Mya (protein)/80.9±4.4 (72.4–89.6) Mya (rRNA) before the K/T (Cretaceous/Tertiary) boundary, which marked

16 Molecular Phylogeny and Evolution in Primates

Table 16.1 Estimated dates with posterior standard deviations based on amino acid sequences (12H-strand encoded protein gene) and rRNA sequences of mitochondrial DNA

Branching among primates	Estimated dates (million years ago, Mya)		95% credibility interval	
	Protein (amino acid sequences)	rRNA	Protein (amino acid sequences)	rRNA
Haplorhini/Strepsirrhini	76.0±3.3	80.9±4.4	(69.3–82.5)	(72.4–89.6)
Anthropoidea/ Tarsiiformes	70.1±3.4	76.8±4.8	(63.2–76.7)	(67.4–86.4)
Lemuriformes/ Lorisiformes	64.5±3.7	69.3±4.8	(57.2–71.7)	(60.2–79.0)
The radiation of Lemuriformes	55.3±3.9	65.6±5.0	(47.7–63.0)	(56.3–75.5)
The radiation of Lemuridae	26.1±3.3	24.9±3.6	(20.0–32.6)	(18.5–32.5)
The radiation of Lorisiformes	35.4±3.7	31.5±3.9	(28.5–43.1)	(24.4–39.7)
Tarsius bancanus/ Tarsius syrichta	30.8±3.9	20.2±3.3	(23.4–38.6)	(14.5–27.4)
Platyrrhini/Catarrhini	45.3±3.1	47.0±4.3	(39.4–51.3)	(39.1–55.7)
Hominoidea/ Cercopithcoidea	30.5±2.5	33.1±3.3	(25.8–35.3)	(27.0–39.9)
Greater ape/small ape (gibbon)	19.9±1.7	21.3±2.2	(16.7–23.0)	(17.3–26.0)
Orangutan/human, chimpanzee, gorilla	15.8±1.3	14.9±1.3	(13.3–17.9) (13–18[a])	(13.1–17.6) (13–18[a])
Gorilla/human, chimpanzee	8.4±1.0	8.8±1.2	(6.6–10.3)	(6.8–11.4)
Human/chimpanzee	6.2±0.8	6.9±1.0	(4.7–7.8)	(5.1–9.1)
Bornean orangutan/ Sumatran orangutan	4.7±0.7	3.8±0.7	(3.5–6.0)	(2.5–5.3)
Common chimpanzee/ pygmy chimpanzee	3.0±0.5	2.3±0.5	(2.1–4.1)	(1.4–3.5)

For orangutan branching, constraints were given with [a] in parentheses

the extinction of the dinosaurs (Hedges et al. 1996). This result is in accord with the estimated dates of the previous studies concerning the interordinal diversification of placental mammals (Hasegawa et al. 2003; Springer et al. 2003). Hasegawa et al. (2003) and Springer et al. (2003) suggested that the Haplorhini/Strepsirrhini divergence was 73.1±2.7 Mya and approximately 77 Mya, respectively, at least several millions of years before the K/T boundary. Steiper and Young (2006) also suggested the origin of primates to be about 77.5 Mya using a large dataset of genome sequences. Although some recent molecular phylogenies assume that primates originated far earlier than the K/T boundary, the earliest fossil record suggests that primates began to diversify just after the K/T boundary, around 55 Mya (Ni et al. 2004). Divergence dates estimated with molecular data often predate the earliest recognized fossil representatives of the species studied, which has been explained by the incompleteness of the fossil record of primates. The common ancestor of primates should be earlier than the oldest known fossils (Martin 1993), but adequate quantification is needed to interpret possible discrepancies between molecular and paleontological estimates. It is often said that the first small primates might have survived the K/T extinction event. Tavaré et al. (2002) estimated that living primates last shared a common ancestor 81.5 (72.0–89.6) Mya, based on a statistical analysis of the fossil record, which takes into account fossil preservation rate.

Subsequently, the split of Haplorhini in Anthropoidea and tarsiers was estimated to be 70.1 ± 3.4 (63.2–76.7) Mya (protein)/76.8 ± 4.8 (67.4–86.4) Mya (rRNA). Because of this deep branching of tarsiers, the position of tarsiers among primates might have to be resolved by the phylogenetic analyses.

The age of the last common ancestor of lorisiforms and lemuriforms was estimated to be approximately 64.5 ± 3.7 (57.2–71.7) Mya (protein)/69.3 ± 4.8 (60.2–79.0) Mya (rRNA). After the divergence of Strepsirrhini, Lemuriformes separated into Daubentonidae and other lemurs (Indriidae/Lemuridae) at the estimated date of 55.3 ± 3.9 (47.7–63.0) Mya (protein)/65.6 ± 5.0 (56.3–75.5) Mya (rRNA). In comparison with the previous results, Yoder et al. (1996), often referring to the Madagascar lemur study, proposed at least 62 Mya, 54 Mya, and 55 Mya, respectively, for the strepsirrhine, lemuriform, and lorisiform radiations. This analysis was, however, based on the calibration point (63 Mya) for the divergence between Strepsirrhini and Anthropoidea using partial mitochondrial sequences. In other words, the calibration point of the Yoder et al. (1996) analysis was a minimum value, and the actual date should be older, as mentioned earlier. Their recent study with the Bayesian method suggested 50–78 Mya for the diversification of Malagasy primates (Yoder et al. 2003). The radiation of Lemuridae was estimated to be 26.1 ± 3.3 (20.0–32.6) Mya (protein)/24.9 ± 3.6 (18.5–32.5) Mya (rRNA) because of the earliest divergence of *Varecia*.

The lorisiformes diverged at 35.4 ± 3.7 (28.5–43.1) Mya (protein)/31.5 ± 3.9 (24.4–39.7) Mya (rRNA) into Galagidae and Lorisidae. Because there is still no fossil record of the Madagascar lemur (Martin 2003) except for recent subfossils, the radiation of Lemuriformes, estimated to be 55.3 ± 3.9 (47.7–63.0) Mya (protein)/65.6 ± 5.0 (56.3–75.5) Mya (rRNA), cannot be confirmed by the fossil evidence. The estimated divergence date between Galagidae and Lorisidae within Lorisiformes is in agreement with fossils recently discovered by Seiffert et al. (2003) from the late middle Eocene, which suggested that the basal divergence between extant Galagidae and Lorisidae began at least 38–40 Mya.

Concerning these divergence times, Poux et al. (2005) suggested a younger date for the last common ancestor of Lorisiformes and Lemuriformes and the radiation of Lemuriformes, estimated to be about 60 Mya and 50 Mya, respectively using the nuclear DNA data set. On the other hand, Horvath et al. (2008) proposed older dates for these, about 75 Mya and 66 Mya, respectively, and the radiation of the Lorisiformes was estimated to be about 39 Mya using a large data set of genome sequences.

The estimated divergence times of Catarrhini/Platyrrhini and hominoids/OWMs are approximately 45.3 ± 3.1 (39.4–51.3) Mya (protein)/47.0 ± 4.3 (39.1–55.7) Mya (rRNA) and 30.5 ± 2.5 (25.8–35.3) Mya (protein)/33.1 ± 3.3 (27.0–39.9) Mya (rRNA), respectively. The split of Catarrhini/Platyrrhini and hominoids/OWMs, often used as calibration points among primates, were at 35 Mya (Rosenberger et al. 1991) and at 25 Mya (Fleagle 1999), respectively. We, however, estimated both divergence dates to be earlier. Steiper and Young (2006) also suggested older divergence times of Catarrhini/Platyrrhini and hominoids/OWMs to be about 43 Mya and 31 Mya, respectively, using large genome sequences. Concerning NWMs, Poux et al. (2006) proposed that NWMs colonized in South America about 37 Mya and that the radiation of the NWMs occurred about 16.8 Mya. Recent studies also reported a similar estimation for NWMs. Hodgson et al. (2009) estimated that the most common ancestor of NWMs dates to about 19.5 Mya and proposed that extant NWMs are descendants of recent successive radiation in the South American primates.

The estimate for the divergence between humans and orangutans, 15.8 ± 1.3 (13.3–17.9) Mya (protein)/14.9 ± 1.3 (13.1–17.6) Mya (rRNA), was also close to the paleontological estimate, 13 Mya (Ward 1997; Stewart and Disotell 1998), used as calibration (>13 Mya). Steiper and Young (2006), in contrast, suggested the time of divergence between humans and orangutans to be about 18 Mya using large genome sequences. Our estimated dates for the (human, chimpanzee)/gorilla and great ape/ small ape divergences were 8.4 ± 1.0 (6.6–10.3) Mya (protein)/8.8 ± 1.2 (6.8– 11.4) Mya (rRNA) and 19.9 ± 1.7 (16.7–23.0) Mya (protein)/21.3 ± 2.2 (17.3– 26.0) Mya (rRNA), respectively.

The estimate for the human/chimpanzee divergence date was 6.2 ± 0.8 (4.7– 7.8) Mya (protein)/6.9 ± 1.0 (5.1–9.1) Mya (rRNA), in accord with the widely accepted date. The currently accepted date of the chimpanzee–human split at about 6–7 Mya was initially proposed based on molecular estimates (Horai et al. 1995), which were much younger than fossil-based estimates at that time. This estimation is close to the ages of the recently discovered oldest hominid fossils (5.4–7.0 Mya) (Aiello and Collard 2001; Haile-Selassie 2001; Brunet et al. 2002).

By using molecular data accompanied with the fossil and geologic evidence, we should reexamine the biogeographic scenarios that have been proposed for the origin of the main lineages of primates and for their dispersal. The major questions in strepsirrhine evolution are when and how lemurs first arrived in Madagascar, and when and how lorises spread over Asia and Africa. The similar question has arisen for the NWMs in South America (when and how NWMs arrived at South America). Recently, estimation of divergence times and a supermatrix approach have been carried out with larger data sets than previously for primate phylogenetic analyses (Chatterjee et al. 2009; Fabre et al. 2009; Perelman et al. 2011).We hope to gain a better understanding of the speciation scenarios of primates by further studies.

References

Adkins RM, Honeycutt RL (1994) Evolution of the primate cytochrome c oxidase subunit II gene. J Mol Evol 38:215–231

Aiello LC, Collard M (2001) Palaeoanthropology Our newest oldest ancestor? Nature 410:526–527

Arnason U, Adegoke JA, Bodin K et al (2002) Mammalian mitogenomic relationships and the root of the eutherian tree. Proc Natl Acad Sci USA 99:8151–8156

Barroso CML, Schneider H, Schneider MPC et al (1997) Update on the phylogenetic systematics of new world monkeys: further DNA evidence for placing the pygmy marmoset (*Cebuella*) within the genus *Callithrix*. Int J Primatol 18:651–674

Begun DR, Ward CV, Rose MD (1997) Events in hominoid evolution. In: Begun DR, Ward CV, Rose MD (eds) Function, phylogeny and fossils: miocene hominoid evolution and adaptation. Plenum, New York

Bigoni F, Stanyon R, Wimmer R et al (2003) Chromosome painting shows that the proboscis monkey (*Nasalis larvatus*) has a derived karyotype and is phylogenetically nested within Asian colobines. Am J Primatol 60:85–93

Bigoni F, Houck M, Ryder O et al (2004) Chromosome painting shows that *Pygathrix nemaeus* has the most basal karyotype among Asian Colobinae. Int J Primatol 25:679–688

Brandon-Jones D, Eudey AA, Geissmann T et al (2004) An Asian primate classification. Int J Primatol 25:97–164

Brunet M, Guy F, Pilbeam D et al (2002) A new hominid from the Upper Miocene of Chad, Central Africa. Nature 418:145–151

Canavez FC, Moreira MAM, Ladasky JJ et al (1999a) Molecular phylogeny of new world primates (Platyrrhini) based on beta2-microglobulin DNA sequences. Mol Phylogenet Evol 12:74–82

Canavez FC, Moreira MAM, Simon F et al (1999b) Phylogenetic relationships of the Callitrichinae (Platyrrhini, primates) based on beta2-microglobulin DNA sequences. Am J Primatol 48:225–236

Chan YC, Roos C, Inoue-Murayama M et al (2010) Mitochondrial genome sequences effectively reveal the phylogeny of *Hylobates* gibbons. PLoS One 5:e14419

Chatterjee HJ, Ho SY, Barnes I et al (2009) Estimating the phylogeny and divergence times of primates using a supermatrix approach. BMC Evol Biol 9:259

Chaves R, Sampaio I, Schneider MP et al (1999) The place of *Callimico goeldii* in the callitrichine phylogenetic tree: evidence from von Willenbrand factor gene intron II sequences. Mol Phylogenet Evol 13:392–404

Chivers DJ, Hladik CM (1980) Morphology of the gastrointestinal tract in primates: comparisons with other mammals in relation to diet. J Morphol 166:377–386

Collins AC (2004) Atelinae phylogenetic relationships: the trichotomy revived? Am J Phys Anthropol 124:285–296

Collura RV, Auerbach MR, Stewart CB (1996) A quick, direct method that can differentiate expressed mitochondrial genes from their nuclear pseudogenes. Curr Biol 6:1337–1339

Davenport TR, Stanley WT, Sargis EJ et al (2006) A new genus of African monkey, *Rungwecebus*: morphology, ecology, and molecular phylogenetics. Science 312:1378–1381

Delson E (1980) Fossil macaques phyletic relationships and a scenario of development. In: Lindburg DG (ed) The macaques: studies in ecology, behavior, and evolution. Van Nostrand Reinhold, New York

Delson E (1992) Evolution of old world monkeys. In: Johns JS, Martin RD, Pilbeam D et al (eds) The Cambridge encyclopedia of human evolution. Cambridge University Press, Cambridge

Delson E (1994) Evolutionary history of the colobine monkeys in paleoenvironmental perspective. In: Oates JF, Davies AG (eds) Colobine monkeys: their ecology, behaviour, and evolution. Cambridge University Press, Cambridge

Delson E (2000) Colobinae. In: Delson E, Tattersall I, Van Couvering JA et al (eds) Encyclopedia of human evolution and prehistory, 2nd edn. Garland, New York

Disotell TR (1994) Generic level relationships of the Papionini (Cercopithecoidea). Am J Phys Anthropol 94:47–57

Disotell TR (2003) Primates: phylogenetics. Nature Publishing Group, London, Encyclopedia of the human genome

Disotell TR, Raaum RL (2002) Molecular timescale and gene tree incongruence in the guenons. In: Glenn ME, Cords M (eds) The guenons: diversity and adaptation in African monkeys. Kluwer, New York

16 Molecular Phylogeny and Evolution in Primates 263

Dutrillaux B, Muleris M, Couturier J (1988) Chromosomal evolution of Cercopithecinae. In: Gautier-Hion A, Bourliere F, Gautier JP et al (eds) A primate radiation: evolutionary biology of the African guenons. Cambridge University Press, New York

Eizirik E, Murphy WJ, O'Brien SJ (2001) Molecular dating and biogeography of the early placental mammal radiation. J Hered 92:212–219

Fa JE (1989) The genus *Macaca*: a review of taxonomy and evolution. Mamm Rev 19:45–81

Fabre PH, Rodrigues A, Douzery EJ (2009) Patterns of macroevolution among Primates inferred from a supermatrix of mitochondrial and nuclear DNA. Mol Phylogenet Evol 53:808–825

Felsenstein J (1978) Cases in which parsimony and compatibility methods will be positively misleading. Syst Zool 27:401–410

Fleagle JG (1999) Primate adaptation and evolution, 2nd edn. Academic, San Diego

Fleagle JG, McGraw WS (1999) Skeletal and dental morphology supports diphyletic origin of baboons and mandrills. Proc Natl Acad Sci USA 96:1157–1161

Fooden J (1976) Provisional classifications and key to living species of macaques (primates: *Macaca*). Folia Primatol (Basel) 25:225–236

Gibson A, Gowri-Shankar V, Higgs PG et al (2005) A comprehensive analysis of mammalian mitochondrial genome base composition and improved phylogenetic methods. Mol Biol Evol 22:251–264

Goodman M, Porter CA, Czelusniak J et al (1998) Toward a phylogenetic classification of primates based on DNA evidence complemented by fossil evidence. Mol Phylogenet Evol 9:585–598

Groves CP (1970) The forgotten leaf-eaters, and the phylogeny of the Colobinae. In: Napier JR, Napier PH (eds) Old World monkeys: evolution, systematics, and behavior. Academic, New York

Groves CP (1978) Phylogenetic and populations systematics of the mangabeys (Primates: Cercopithecoidea). Primates 19:1–34

Groves CP (1989) A theory of human and primate evolution. Oxford University Press, New York

Groves CP (2001) Primate taxonomy. Smithsonian Institution Press, Washington, DC

Groves CP (2005) Order Primates. In: Wilson DE, Reeder DM (eds) Mammal species of the world, 3rd edn. Johns Hopkins University Press, Baltimore

Haile-Selassie Y (2001) Late Miocene hominids from the Middle Awash, Ethiopia. Nature 412:178–81

Harada ML, Schneider H, Schneider MP et al (1995) DNA evidence on the phylogenetic systematics of New World monkeys: support for the sister-grouping of *Cebus* and *Saimiri* from two unlinked nuclear genes. Mol Phylogenet Evol 4:331–349

Harris EE, Disotell TR (1998) Nuclear gene trees and the phylogenetic relationships of the mangabeys (Primates: Papionini). Mol Biol Evol 15:892–900

Hasegawa M, Thorne JL, Kishino H (2003) Time scale of eutherian evolution estimated without assuming a constant rate of molecular evolution. Genes Genet Syst 78:267–283

Hayasaka K, Fujii K, Horai S (1996) Molecular phylogeny of macaques: implications of nucleotide sequences from an 896-base pair region of mitochondrial DNA. Mol Biol Evol 13:1044–1053

Hedges SB, Parker PH, Sibley CG et al (1996) Continental breakup and the ordinal diversification of birds and mammals. Nature (Lond) 381:226–229

Hodgson JA, Sterner KN, Matthews LJ et al (2009) Successive radiations, not stasis, in the South American primate fauna. Proc Natl Acad Sci USA 106:5534–5539

Horai S, Hayasaka K, Kondo R et al (1995) The recent African origin of modern humans revealed by complete sequences of hominoid mitochondrial DNAs. Proc Natl Acad Sci USA 92:532–536

Horovitz I, Meyer A (1995) Systematics of New World monkeys (Platyrrhini, primates) based on 16 S mitochondrial DNA sequences: a comparative analysis of different weighting methods in cladistic analysis. Mol Phylogenet Evol 4:448–456

Horovitz I, Zardoya R, Meyer A (1998) Platyrrhine systematics: a simultaneous analysis of molecular and morphological data. Am J Phys Anthropol 106:261–281

Horvath JE, Weisrock DW, Embry SL et al (2008) Development and application of a phylogenomic toolkit: resolving the evolutionary history of Madagascar's lemurs. Genome Res 18:489–499

Hudelot C, Gowri-Shankar V, Jow H et al (2003) RNA-based phylogenetic methods: application to mammalian mitochondrial RNA sequences. Mol Phylogenet Evol 28:241–252

Israfil H, Zehr SM, Mootnick AR et al (2011) Unresolved molecular phylogenies of gibbons and siamangs (family: Hylobatidae) based on mitochondrial, Y-linked, and X-linked loci indicate a rapid Miocene radiation or sudden vicarance event. Mol Phylogenet Evol 58:447–455

Jablonski N (1998) The evolution of the doucs and snub-nosed monkeys and the question of the phyletic unity of the odd-nosed colobines. In: Jablonski N (ed) Natural history of the doucs and snub-nosed monkeys. World Scientific Publishing, New Jersey

Jablonski NG, Peng YZ (1993) The phylogenetic relationships and classification of the doucs and snub-nosed langurs of China and Vietnam. Folia Primatol 60:36–55

Jow H, Hudelot C, Rattray M et al (2002) Bayesian phylogenetics using an RNA substitution model applied to early mammalian evolution. Mol Biol Evol 19:1591–1601

Karanth KP, Singh L, Collura RV et al (2008) Molecular phylogeny and biogeography of langurs and leaf monkeys of South Asia (Primates: Colobinae). Mol Phylogenet Evol 46:683–694

Karanth KP, Delefosse T, Rakotosamimanana B et al (2005) Ancient DNA from giant extinct lemurs confirms single origin of Malagasy primates. Proc Natl Acad Sci USA 102:5090–5095

Kelley J (2002) The hominoid radiation in Asia. In: Hartwig WC (ed) The primate fossil record. Cambridge University Press, Cambridge

Li J, Han K, Xing J et al (2009) Phylogeny of the macaques (Cercopithecidae: *Macaca*) based on Alu elements. Gene (Amst) 448:242–249

Martin RD (1993) Primate origins: plugging the gaps. Nature (Lond) 363:223–234

Martin RD (2003) Combing the primate record. Nature (Lond) 422:388–391

Masters JC, Anthony NM, de Wit MJ et al (2005) Reconstructing the evolutionary history of the Lorisidae using morphological, molecular, and geological data. Am J Phys Anthropol 127:465–480

Masters JC, Boniotto M, Crovella S et al (2007) Phylogenetic relationships among the Lorisoidea as indicated by craniodental morphology and mitochondrial sequence data. Am J Primatol 69:6–15

Matsudaira K, Ishida T (2010) Phylogenetic relationships and divergence dates of the whole mitochondrial genome sequences among three gibbon genera. Mol Phylogenet Evol 55:454–459

Matsui A, Rakotondraparany F, Hasegawa M et al (2007) Determination of a complete lemur mitochondrial genome from feces. Mamm Study 32:7–16

Matsui A, Rakotondraparany F, Munechika I et al (2009) Molecular phylogeny and evolution of prosimians based on complete sequences of mitochondrial DNAs. Gene (Amst) 441:53–66

Meireles CM, Czelusniak J, Schneider MPC et al (1999) Molecular phylogeny of ateline new world monkeys (Platyrrhini, Atelinae) based on gamma-globin gene sequences: evidence that *Brachyteles* is the sister group of *Lagothrix*. Mol Phylogenet Evol 12:10–30

Messier W, Stewart CB (1997) Episodic adaptive evolution of primate lysozymes. Nature (Lond) 385:151–154

Mootnick A, Groves CP (2005) A new generic name for the hoolock gibbon (Hylobatidae). Int J Primatol 26:971–976

Morales JC, Melnick DJ (1998) Phylogenetic relationships of the macaques (Cercopithecidae: *Macaca*), as revealed by high resolution restriction site mapping of mitochondrial ribosomal genes. J Hum Evol 34:1–23

Murphy WJ, Eizirik E, Johnson WE et al (2001) Molecular phylogenetics and the origins of placental mammals. Nature (Lond) 409:614–618

Ni X, Wang Y, Hu Y et al (2004) A euprimate skull from the early Eocene of China. Nature (Lond) 427:65–68

Opazo JC, Wildman DE, Prychitko T et al (2006) Phylogenetic relationships and divergence times among New World monkeys (Platyrrhini, Primates). Mol Phylogenet Evol 40:274–280

Osterholz M, Walter L, Roos C (2008) Phylogenetic position of the langur genera *Semnopithecus* and *Trachypithecus* among Asian colobines, and genus affiliations of their species groups. BMC Evol Biol 8:58

Oxnard CE (1981) The uniqueness of *Daubentonia*. Am J Phys Anthropol 54:1–21

Page SL, Goodman M (2001) Catarrhine phylogeny: noncoding DNA evidence for a diphyletic origin of the mangabeys and for a human-chimpanzee clade. Mol Phyl Evol 18:14–25

Page SL, Chiu C, Goodman M (1999) Molecular phylogeny of Old World monkeys (Cercopithecidae) as inferred from gamma-globin DNA sequences. Mol Phylogenet Evol 13:348–359

Pastorini J, Forstner MR, Martin RD (2002) Phylogenetic relationships among Lemuridae (Primates): evidence from mtDNA. J Hum Evol 43:463–478

Pastorini J, Thalmann U, Martin RD (2003) A molecular approach to comparative phylogeography of extant Malagasy lemurs. Proc Natl Acad Sci USA 13:5879–5884

Peng YZ, Pan RL, Jablonski NG (1993) Classification and evolution of Asian colobines. Folia Primatol (Basel) 60:106–117

Perelman P, Johnson WE, Roos C et al (2011) A molecular phylogeny of living primates. PLoS Genet 7:e1001342

Phillips MJ, Penny D (2003) The root of the mammalian tree inferred from whole mitochondrial genomes. Mol Phylogenet Evol 28:171–185

Pilbeam D, Rose MD, Barry JC et al (1990) New *Sivapithecus humeri* from Pakistan and the relationship of *Sivapithecus* and *Pongo*. Nature (Lond) 348:237–239

Porter CA, Sampaio I, Schneider H et al (1995) Evidence on primate phylogeny from ε-globin gene sequences and flanking regions. J Mol Evol 40:30–55

Porter CA, Page SL, Czelusniak J et al (1997a) Phylogeny and evolution of selected primates as determined by sequences of the ε-globin locus and 50 flanking regions. Int J Primatol 18:261–295

Porter CA, Czelusniak J, Schneider H et al (1997b) Sequence of the primate epsilon-globin gene: implication for systematics of the marmosets and other new world primates. Gene (Amst) 205:59–71

Porter CA, Czelusniak J, Schneider H et al (1999) Sequence from the 5′ flanking region of the epsilon-globin gene support the relationship of *Callicebus* with the Pitheciins. Am J Primatol 48:69–75

Poux C, Douzery EJ (2004) Primate phylogeny, evolutionary rate variations, and divergence times: a contribution from the nuclear gene IRBP. Am J Phys Anthropol 124:1–16

Poux C, Madsen O, Marquard E et al (2005) Asynchronous colonization of Madagascar by the four endemic clades of primates, tenrecs, carnivores, and rodents as inferred from nuclear genes. Syst Biol 54:719–730

Poux C, Chevret P, Huchon D et al (2006) Arrival and diversification of caviomorph rodents and platyrrhine primates in South America. Syst Biol 55:228–244

Prychitko T, Johnson RM, Wildman DE et al (2005) The phylogenetic history of New World monkey beta globin reveals a platyrrhine beta to delta gene conversion in the atelid ancestry. Mol Phylogenet Evol 35:225–234

Raaum RL, Sterner KN, Noviello CM et al (2005) Catarrhine primate divergence dates estimated from complete mitochondrial genomes: concordance with fossil and nuclear DNA evidence. J Hum Evol 48:237–257

Ray DA, Xing J, Hedges DJ et al (2005) Alu insertion loci and platyrrhine primate phylogeny. Mol Phylogenet Evol 35:117–126

Roos C, Geissmann T (2001) Molecular phylogeny of the major hylobatid divisions. Mol Phylogenet Evol 19:486–494

Roos C, Schmitz J, Zischler H (2004) Primate jumping genes elucidate strepsirrhine phylogeny. Proc Natl Acad Sci USA 101:10650–10654

Rosenberger AL, Hartwig WC, Wolff RG (1991) *Szalatavus attricuspis*, an early platyrrhine primate. Folia Primatol (Basel) 56:225–233

Rumpler Y, Warter S, Petter JJ et al (1988) Chromosomal evolution of Malagasy lemurs. XI. Phylogenetic position of Daubentonia madagascariensis. Folia Primatol (Basel) 50:124–129

Ruvolo M (1988) Genetic evolution in the African guenons. In: Gautier-Hion A, Bourliere F, Gautier JP et al (eds) A primate radiation: evolutionary biology of the African guenons. Cambridge University Press, New York

Schmitz J, Ohme M, Zischler H (2001) SINE insertions in cladistic analyses and the phylogenetic affiliations of *Tarsius bancanus* to other primates. Genetics 157:777–784

Schmitz J, Ohme M, Zischler H (2002) The complete mitochondrial sequence of *Tarsius bancanus*: evidence for an extensive nucleotide compositional plasticity of primate mitochondrial DNA. Mol Biol Evol 19:544–553

Schneider H (2000) The current status of the New World monkey phylogeny. An Acad Bras Cienc 72:165–172

Schneider H, Schneider MPC, Sampaio I et al (1993) Molecular phylogeny of the new world monkeys (Platyrrhini, Primates). Mol Phylogenet Evol 2:225–242

Schneider H, Sampaio I, Harada ML et al (1996) Molecular phylogeny of the New World monkeys (Platyrrhini, primates) based on two unlinked nuclear genes: IRBP intron 1 and epsilon-globin sequences. Am J Phys Anthropol 100:153–179

Schwartz JH (1992) Topics in primatology. In: Matano S, Tuttle RH, Ishida H et al (eds) Evolutionary biology, reproductive endocrinology, and virology. University of Tokyo Press, Tokyo

Seiffert ER, Simons EL, Attia Y (2003) Fossil evidence for an ancient divergence of lorises and galagos. Nature (Lond) 422:421–424

Singer SS, Schmitz J, Schwiegk C et al (2003) Molecular cladistic markers in the new world monkey phylogeny (Platyrrhini, Primates). Mol Phylogenet Evol 26:490–501

Springer MS, Murphy WJ, Eizirik E et al (2003) Placental mammal diversification and the Cretaceous–Tertiary boundary. Proc Natl Acad Sci USA 100:1056–1061

Steiper ME, Ruvolo M (2003) New world monkey phylogeny based on X-linked G6PD DNA sequences. Mol Phylogenet Evol 27:121–130

Steiper ME, Young NM (2006) Primate molecular divergence dates. Mol Phylogenet Evol 41:384–394

Sterner KN, Raaum RL, Zhang YP et al (2006) Mitochondrial data support an odd-nosed colobine clade. Mol Phylogenet Evol 40:1–7

Stewart C, Disotell T (1998) Primate evolution in and out of Africa. Curr Biol 8:R582–R588

Strasser E, Delson E (1987) Cladistic analysis of cercopithecid relationships. J Hum Evol 16:81–99

Szalay FS, Delson E (1979) Evolutionary history of the primates. Academic, New York

Takacs Z, Morales JC, Geissmann T et al (2005) A complete species-level phylogeny of the Hylobatidae based on mitochondrial ND3-ND4 gene sequences. Mol Phylogenet Evol 36:456–467

Tavaré S, Marshall CR, Will O et al (2002) Using the fossil record to estimate the age of the last common ancestor of extant primates. Nature (Lond) 416:726–729

Thinh VN, Mootnick AR, Geissmann T et al (2010) Mitochondrial evidence for multiple radiations in the evolutionary history of small apes. BMC Evol Biol 10:74

Thorne JL, Kishino H (2002) Divergence time and evolutionary rate estimation with multilocus data. Syst Biol 51:689–702

Thorne JL, Kishino H, Painter IS (1998) Estimating the rate of evolution of the rate of molecular evolution. Mol Biol Evol 15:1647–1657

Ting N (2008) Mitochondrial relationships and divergence dates of the African colobines: evidence of Miocene origins for the living colobus monkeys. J Hum Evol 55:312–325

Ting N, Tosi AJ, Li Y et al (2008) Phylogenetic incongruence between nuclear and mitochondrial markers in the Asian colobines and the evolution of the langurs and leaf monkeys. Mol Phylogenet Evol 46:466–474

Tosi AJ, Morales JC, Melnick DJ (2000) Comparison of Y chromosome and mtDNA phylogenies leads to unique inferences of macaque evolutionary history. Mol Phylogenet Evol 17:133–144

Tosi AJ, Buzzard PJ, Morales JC et al (2002) Y-chromosome data and tribal affiliations of *Allenopithecus* and *Miopithecus*. Int J Primatol 23:1287–1299

Tosi AJ, Disotell TR, Morales JC et al (2003) Cercopithecine Y-chromosome data provide a test of competing morphological evolutionary hypotheses. Mol Phylogenet Evol 27:510–521

Tosi AJ, Melnick DJ, Disotell TR (2004) Sex chromosome phylogenetics indicate a single transition to terrestriality in the guenons (tribe Cercopithecini). J Hum Evol 46:223–237

Tosi AJ, Detwiler KM, Disotell TR (2005) X-chromosomal window into the evolutionary history of the guenons (Primates: Cercopithecini). Mol Phylogenet Evol 36:58–66

von Dornum M, Ruvolo M (1999) Phylogenetic relationships of the New World monkeys (primates, Platyrrhini) based on nuclear G6PD DNA sequences. Mol Phylogenet Evol 11:459–476

Waddell P, Penny D (1996) Evolutionary trees of apes and humans from DNA sequences. In: Lock AJ, Peters CR (eds) Handbook of human symbolic evolution. Oxford University Press, Oxford

Wang W, Su B, Lan H et al (1995) Phylogenetic relationships among two species of golden monkey and three species of leaf monkey inferred from rDNA variation. Folia Primatol (Basel) 65:138–143

Ward S (1997) The taxonomy and phylogenetic relationships of *Sivapithecus* revisited. In: Begun DR, Ward CV, Rose MD (eds) Function, phylogeny, and fossils. Plenum, New York

Xing J, Wang H, Han K et al (2005) A mobile element based phylogeny of Old World monkeys. Mol Phylogenet Evol 37:872–880

Yoder AD (1994) Relative position of the Cheirogaleidae in strepsirrhine phylogeny: a comparison of morphological and molecular methods and results. Am J Phys Anthropol 94:25–46

Yoder AD, Cartmill M, Ruvolo M et al (1996) Ancient single origin for Malagasy primates. Proc Natl Acad Sci USA 93:5122–5126

Yoder AD, Irwin JA, Payseur BA (2001) Failure of the ILD to determine data combinability for slow loris phylogeny. Syst Biol 50:408–424

Yoder AD, Burns MM, Zehr S et al (2003) Single origin of Malagasy Carnivora from an African ancestor. Nature (Lond) 421:734–737

Zhang Y, Ryder O (1998) Mitochondrial cytochrome *b* gene sequences of Old World monkeys: with special reference on evolution of Asian colobines. Primates 39:39–49

Chapter 17
Origins and Evolution of Early Primates

Masanaru Takai

17.1 Introduction

It has long been believed that primates originated in North America around the time boundary between the latest Cretaceous and early Tertiary (K/T boundary), mainly because of the rich fossil records of early primates and possible close relatives – such as the Plesiadapiformes – from the Paleocene of North America and Europe (Fleagle 1999). In particular, *Purgatorius*, which has been discovered from the latest Cretaceous to early Paleocene sediments of North America, has been regarded as the oldest fossil primate in the world, because it was not specialized but still sufficiently generalized in dental morphology to be a possible ancestor of Eocene euprimates. *Purgatorius* has been included in the Plesiadapiformes, as the "wastebasket" grouping of the Paleocene mammals, but it is now often treated as belonging to a different taxonomic group (Fleagle 1999).

Most plesiadapiform mammals were so specialized as to resemble modern rodents, in having large incisors, extremely reduced or absent canines, and a large diastema between the incisors and cheek teeth. For these reasons, plesiadapiforms have been regarded as possible close relatives to primates. However, nearly complete skeletal specimens of tiny plesiadapiforms, such as *Carpolestes*, have now been reported from the Paleocene sediments of North America (Bloch and Boyer 2002, 2007; Bloch et al. 2007). Although *Carpolestes* had been described originally based on dental fragments in the 1970s, not only cranial but also postcranial materials were discovered together, and ultrahigh-resolution X-ray computed tomography (CT) has made possible the detailed analysis of minute features of these specimens. In addition, the detailed observation of the cranial materials has confirmed the definitive features of primates for *Carpolestes* (Silcox et al. 2009). However, the

---

M. Takai (✉)
Primate Research Institute, Kyoto University, 41-2 Kanrin, Inuyama, Aichi 484-8506, Japan
e-mail: takai@pri.kyoto-u.ac.jp

H. Hirai et al. (eds.), *Post-Genome Biology of Primates*, Primatology Monographs,
DOI 10.1007/978-4-431-54011-3_17, © Springer 2012

detailed analyses of *Carpolestes* and other plesiadapiforms still invoke controversies on evolutionary scenarios for the earliest primates.

On the other hand, recent molecular biological research revealed the phylogenetic relationships among the "Archonta," which was originally defined to include Primates, Chiroptera, Scandentia, and Dermoptera. According to recent molecular studies, the chiropterans are not closely related to primates but to ungulates and carnivores (Murphy et al. 2001a, b; Hasegawa et al. 2003; Springer et al. 2003, 2007; but see Sargis 2007). In addition, molecular biologists have estimated the divergence date for primates as early as the middle Cretaceous. These two findings based on molecular biology surprised paleontologists working on primates and led to revisions of their traditional views. In addition, the topology of the phylogenetic tree of extant placental mammals indicates that the major mammalian groups, including primates, might have originated not in the Northern Hemisphere continents such as North America and Eurasia, but in the continents of the Southern Hemisphere. Although some paleontologists have already advocated hypotheses of an origin in the Indian subcontinent or East Asia based on the Paleocene/early Eocene fossil evidence in Asia (Kraus and Maas 1990; Beard 1998), the southern continent origin hypothesis supported by molecular biology has been a revolutionary proposal from a new research field.

17.2 Recent Consensus on Primate Evolution from Molecular Biology

17.2.1 Phylogenetic Relationships

Recent molecular biological studies have revealed the phyletic, chronological, and geographic origins of primates (Madsen et al. 2001; Murphy et al. 2001a, b; Hasegawa et al. 2003; Springer et al. 2003). According to these studies, although the detailed topography of the phylogenetic trees differs slightly (Asher et al. 2009), the living placental mammals are divided into four major clades: the Afrotheria, Edentata, Laurasiatheria, and Euarchontoglires (Fig. 17.1). Primates are included in the Euarchontoglires, together with the Scandentia, Dermoptera, Rodentia, and Lagomorpha. Although Chiroptera has long been regarded as close relatives of primates by morphologists (Beard 1993; MacKenna and Bell 1997), it is now included not in the Euarchontoglires but in the Laurasiatheria together with artiodactyls, perissodactyls, and carnivores. Therefore, primates, scandentians, and dermopterans are now combined in the Euarchonta (Waddell et al. 1999; Silcox et al. 2005), excluding chiropterans from the "Archonta," and the scandentian/dermopteran clade is regarded as being most closely related to the primates (Springer et al. 2003, 2007; Kriegs et al. 2006; Sargis 2007).

Among morphologists and paleontologists, some have regarded scandentians as the closest group to primates (Wible and Covert 1987), whereas others have considered

17 Origins and Evolution of Early Primates

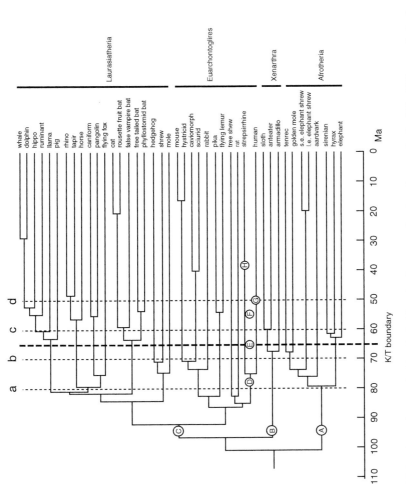

Fig. 17.1 Phylogenetic trees of living placental mammals. Four *dashed lines* (*a–d*) correspond to the paleomaps (*a–d*) of Fig. 17.2, respectively. *A*, settlement toward African continent; *B*, settlement toward South American continent; *C*, transverse event across the circum-Tethys region; *D*, the divergence time of primates estimated by molecular biology; *E*, the age of *Purgatorius*, the possible oldest fossil record of primates; *F*, the age of *Altiatlasius*, the oldest euprimate fossil from North Africa (Sigé et al. 1990); *G*, the age of the Vastan primates, the possible oldest fossil record of anthropoid primates (Bajpai et al. 2008; Rose et al. 2009); *H*, the oldest fossil record of crown strepsirrhine (Seiffert et al. 2005). (Modified from Springer et al. 2005)

dermopterans to be closer to primates than to the scandentians (Beard 1993; MacKenna and Bell 1997). Recent cladistic analyses by several paleontologists support the scandentian/dermopteran clade but insist that the fossil taxon Plesiadapiformes was more closely related to primates (Bloch and Boyer 2007; Silcox et al. 2007, 2009).

17.2.2 Estimates of Divergence Date

Molecular biological studies have estimated the divergence times of the members of placental mammals, including the primates. One of the most remarkable points of these estimations is that the major placental mammalian groups originated in the Cretaceous. Arnason et al. (1996, 1998) estimated the splitting date between strepsirrhines and haplorrhines to be 80 million years ago (Ma). Springer et al. (2003) suggested 77 Ma and Hasegawa et al. (2003) proposed 67.7–78.5 Ma as the divergence date for the base group of primates (Janke et al. 1994; Hedges et al. 1996; Kumar and Hedges 1998; Murphy et al. 2001a). On the other hand, the oldest primate fossil is possibly *Purgatorius* at 65 Ma (Fleagle 1999), and the earliest certain fossil record of the euprimates is *Altiatlasius* from the late Paleocene, 55 Ma, from Morocco, North Africa (Sigé et al. 1990; Fleagle 1999; but see Hooker et al. 1999). Consequently, fiery controversies have arisen on the divergence date of modern mammalian clades, mostly from paleontologists (Benton 1999; Archibald and Deutschman 2001). Aside from the technical problems in molecular analysis, the most important point of this controversy is the question of how robust are the Cretaceous fossil records. Molecular biologists assert that the fossil evidence is unreliably incomplete, especially in the Cretaceous, whereas paleontologists insist that the Cretaceous fossil records are not any more scarce than those of the Tertiary (Benton 1999). As paleontologists insist, the fossil records of North America and Europe are relatively well documented, but they are rather scarce or even poor in Asia and Africa. In any case, if the estimated dates of primate origins of about 80 Ma given by molecular biology are correct, it is obvious that no definitive early primate fossils have been discovered from the late Cretaceous. Martin (1993) predicted an old Cretaceous origin of primates, based on the incompleteness of fossil specimens. According to his argument, only 3% of the fossil primate taxa have been discovered to date, and he has estimated that primates first appeared about 80 Ma. His prediction is now being supported by statistical analyses (Tavaré et al. 2002) and by molecular evidence (see also Martin 2007).

The discrepancies between the estimated divergence times given by molecular clock dating and the oldest fossil records in early primates could be explained by two alternative hypotheses. One is the "Garden of Eden" hypothesis, which predicts discoveries of fossil evidence from unknown Cretaceous localities in the Southern Hemisphere continents, such as Africa, South Africa, Antarctica, or Madagascar (Kraus and Maas 1990; Madsen et al. 2001). The other hypothesis assumes either acceleration or heterogeneous rates of molecular evolution in some lineages of placental mammals (Foote et al. 1999). Molecular biologists believe the former prediction, whereas most paleontologists prefer the latter. Improvements in the analytical methods used in molecular paleontology and further excavations in the southern continents would help solve this question.

17.2.3 Place of Origins

As already mentioned, the pattern of distribution of fossil and living placental mammals indicates that the place of origin of two major basal clades, afrotherians and xenarthrans, is likely to be Africa and South America, respectively. Both these places formed part of the ancient Southern Hemisphere continent, Gondwanaland (Madsen et al. 2001). In one cladogram of placental mammals, afrotherians and xenarthrans are located as an outgroup to the laurasiatherian/euarchontoglires clade (Murphy et al. 2001a, b), so it is suggested that these two groups could have originated in Gondwanaland and then invaded northern Laurasialand (Eizirik et al. 2001; Soligo et al. 2007). If the initial divergence of placental mammals had actually occurred about 100–80 Ma, as molecular biological studies suggest, southern Gondwanaland and northern Laurasialand were not yet fully separated from each other and the intercontinental migration of terrestrial mammals could have been much easier than during the later age (Fig. 17.2). However, most paleoprimatologists are still opposed to the southern continent origin hypothesis because of the rich fossil records of euprimates (adapiforms and omomyiforms) from the Paleocene to Eocene sediments of North America and Europe.

17.3 Adaptive Explanation for Primate Origins

At present, molecular biology provides the most powerful and credible methods for phylogenetic studies. It has elucidated phylogenetic relationships among placental mammals, including primates and close relatives, estimating the divergence time among the clades and suggesting the geographic origins of main mammalian clades. However, theories on the evolutionary adaptation of primate origins are still constructed using morphological analyses of fossil and living primates and close relatives.

Several major hypotheses have been advocated on the initial divergence of primates since the early twentieth century. First, Smith (1913) considered that the stereoscopic vision and grasping hands and feet seen in early primates were adaptations to a three-dimensional arboreal habitat (the "Arboreal Adaptation" hypothesis). Although this hypothesis has long been treated as the pivotal theory for the adaptation of initial primates by many workers, Cartmill (1972) pointed out that it does not explain why other arboreal mammals did not evolve stereoscopic vision and grasping hands and feet, and instead argued that the earliest primates were specialized visual predators using stereoscopic view and grasping hands and feet to capture active arthropods or insects (the "Visual Predation" hypothesis).

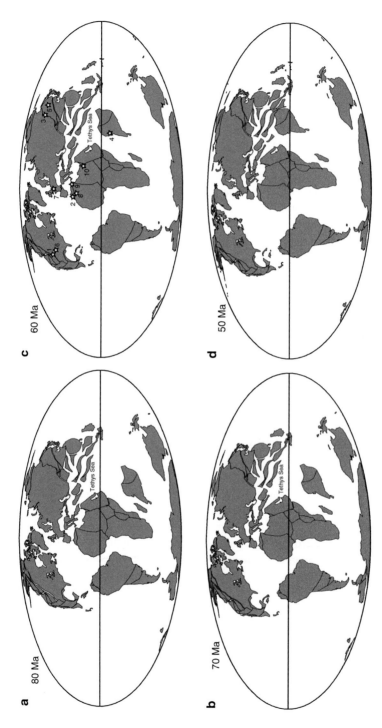

Fig. 17.2 Paleomap of from the late Cretaceous through early Eocene (*a*, 80 million years ago, Ma; *b*, 70 Ma; *c*, 60 Ma; *d*, 50 Ma). *Stars* indicate the localities of several important primate fossils. *1*, Purgatory Hill (*Purgatorius*); *2*, Adrar Mgorn (*Altiatlasius*); *3*, (*Altanius*); *4*, Vastan Mine (the Vastan primates); *5*, Hengyang Basin (*Teilhardina*); *6*, Bighorn Basin (*Teilhardina*); *7*, Dormaal (*Teilhardina*); *8*, Glib Zegdou (*Azgeripithecus*); *9*, Chambi (Djebelemur); *10*, Fayum (Karanisia and Saharagalago). (All maps modified from Scotese 2001)

Many workers have accepted these two hypotheses widely, but there has been no explanation of why the earliest primates occurred in the Paleocene/Eocene ages. Considering the paleoenvironmental background of the Paleocene, Sussman (1991) (also see Rasmussen 1990) proposed the "Angiosperm Exploitation" hypothesis for the adaptive radiation of the earliest primates. According to this hypothesis, the earliest primates were adapted to the environment in the fine terminal branches of angiosperm plants, where rich products, such as flowers, buds, and nectar, were to be found and where many arthropods came together to feed on them. The earliest primates would also have fed on such angiosperm products and insects and needed grasping hands and stereoscopic vision to capture insects. They would also have needed grasping feet to support their body firmly in mobile substrates. This hypothesis likely explains the adaptive radiation of the earliest primates, redeeming the weak points of the previous two hypotheses, and it seems to be most widely accepted by many workers today.

The angiosperm exploitation hypothesis was constructed based on a deduced history of coevolution between angiosperms and insects, but the evolutionary history of the angiosperms is now being revised according to new paleontological evidence (Crane et al. 1986, 1995). Although Sussman (1991) noted that many modern families and genera of flowering plants first appeared at the Paleocene–Eocene boundary and that modern evergreen tropical rainforests became widespread at that time, the earliest fossil specimens of flowering plants have been discovered from 140 Ma and became widespread as early as 100 Ma (Crane et al. 1995). The coevolution of flowering plants, insects, and early mammals (probably including stem primates) most likely occurred in the late Cretaceous, which is much earlier than the Paleocene–Eocene boundary that Sussman (1991) assumed. However, the estimated divergence time of primate origins has now been pushed back to the middle Cretaceous by molecular biology studies, corresponding to the evolutionary history of flowering plants.

17.4 Impact of "Grasping Primate Origins" Hypothesis

Early primates have been defined traditionally mainly by two morphological features: grasping hands and feet and orbital convergence. However, most of the fossil materials of the late Cretaceous to Paleocene consist of gnathodental fragments, and only a few skeletal specimens articulated with crania have been discovered. Fossil mammals including primates are usually identified from their dental morphology, and it is often very hard to identify postcranial specimens. Recent discoveries of relatively well preserved, articulated skeletons of plesiadapiforms from the Paleocene of North America have led to detailed observations and analysis of the cranial, dental, postcranial features of fossil taxa using ultrahigh-resolution CT analysis (Bloch and Boyer 2002, 2007; Silcox et al. 2009). According to these studies, the following morphological features of *Carpolestes* (Carpolestidae, Plesiadapiformes) were revealed: (1) orbital convergence had not yet occurred, but small and divergent orbits such as those of *Plesiadapis* were present; (2) the grasping abilities of hands and feet had already been acquired in well-preserved skeletal specimens, as in some fossil euprimates;

and (3) the low-crowned molar teeth indicate that they were not insectivorous but frugivorous (Bloch and Boyer 2002, 2007). In addition, based on a heuristic cladistic analysis, these authors advocated that plesiadapiforms are not closely related to dermopterans but are confined to the euprimate clade. Based on these findings, they concluded that *Carpolestes* and other plesiadapiforms evolved grasping first and convergent orbits later. This conclusion supports the "Terminal Branch Feeding" hypothesis (= "Angiosperm Exploitation" hypothesis of Sussman 1991; see also Rasmussen 1990) rather than the "Visual Predator" hypothesis (Cartmill 1972) as an adaptational explanation for primate origins (Bloch and Boyer 2002, 2007).

However, the most critical point of the "Grasping Primate Origins" hypothesis is the reliability of the proposed phylogenetic relationships between plesiadapiforms, especially the carpolestids and euprimates. As Kirk et al. (2003) pointed out, the cladistic analyses of Bloch and Boyer (2002, 2007) were mainly based on postcranial features, although most early mammals have been classified from dental morphology. Moreover, the morphological analysis of the dentition of carpolestids suggests a faunivorous diet rather than a frugivorous one (Rose 1975), and their anterior dentition is too specialized. Silcox et al. (2007), the advocator of the "Grasping Primate Origins" hypothesis, also mentioned that there is an apparent morphological gap, especially in dental morphology, between known plesiadapiforms and the oldest euprimates, and that the fossil taxa filling this morphological gap could be discovered not from North America but from Europe or Asia in the future.

Although the "Grasping Primate Origins" hypothesis looks fascinating, the weakest points of the discussion on this hypothesis are that all fossil materials of plesiadapiforms used for analysis have been discovered from North America. In other words, the evolutionary scenario of early primates – from primitive plesiadapiforms through early euprimates – of this hypothesis is constructed only by fossil taxa discovered from North America and Europe. In this model the fossil taxa from other continents, such as Asia and Africa, look unnecessary for the adaptive evolution of early primates and their close relatives. The origins and evolution of primates are likely to be more complicated than the scenario of the "Grasping Primate Origins" hypothesis.

17.5 Acquisition of Convergent Orbits and a Circum-Tethys Origin

Molecular biologists and paleontologists have long disputed the divergence dates of many organisms, including primates. The divergence dates estimated by molecular biology are usually much older than those of the fossil record, because morphological differentiation between any two groups arise from genetic isolation and the initial members of the diverging groups were too similar to allow discrimination. In addition, the incompleteness of fossil records produces inevitable discrepancies between the estimated divergence time by molecular estimation and the age of the first fossil specimens.

To resolve this discrepancy, it is necessary to discover late Cretaceous fossil specimens preserving the definitive features of primates, such as grasping hands and feet and convergent orbits. As the "Grasping Primate Origins" hypothesis indicates, the acquisition of the latter feature is attracting particular attention because among the extant primates, both strepsirrhines and haplorrhines retain well-converged orbits, suggesting that such morphology was acquired at the taxonomic base of the living crown primates. As molecular biological analysis has demonstrated that the divergence date of the base group of primates was in the late Cretaceous, one presumes that convergent orbits were well evolved at that time. However, the proponents of the "Grasping Primate Origins" hypothesis point out that no primate fossils preserving well-converged orbits have been discovered from earlier than the late Paleocene from North America (Bloch and Boyer 2002). The combination of the late Cretaceous origins of the crown primates and the lack of fossil taxa preserving well-convergent orbits from the Paleocene of North America may lead to the expectation that early primate fossils preserving well-converged orbits would be discovered from the late Cretaceous sediments not from North America but from other continents, probably from Asia or Africa. In fact, a nearly complete skull of *Teilhardina*, which preserves well-converged orbits, has been discovered from the early Eocene of China, likely supporting the Asian hypothesis (Ni et al. 2004). Moreover, *Teilhardina* has also been discovered from a number of early Eocene localities in North America and Europe, suggesting the intercontinental dispersal event among the three continents, even in the early Eocene (Smith et al. 2006; Beard 2008).

Many primitive primate fossils are now being reported from the early Eocene Vastan locality of India (Bajpai et al. 2008; Rose et al. 2009). Aside from their phyletic positions (early euprimates or primitive anthropoids), the discoveries of these Vastan primates from southern Asia suggest the importance of this area as the stage of the initial adaptive radiation of early primates. On the other hand, several primate taxa have already been discovered from a number of localities in northern Africa (Sigé et al. 1990; Godinot and Mahboubi 1992; Hartenberger and Marandat 1992; Seiffert et al. 2005). Although the geologic age of these north African localities is slightly younger than those of southern Asia, the presence of the middle Eocene anthropoid primates in this area also suggest the importance for the evolution of early primates here as well as in southern Asia.

The combination of these discoveries suggests that the circum-Tethys area, situated between southern Gondwanaland and northern Laurasialand during the late Cretaceous, could have been the stage for the evolution of the first primates. The phylogenetic tree of placental mammals proposed by molecular biologists also indicates a dispersal event between the two primitive southern continent clades (afrotherians and xenarthrans) toward the northern continent clades (euarchontoglires and laurasiatherians). Members of the latter groups could have moved across the Tethys Sea during the late Cretaceous. The much greater taxonomic variation seen in groups from northern rather than southern continents might indicate that such a sea-crossing event could have acted as a filter, leading to the genetic isolation of the late Cretaceous mammals and the consequent explosive divergence of those in the northern continents. The role of the Tethys sea and the circum-Tethys area might have been more important for the evolutionary history, not only of primates but also of other mammals, than ever imagined.

References

Archibald JD, Deutschman DH (2001) Quantitative analysis of the timing of the origin and diversification of extant placental orders. J Mamm Evol 8(2):107–124

Arnason U, Gullberg A, Janke A et al (1996) Pattern and timing of evolutionary divergences among hominoids based on analyses of complete mtDNAs. J Mol Evol 43:650–661

Arnason U, Gullberg A, Janke A (1998) Molecular timing of primate divergences as estimated by two nonprimate calibration points. J Mol Evol 47:718–727

Asher RJ, Bennett N, Lehmann T (2009) The new framework for understanding placental mammal evolution. Bioessays 31:853–864

Bajpai S, Kay RF, Williams BA et al (2008) The oldest Asian record of Anthropoidea. Proc Natl Acad Sci USA 105:11093–11098

Beard KC (1993) Phylogenetic systematics of the Primatomorpha, with special reference to Dermoptera. In: Szalay FS, Novacek MJ, McKenna MC (eds) Mammal phylogeny: placentals. Springer, New York, pp 129–150

Beard KC (1998) East of Eden: Asia as an important center of taxonomic origination in mammalian evolution. Bull Carnegie Mus Nat Hist 34:5–39

Beard KC (2008) The oldest North American primate and mammalian biogeography during the Paleocene-Eocene thermal maximum. Proc Natl Acad Sci USA 105:3815–3818

Benton MJ (1999) Early origins of modern birds and mammals: molecules vs. morphology. Bioessays 21(12):1043–1051

Bloch JI, Boyer DM (2002) Grasping primate origins. Science 298:1606–1610

Bloch JI, Boyer DM (2007) New skeletons of Paleocene-Eocene Plesiadapiformes: a diversity of arboreal positional behaviors in early primates. In: Rovosa MJ, Dagosto M (eds) Primate origins: adaptations and evolution. Springer, New York, pp 535–581

Bloch JI, Silcox MT, Boyer DM et al (2007) New Paleocene skeletons and the relationship of plesiadapiforms to crown-clade primates. Proc Natl Acad Sci USA 104:1159–1164

Cartmill M (1972) Arboreal adaptations and the origin of the order Primates. In: Tuttle R (ed) Functional and evolutionary biology of primates. Aldine-Atherton, Chicago, pp 97–122

Crane PR, Friis EM, Pedersen KR (1986) Lower Cretaceous angiosperm flowers: fossil evidence on early radiation of dicotyledons. Science 232:852–854

Crane PR, Friis EM, Pedersen KR (1995) The origin and early diversification of angiosperms. Nature (Lond) 374:27–33

Eizirik E, Murphy WJ, O'Brien SJ (2001) Molecular dating and biogeography of the early placental mammal radiation. J Hered 92:212–219

Fleagle JG (1999) Primate adaptation and evolution. Plenum, San Diego

Foote M, Hunter JP, Janis CM et al (1999) Evolutionary and preservational constraints on origins of biologic groups: divergence times of eutherian mammals. Science 283:1310–1314

Godinot M, Mahboubi M (1992) Earliest known simian primate found in Algeria. Nature (Lond) 357:324–326

Hartenberger J-L, Marandat B (1992) A new genus and species of an early Eocene primate from North Africa. Hum Evol 7:9–16

Hasegawa M, Thorne JL, Kishino H (2003) Time scale of eutherian evolution estimated without assuming a constant rate of molecular evolution. Genes Genet Syst 78:267–283

Hedges SB, Parker PH, Sibley CG et al (1996) Continental breakup and the ordinal diversification of birds and mammals. Nature (Lond) 381:226–229

Hooker JJ, Russell DE, Phélizon A (1999) A new family of Plesiadapiformes (Mammalia) from the Old World lower Paleogene. Palaeontology 42(3):377–407

Janke A, Feldmaier-Fuchs G, Thomas WK et al (1994) The marsupial mitochondrial genome and the evolution of placental mammals. Genetics (Dordr) 137:243–256

Kirk EC, Cartmill M, Kay RF et al (2003) Comment on "Grasping Primate Origins.". Science 300:741b

Kraus DW, Maas MC (1990) The biogeographic origins of late Paleocene-early Eocene mammalian immigrants to the Western Interior of North America. Geol Soc Am Special Pap 243:71–105

Kriegs JO, Chrakov G, Jurka J et al (2006) Evolutionary history of 7SL RNA-derived SINEs in supraprimates. Trends Genet 23:158–161

Kumar S, Hedges B (1998) A molecular timescale for vertebrate evolution. Nature (Lond) 392:917–920

MacKenna MC, Bell SK (1997) Classification of mammals above the species level. Columbia University Press, New York

Madsen O, Scally M, Douady CJ et al (2001) Parallel adaptive radiations in two major clades of placental mammals. Nature (Lond) 409:610–614

Martin RD (1993) Primate origins: plugging the gaps. Nature (Lond) 363:223–234

Martin RD (2007) Primate origins: implications of a Cretaceous ancestry. Folia Primatol (Basel) 78:277–296

Murphy WJ, Eizirik E, Johnson WE et al (2001a) Molecular phylogenetics and the origins of placental mammals. Nature (Lond) 401:614–618

Murphy WJ, Eizirik E, O'Brien SJ et al (2001b) Resolution of the early placental mammal radiation using Bayesian phylogenetics. Science 294:2348–2351

Ni X, Wang Y, Hu Y et al (2004) A euprimate skull from the early Eocene of China. Nature (Lond) 427:65–68

Rasmussen DT (1990) Primate origins: lessons from a neotropical marsupial. Am J Primatol 22:263–277

Rose KD (1975) The Carpolestidae, early Tertiary primates from North America. Bull Mus Comp Zool 147:1–74

Rose KD, Rana RS, Sahni A et al (2009) Early Eocene primates from Gujarat, India. J Hum Evol 56:366–404

Sargis EJ (2007) The postcranial morphology of *Ptilocercus lowii* (Scandentia, Tupaiidae) and its implications for primate supraordinal relationships. In: Rovosa MJ, Dagosto M (eds) Primate origins: adaptations and evolution. Springer, New York, pp 51–82

Scotese CR (2001) Digital paleogeographic map archive on CD-ROM. PALEOMAP Project, Arlington, Texas

Seiffert ER, Simons EL, Clyde WC et al (2005) Basal anthropoids from Egypt and the antiquity of Africa's higher primate radiation. Science 310:300–304

Sigé B, Jaeger J-J, Sudre J et al (1990) *Altiatlasius koulchii* n. gen. et sp., Primate omomyidé du Paléocène Supérieur du Maroc, et les origins des Euprimates. Palaeontographica A 214:31–56

Silcox MT, Bloch JI, Sargis EJ et al (2005) Euarchonta (Dermoptera, Scandentia, Primates). In: Rose KD, Archibald JD (eds) The rise of placental mammals. Johns Hopkins University Press, Baltimore, pp 127–144

Silcox MT, Sargis EJ, Bloch JI et al (2007) Primate origins and supraordinal relationships: morphological evidence. In: Henke W, Tattersall I (eds) Handbook of paleoanthropology, vol 2, Primate evolution and human origins. Springer, Berlin, pp 831–859

Silcox MT, Dalmyn CK, Bloch JI (2009) Virtual endocast of *Ignacius graybullianus* (Paromomyidae, Primates) and brain evolution in early primates. Proc Natl Acad Sci USA 106:10987–10992

Smith GE (1913) The evolution of man. Annu Rep Board Regents Smithson Inst 1912:553–572

Smith T, Rose KD, Gingerich PD (2006) Rapid Asia-Europe-North America geographic dispersal of earliest Eocene primate *Teilhardina* during the Paleocene-Eocene thermal maximum. Proc Natl Acad Sci USA 103:11223–11227

Soligo C, Will O, Tavaré S et al (2007) New light on the dates of primate origins and divergence. In: Ravosa MJ, Dagosto M (eds) Primate origins: adaptations and evolution. Springer, New York, pp 29–49

Springer MS, Murphy WJ, Eizirik E et al (2003) Placental mammal diversification and the Cretaceous–Tertiary boundary. Proc Natl Acad Sci USA 100:1056–1061

Springer MS, Murphy WJ, Eizirik E et al (2005) Molecular evidence for major placental clades. In: Rose KD, Archibald JD (eds) The rise of placental mammals. Johns Hopkins University Press, Baltimore, pp 37–49

Springer MS, Murphy WJ, Eizirik E et al (2007) A molecular classification for the living orders of placental mammals and the phylogenetic placement of primates. In: Rovosa MJ, Dagosto M (eds) Primate origins: adaptations and evolution. Springer, New York, pp 1–28

Sussman RW (1991) Primate origins and evolution of angiosperms. Am J Primatol 23:209–223

Tavaré S, Marshall CR, Will O et al (2002) Using the fossil record to estimate the age of the last common ancestor of extant primates. Nature (Lond) 416:726–729

Waddell PJ, Okada N, Hasegawa M (1999) Towards resolving the interordinal relationships of placental mammals. Syst Biol 48:1–5

Wible JR, Covert HH (1987) Primates: cladistic diagnosis and relationships. J Hum Evol 16:1–22

Index

A

ABO blood group, 187–189
Accelerated evolution, 135
Accessory olfactory bulb, 55, 56, 64, 65
Accessory olfactory system, 64
Accumulation, 24, 34, 35, 58, 150, 156, 166, 231, 235–238
Acquired immunodeficiency syndrome (AIDS), 133, 143
Activating receptor, 128, 133, 137
Adaptation, 25, 30, 37, 67, 80, 85, 88, 96, 130, 136, 273, 276
Adaptive evolution, 72, 136, 137, 276
ADH. *See* Alcohol dehydrogenase (ADH)
African apes, 201, 221, 227, 228, 231–234, 258
Aging, 32–37
    antagonistic pleiotropy hypothesis of, 36
    disposable soma hypothesis of, 34
    gene expression changes during, 28, 30, 32, 34–37
    molecular mechanism of, 33
    oxidative damage theory of, 34, 35
AIDS. *See* Acquired immunodeficiency syndrome (AIDS)
Alcohol, 3, 25, 149–150, 157
α2-6-linked sialic acids, 139, 140, 142
Alcohol dehydrogenase (ADH), 3, 149–158
Alternatively spliced transcript, 12
Alu, 129, 164, 206, 223, 248
Alzheimer's disease, 138, 143
Anatomy, 24, 64–65, 194
Ancestral, 4, 26, 27, 72, 73, 100, 130, 137, 151, 186, 187, 189, 193–214, 220–221, 238
Androstenone, 47, 67, 70–73
Anthropoid (Anthropoidea), 157, 212, 243, 247, 248, 259, 260, 271, 277

B

Ape, 3, 4, 14, 16, 24–29, 32, 48, 58, 65, 67–69, 88, 97, 99, 100, 113, 126, 130, 131, 133–136, 139, 141, 143, 154–157, 184–186, 188, 189, 197, 199–205, 207, 210, 217, 218, 220–222, 227, 238, 244, 255, 256, 259, 261
Ardipithecus ramidus, 157
Arms race, 17, 142
"Arms race" between hosts and pathogens ("Red Queen" effects), 130, 142
Array painting, 223
Australopithecus, 129, 130

Baboon, 1, 14, 133, 134, 139, 141, 153–156, 196, 218, 251
Bacterial artificial clone (BAC), 223
Balancing selection, 87, 89, 110–113
Behavioral experiments, 87, 101, 106, 107
Biased gene conversion, 18
Biological diversity, 228
Birth-and-death evolution, 54
BLAST search, 11, 213, 234
Bonobo, 71, 126, 137, 154–156, 217, 227, 229–235
Bootstrap, 153, 156, 170, 184
Bouquet
    frequency, 234
    stage, 234, 236
Brain
    evolution, 138
    expansion, 19, 129, 138
    microglia, 137, 138
    size, 25, 29, 33, 129, 138
Bronchial asthma, 133

C

Campylobacter jejuni, 136
Capuchin monkey, 98, 108–113, 232, 249, 250
Catalytic domain, 157
Catarrhine (Catarrhini), 52, 53, 55, 67, 97,
     99–102, 104, 113, 164–167, 170,
     200, 204, 205, 244, 251, 259–261
C-band patterns, 229
cDNA library, 11, 12, 16
CD33/Siglec-3-related Siglec, 128, 132,
     134–137, 139, 141
Cell–cell communication, 124, 128
Centromeric regions, 233, 234
Character states, 181, 183–185
Chemosensation, 43, 44
Chemosensory receptor, 2, 43–58, 69–70, 94
Chimpanzee, 1–3, 5, 10, 12, 13, 15–19, 23–31,
     33, 36, 37, 46, 51, 52, 55, 67–69,
     72, 82, 84–88, 99, 124, 126, 128,
     130–141, 154–156, 165–171, 182,
     184–186, 194, 195, 201, 207, 210,
     217, 222, 223, 227–238, 255, 256,
     259, 261
Chromaticity, 105
Chromatin, 224, 228
Chromophore, 94, 95
Chromosome
     association, 102, 199, 202–205, 221, 234
     breakpoints, 171, 198, 200, 207, 213, 220,
          222–224
     painting, 4, 194, 197–205, 219–221, 223, 234
     sorting, 198, 222
Circum-Tethys area, 277
Cladistic, 221, 272, 276
Clone-by-clone method, 154
CMAH. *See* CMP-*N*-acetylneuraminic acid
     hydroxylase (CMAH)
CMP-*N*-acetylneuraminic acid hydroxylase
     (CMAH), 126, 127, 129–131, 135,
     140, 142, 143, 164
     inactivation, 129–131, 142
Coenzyme binding domain, 157
Cognitive abilities, human-specific, 15, 17, 24,
     26, 29, 31, 33, 37
Colinear chromosomes, 231, 238
Color vision, 3, 68–69, 94–114, 166, 171
     polymorphism, 93–114
     priority hypothesis, 52, 53, 58, 69
Combinatorial coding, 47
Constitutive heterochromatin, 229, 230
Copy number variations (CNV), 3, 50, 73, 85,
     165–166, 171
Crossing-over, 236–238
Cultural transmission, 25

D

Deletion, 3–5, 10, 25, 49, 54, 81, 82, 99, 129,
     133, 134, 138, 139, 141, 164, 175,
     179, 211, 238
De novo transcriptome sequencing, 14
Dichromacy, 3, 95, 97, 99, 105, 106, 108, 113
Dichromatic vision, 68, 106
Dietary adaptation, 157
Distance matrix, 183
Diversification of sialic acid-containing glycan
     chains, 128
D-loop, 182, 256
DNA-based transposable element, 175–179,
     229
DNA microarray, 12–15, 18, 50
     cDNA microarray, 12, 13, 15
     oligo-DNA microarray, 12, 13
     tiling array, 13

E

Echolocation, 53
Ectopic recombination, 228, 235–237
Encephalization, 24, 27, 33
Endophenotype, 10
Enterotoxigenic *Escherichia coli* K99, 129
Enzyme, 11, 127, 129, 140, 149–152, 157,
     175, 177
Epigenetic, 2, 223, 228
"Essential" arginine, 128, 133, 138, 139, 141
Euchromatic region, 124, 230, 236–238
Expressed sequence tag (EST), 11, 12, 16

F

Fluorescence in situ hybridization (FISH),
     4, 171, 194, 197–199, 206, 207,
     211–214, 219, 223, 233
Formyl peptide receptors (FPR), 44, 45, 57,
     58, 69, 70
Frontal cortex, 29–31
Frugivorous anthropoid, 157
Frugivorous behavior, 157
Functional variance, 73–74
Fusion, 108, 133, 134, 195, 196, 203, 205,
     210, 212, 213, 221–223, 227, 237

G

Gene
     conversion, 4, 18, 100, 132, 133, 137, 138,
          142, 155, 156, 158, 167, 169, 170
     conversion events, 132, 238
     dense, 224

Index

duplication, 48, 50, 54, 95, 97, 99–102, 113, 132, 135, 137, 139, 151, 152, 155, 165
losses, 51, 139
network, 17, 19
poor, 224
shuffling, 48, 236
silencing, 238
Genetically inert, 228–231, 236
sequences, 229–231
Genetic recombination, 238
Genome sequence, 1, 3, 5, 12, 13, 14, 15, 16, 19, 44, 46, 49–52, 85, 86, 88, 124, 164, 172, 176, 184, 207, 212, 228, 231, 234, 236, 238, 256, 259–261
Genomic
drift, 58
reactions, 228
Genomic wastelands, 5, 227–238
Gibbons, 1, 4, 24, 187–189, 199, 202, 203, 213, 217–224, 232, 255, 256, 259
Gorilla, 1, 15, 126, 133, 135, 137, 141, 154–156, 165, 182, 184–186, 201, 217, 227, 229, 230, 232–236, 255, 259, 261
Gould, S.J., 24, 26, 27, 29, 30
G-protein
coupled receptors, 44, 47, 48, 56, 57, 65, 69, 80, 81, 94, 186
coupling, 83, 88
Grasping abilities, 275
Gray matter, 30
Guanylyl cyclase type D (GC-D), 69

H
Haplorhines (Haplorhini), 52, 64, 65, 68, 244, 247, 248, 258, 259, 260, 272, 277
Heterochromatinization, 231, 234, 237
Heterochrony, 2, 26–28, 31
of gene expression, 28–32, 36
Heterodimer, 57, 150, 157
Heterozygous, 47, 52, 68, 89, 103, 112
Hominoid (Hominoidea), 50, 52, 164, 165, 169, 202, 212, 217, 221, 232, 244, 255–256, 259–261
Homo, 33, 129, 130, 255
Homodimer, 150, 157
Homozygote, 87
Hoolock, 217–219, 221, 256
Host–pathogen interaction, 124, 143
Human
color vision, 105, 113, 114
evolution, 3, 4, 15, 23–37, 58, 139, 227

inactivation of CMAH, 129–131, 142
influenza virus A and B, 140
Human longevity, 32, 33, 36, 37
embodied capital hypothesis of human longevity, 33
grandmother hypothesis of, 33
Human malaria parasite, 130
Human-specific Alu element, 129
Human-specific change, 2, 123–143
Human-specific genomic changes, 126
Human-specific inactivation of CMAH (human-specific loss of Neu5Gc), 129–131, 142
Human-specific loss of Neu5Gc, 126, 130, 131, 135–137, 139, 142
Human-specific paired receptors, 137
Human-specific pathogen group B *Streptococcus*, 136
Human-specific pathogens, 142
Human uniqueness, 3, 24, 29, 124, 126, 128, 134, 138, 140, 142, 143, 167
Hylobates, 187, 203, 217–221, 256

I
Immune systems, 17, 18, 57, 127
Immunity, 132, 143
Inactivation of CMAH, 129–131, 142
Incompatible phylogenetic information, 181, 182
Inhibitory receptors, 127, 132, 135, 137
Insectivore, 65, 157, 158, 200
Insertion, 3, 5, 10, 25, 82, 129, 164, 175, 176, 178, 179, 231, 234, 235, 238, 247, 248, 250
Intercalary insertions, 231
Interphase nucleus, 224
Inversion, 4, 10, 167, 175, 201, 206, 207, 213, 219–221, 227, 235, 237
Isovaleric acid, 47, 73

K
Ka/Ks, 87
Karyotype, 4, 195, 196, 198, 200–205, 208, 211–212, 220–221, 227, 238, 244
Karyotypic differentiation, 238
King, M.C., 18, 28, 29
K/T (cretaceous/tertiary) boundary, 258, 259, 269

L
LCR.*See* Locus control region (LCR)
Leaf-eater, 157, 158

Lemur (Lemuriformes), 1, 52, 65, 68–70, 154, 200, 204, 232, 243–246, 259, 260, 261
Ligand(s), 2, 47, 49, 54–55, 65, 66, 72–74, 82–83, 85, 88, 131, 134–136
binding, 44, 80, 83, 84, 133
Locus control region (LCR), 101, 102
Lorise (Lorisiformes), 52, 68, 97, 204, 243, 244, 246–247, 259–261
Loss of Neu5Gc, 126, 130, 131, 135–137, 139, 142

M

Macaque, 12–14, 16–18, 28, 29, 32, 35, 52, 55, 71, 99, 154, 195, 196, 201, 203, 207, 209–213, 218, 236, 251, 252
Main olfactory
  bulb, 51, 64, 65
  epithelium, 44, 46, 50, 52, 56, 64, 69, 70
  system, 54, 55, 64, 65, 68
Malaria parasite *Plasmodium reichenowi*, 130
Mammalian radiation, 153, 155, 176
Medaka fish, 177, 179
Meiosis, 234, 236
Metabolism, 3, 34, 149–150
Microarray, 12, 13, 18–19, 28, 34
Microglia, 138
Microglial cells, 138
Micro RNA (miRNA), 9, 13, 31, 32, 35
Mitochondrial data, 185, 244, 250, 254–261
Mitochondrial dataset, 246, 254–256
Mitochondrial DNA (mtDNA), 183, 246–248, 256–259
Mitochondrial gene, 16, 248, 254
Mitochondrial genome, 182, 246–249, 254, 256–261
Mitochondrial protein, 258
Mitochondrial sequence, 249, 254, 255, 260
Model organism, 10, 28, 30, 35, 58
Molecular coevolution, 136
Molecular dissection analysis, 230, 232
Motif sequence, 232
Multigene families, 2, 44, 46

N

N-acetylneuraminic acid (Neu5Ac), 124, 129–132, 135, 136, 139, 142
Natural selection, 15, 16, 72, 87, 95, 107, 110, 114, 171, 176, 178, 193, 223
Neanderthal, 1, 29
Nearly neutral theory, 17
Neighbor-joining method, 182–184, 248

Neighbor-net method, 182, 183
Neocentromeres, 4, 208–211
Neoteny, 2, 27, 33, 36, 37
  definition and examples of, 26
  in human development, 26, 30, 33, 36
  of human encephalization, 27, 33
  of human gene expression, 28–32
  of human morphology, 24, 26, 27, 32
  theory of human evolution, 23–37
Neutral evolution, 25, 58
Neutral fixation, 156
Neutral theory, 16, 17
New World monkey (NWM), 3, 14, 50, 52, 55, 65, 68–70, 88, 95–101, 104, 106–114, 141, 143, 154, 155, 157, 158, 165–167, 170, 177, 200, 204, 207, 243, 244, 249–251, 261
Next-generation sequencer, 13–14, 19
N-glycolylneuraminic acid (Neu5Gc), 124–127, 129–132, 135–137, 139, 142
Nomascus, 199, 207, 213, 218, 220–223, 256
Nonsynonymous, 15, 47, 67, 71, 72, 84, 86, 87
  substitutions, 15, 16, 51, 71, 87, 135, 157
Nuclear architecture, 224
Nuclear compartment, 224
Nuclear data, 250
Nuclear dataset, 255
Nuclear DNA, 244, 246–249, 260
Nuclear gene, 5, 16, 177, 246

O

Odorant receptors (ORs)
  OR7D4, 47, 66, 67, 71–73
  OR repertoire, 67–69, 73
Old World monkey (OWMs), 3, 14, 16, 18, 50, 52, 65–69, 88, 97, 99, 100, 104, 141, 154, 155, 157, 158, 164, 165, 169, 199, 200, 202, 207, 210, 211, 218, 243–244, 251–255, 260, 261
Olfaction, 2, 43, 46, 51, 52, 54, 64, 68–69, 71, 73
Olfactory epithelium, 64–66, 69, 70
Olfactory receptors, 2, 44–54, 66
One neuron–one receptor rule, 46
Opsins, 3, 47, 52, 53, 68, 72, 73, 94–103, 107–114
Orangutan, 1, 12, 52, 69, 126, 133, 137, 141, 154–156, 169, 182, 184–186, 199, 207, 217, 220, 221, 227, 232, 233, 255, 257–259, 261
Orbital convergence, 275
Orthologous, 51, 72, 80, 81, 136, 153, 168, 248
  genes, 46, 51, 52, 71, 238

Index 285

Orthologues, 46, 51, 56, 66–68, 71, 72, 74, 134, 137, 156, 223
Outgroup, 27, 30, 151, 169, 187, 199, 220, 246, 258, 273
Oxytocin, 184
Oxytocin receptor (*OXTR*), 184–186

P

Paired receptors, 133, 137, 138
Pan troglodytes verus, 86
Parsimony informative sites, 182
Pathogens, 17, 57, 124, 129, 130, 132, 134, 136, 140, 142
Phenylthiocarbamide (PTC), 2, 86, 87
Pheromones, 44, 47, 54–56, 64, 70–71, 94
Phylogenetic network, 4, 181–189
Phylogenetic tree, 4, 48, 98, 137, 139, 151, 156, 181–184, 189, 218, 270, 271, 277
Plasmodium falciparum, 130
Plasmodium reichenowi, 130
Plasticity, neural, 31
Platypus, 53, 56, 258
Platyrrhine (Platyrrhini), 204, 205, 212, 244, 249, 259–261
Polymorphisms, 2, 16, 17, 84–89, 97, 98, 100, 103, 107, 108, 110–113, 150, 166, 177, 219–220
Position effect, 223, 238
Positive selection, 17, 18, 25, 51, 71–74, 130, 150, 155, 156, 189
  on gene expression, 18, 30
Post-genomic research, 237
Prosimian (Prosimii), 69, 70, 88, 97, 100, 103, 104, 154, 155, 157, 158, 166, 177, 200, 204, 243, 247, 248
Pseudogene, 45, 46, 49–55, 58, 65, 68, 69, 73, 81, 82, 85–87, 99, 136, 138, 169, 171, 176
Pseudogenization, 51, 55, 68, 69, 129

R

Rapid evolution, 16, 18, 141–142
Receptors type 1 and 2, 44
Recombination, 3, 4, 99, 164, 166, 175, 182, 183, 186–189, 223, 228, 234–238
Reticulations, 181–183, 187
Retrotransposable compound repeated DNA organization (RCRO), 227–238
Retrotransposable elements, 234
Reverse transcription polymerase chain reaction (RT-PCR), 81

Rhesus macaque, 1, 2, 12, 13, 30, 32, 33, 35, 69, 155, 195, 229, 230, 232–234, 236
Rheumatoid arthritis, 133
Rhinarium, 52, 65
Rhodopsin-like GPCR, 47, 48, 56, 57
Ripe fruit, 82, 103, 112, 157
RNA-mediated transposable element (RTE), 176

S

Secondary "Red Queen" effect, 142
Segmental duplications, 3, 164, 165, 208, 210, 212, 222, 223
Selection, 18, 33, 34, 36, 51, 72, 87, 89, 95, 103, 107, 110–114, 129, 157
  negative, 15, 16, 156, 178
  positive, 17, 18, 25, 30, 51, 71–74, 130, 150, 155, 156, 189
Selective constraint, 18, 72, 74, 87
Selective pressure
  dN, 71–72
  dS, 71–72
  positive selection, 73–74
Serial analysis of gene expression (SAGE), 11, 12, 14
Shotgun sequence, 153–155
Short interspersed element (SINE), 129, 164, 244, 247, 249, 250, 252–254
Sialic acid, 123–143
  binding, 127, 129–131, 133–137, 139, 141
  binding ability, 134, 137, 139, 142
  binding domain, 133, 135–138, 142
  binding immunoglobulin superfamily lectins (Siglecs), 127, 128, 131–143
  binding pathogens (primary "Red Queen" effect), 142
  binding specificity, 136, 137
  biology, 123–143
  mediated interactions, 128
  preference, 128–131, 135, 136, 139, 140
  recognition molecules, 130, 141–143
  recognition proteins, 127
Sialome, 124, 128, 142, 143
  evolution, 142
Sialyltransferase, 126, 128, 140, 142
Siamang, 187, 188, 202, 203, 217–219, 221, 223, 230, 232–235
Siglec paired receptors, 138
Siglecs, 127, 128, 131–143
Signal transduction, 55, 65–67, 80, 127, 131
SINE. *See* Short interspersed element (SINE)
Single nucleotide polymorphism, 47, 50, 73, 84, 86
Specific anosmia, 47

Spider monkey, 53, 71, 97–100, 108–111, 232, 249
ST6GAL1, 127, 128, 139–140
ST6GALI, 139
ST6Gal-I, 140, 143
Strepsirrhines (Strepsirrhini), 52, 64, 65, 68, 204, 205, 244, 245, 248, 258, 260, 261, 271, 272, 277
Substitution, 10, 18, 25, 72, 84, 87, 110, 170, 182, 185–187, 246
  nonsynonymous, 15, 16, 71–72, 87, 135, 157
  synonymous, 15, 16, 51, 71, 86, 87, 135, 170
Subtelomeric regions, 5, 227, 228, 234–237
Subterminal heterochromatin, 232
Subterminal satellite, 229, 231, 233–235
S40 virus, 130
Symphalangus, 187, 217, 218, 221, 245, 256
Synonymous, 72, 86, 166
  substitutions, 15, 16, 51, 71, 86, 87, 135, 170
Synteny, 153, 199–202, 205, 210, 213, 214, 221

T
Tajima's D, 87, 111,
Tandem gene duplications, 48
Tarsier (Tarsiiformes), 1, 3, 6, 7, 19, 52, 65, 154, 243, 244, 247–248, 259, 260
Tastant
  bitter, 2, 57, 80, 81, 83–88
  sweet, 2, 57, 80–84
  umami, 2, 57, 80–84
Taste, 2, 3, 43, 44, 46, 57, 79–89, 94
Taste receptor cell (TRC), 80, 81
Taste receptors, 2, 44, 57–58, 80–89, 157
  type 1 and 2, 44
  type 1, T1R (TAS1R), 44, 57, 80–85
  type 2, T2R (TAS2R), 44, 45, 57, 58, 80–82, 84–89
Telomere sequences, 229, 231, 234
The loss of Neu5Gc, 126, 130, 131, 135–137, 139, 142
Topology, 44, 156, 182–186, 244, 245, 248, 249, 254, 258, 270

Toxico-genomics, 14
Trace amine-associated receptors (TAARs), 44, 56, 57, 69
Transcription factor, 19, 25, 31, 35, 101
Transcriptome, 1, 2, 9–19, 30, 126
Transition/transversion, 129, 130, 182, 183, 205
Translocations, 4, 10, 170, 175, 200, 201, 203, 206, 213, 219–221, 235
Transmissible gastoenteritis coronavirus, 130
Transposition, 167, 175, 176, 178, 179, 231, 235
Trichromacy, 3, 52, 53, 95–110, 112, 113
Trichromatic vision, 52, 53, 55, 68, 69, 95–99, 101, 104–107, 109, 110, 112–114
Type I diabetes, 133

U
Umami, 2, 57, 80–84
Unweighted pair group method with arithmetic means (UPGMA), 183

V
Visual opsin, 72, 97–100
Visual pigment, 94, 95
Visual system
  color vision priority hypothesis, 52, 53, 58, 69
  opsins, 68, 72, 97–100
Vomeronasal, 54, 55, 70
  organ, 46, 54–58, 64, 65, 69, 70
  receptor, 54–57, 70
  system, 54, 55
  type II receptors, 44, 54–58, 64, 69, 70
  type I receptors, 44, 54–58, 69, 70

W
Whole genome sequencing (WGS), 13, 44, 49, 153–155, 231
Wilson, A., 15, 18, 19, 28, 29